Quantum Untangling

Quantum Untangling

An Intuitive Approach to Quantum Mechanics from Einstein to Higgs

Simon Sherwood

Registered Offices
John Wiley & Sons, Inc., 111 River Street, Hoboken, NJ 07030, USA
John Wiley & Sons Ltd, The Atrium, Southern Gate, Chichester, West Sussex, PO19 8SQ, UK

For details of our global editorial offices, customer services, and more information about Wiley products visit us at www.wiley.com.

Wiley also publishes its books in a variety of electronic formats and by print-on-demand. Some content that appears in standard print versions of this book may not be available in other formats.

A catalogue record for this book is available from the Library of Congress

Paperback ISBN: 9781394190577; ePDF: 9781394190638; epub: 9781394190621

Cover Design: Wiley
Cover Image: © David Malan/Getty Images

Set in 9.5/12.5pt STIXTwoText by Integra Software Services Pvt. Ltd, Pondicherry, India
Printed and bound by CPI Group (UK) Ltd, Croydon, CR0 4YY

C9781394190577_050924

Contents

Introduction   *xii*
Acknowledgements   *xiii*

Module I   Special Relativity   *1*

1   **Special Relativity**   *3*
1.1   Special Relativity: Simple, Yet Baffling   *3*
1.2   The Speed of Light Is Constant: So What?   *4*
1.3   The Invariant Interval Equation   *5*
1.4   Time Distortion Quantified   *6*
1.5   Length Distortion   *8*
1.6   Leading Clocks Lag   *9*
1.7   Lorentz Transformations and Invariance   *10*
1.8   Summary: Are You Joking Mr Einstein?   *11*

2   **Paradoxes of Special Relativity**   *13*
2.1   Journey to a Distant Planet (1)   *13*
2.2   Journey to a Distant Planet (2)   *14*
2.3   The Twin Paradox   *16*
2.4   Experimental Proof   *18*

3   **Einstein's Famous Equation**   *20*
3.1   Mass, Energy, Momentum – and Particle Time   *20*
3.2   How Did Albert Figure It Out?   *21*
3.2.1   The Ingredients   *21*
3.2.2   The Calculation   *21*
3.2.3   The Intuition   *22*
3.3   Three Beautiful Equations   *23*
3.4   How Wrong Were We?   *24*
3.5   One Further Equation   *25*
3.6   Summary   *26*

Module II   Essential Quantum Mechanics   *27*

4   **Wave-particle Duality**   *29*
4.1   Classical Physics Cannot Explain...   *29*
4.2   Quanta of Light and the Photoelectric Effect   *30*
4.3   De Broglie's Crazy Idea   *31*

4.4   The Double-slit Experiment   *32*
4.5   Schrödinger's Mistreated Cat   *34*
4.6   Summary   *35*

5   Superpositions and Uncertainty   *37*
5.1   The Free Particle Wave Function   *37*
5.1.1   The Phase of the Wave   *38*
5.1.2   Derivatives of the Free Particle Wave Function   *38*
5.1.3   Linking Back to Special Relativity   *39*
5.1.4   Consider a Rocket...   *40*
5.2   From Sinusoid to Uncertainty   *41*
5.3   Superposition   *42*
5.3.1   Superposition Saves the Day   *42*
5.3.2   Combining Eigenstates   *43*
5.4   Heisenberg's Uncertainty Principle   *44*
5.5   In Praise of Fuzziness   *45*
5.6   God Plays Dice: The Role of Probability   *46*
5.7   Summary   *47*
5.8   What Is This Wave Function?   *47*
5.9   The Role of Rest Mass   *48*

6   Everything Happens ... Kind of   *49*
6.1   The Feynman Path Integral   *49*
6.2   Change in Phase of the Wave Function   *50*
6.3   Simplified Path Integral Model   *51*
6.4   The Principle of Stationary Action   *53*
6.5   Action and the Lagrangian   *54*
6.6   From the Lagrangian to the Equations of Motion   *55*
6.7   The Uncertainty Relationship: A Different Perspective   *56*
6.8   Feynman Diagrams   *57*
6.9   Summary   *58*

7   Measurement and Interaction   *60*
7.1   What Can You Know about a Quantum System?   *60*
7.2   Collapse of the Wave Function   *61*
7.3   When a Body Meets a Body ...   *63*
7.4   An Electron in a Box   *63*
7.5   Collapse of the Wave Function – a Twist   *65*
7.6   Decoherence and the Measurement Problem   *66*
7.7   When a Body Leaves a Body – Entanglement at a Distance   *67*
7.8   Summary   *68*

8   Module Summary and Schrödinger   *70*
8.1   Module Summary   *70*
8.2   Adding up the Implications...   *73*
8.3   The Path to Schrödinger's Equation   *73*
8.3.1   The Klein-Gordon Equation   *74*
8.3.2   A Taste of Schrödinger's Equation   *75*
8.3.3   Incorporating Potential Energy   *76*
8.4   Module Memory Jogger   *78*

Module III   Complex Quantum Mechanics   *79*

9   Introducing Complex Numbers   *81*
9.1   Welcome to Complex Numbers   *81*
9.1.1   We Have a Problem   *82*
9.1.2   Complex Notation for Phase   *82*
9.1.3   Interference Calculations   *83*
9.1.4   A Friend with Benefits   *84*
9.1.5   Not a Free Lunch   *84*
9.2   Representing the Wave Function with Complex Notation   *85*
9.3   Summary   *85*

10   Superpositions and Fourier Transforms   *86*
10.1   The Maths of Fourier Transforms   *87*
10.1.1   Example 1: Fourier Transform of a Position Eigenstate   *88*
10.1.2   Example 2: Fourier Transform of $\frac{\partial \Psi}{\partial x}$ *88*
10.2   Heisenberg's Uncertainty Principle and the Gaussian Distribution   *89*
10.3   The Quantum Footprint   *90*
10.4   Time and Energy   *92*
10.5   Summary   *93*

11   Schrödinger's Equation   *95*
11.1   Understanding Schrödinger's Equation   *95*
11.1.1   Incorporating Potential Energy   *96*
11.1.2   Superpositions   *96*
11.1.3   Schrödinger's Equation in Words   *96*
11.2   Operators, Eigenstates and Eigenvalues   *97*
11.3   Commutation Relations   *100*
11.4   Expectation Values and Dirac Notation   *101*
11.5   Energy Eigenstates are Stationary   *102*
11.6   Time-independent Schrödinger Equation   *102*

12   Schrödinger's Equation in Action   *104*
12.1   Free Particle Wave Function (E > V)   *104*
12.2   Creeping into Forbidden Places (E < V)   *105*
12.3   The Finite Potential Well   *106*
12.4   Quantum Tunnelling and the Sun   *106*
12.5   Dodging Potential Obstacles (E > V)   *108*
12.6   Quantum Biology   *110*
12.7   Wave Packets: A Model for Localised Particles   *110*
12.8   Summary   *113*

13   Quantum Harmonic Oscillator   *114*
13.1   Introduction   *114*
13.1.1   The Simple Harmonic Oscillator   *114*
13.1.2   The SHO and QHO: Why Do We Care?   *115*
13.2   Penetration Model for the QHO   *116*
13.3   Schrödinger's Equation for the QHO   *117*
13.3.1   Ground State of the QHO   *118*
13.3.2   A Trick to Find the Other Energy Eigenstates of the QHO   *119*
13.3.3   The QHO Energy Eigenstate Ladder   *120*
13.3.4   QHO Superpositions   *121*

13.4   The QHO in Three Dimensions   *122*

13.5   Formal Definition of the Creation and Annihilation Operators   *123*

13.6   The Path to Quantum Field Theory (QFT)   *125*

14   Angular Momentum   *126*

14.1   A Primer on Classical Angular Momentum   *126*

14.2   Quanta of Angular Momentum   *128*

14.3   Angular Momentum's Intricate Dance   *128*

14.4   Angular Kinetic Energy and Angular Momentum   *129*

14.5   The Pattern of Angular Momentum Eigenstates   *130*

14.5.1   Ground State: $l = 0$ *131*

14.5.2   First Energy Level: $l = 1$ *131*

14.5.3   Three Distinct First Level States: $l = 1, \quad m = -1, 0, +1$ *131*

14.5.4   Resulting in the Pattern   *132*

14.6   The Angular Momentum Creation Operator   *133*

14.7   Summary   *134*

15   Coulomb Potential   *136*

15.1   The Hydrogen Emission Spectrum   *136*

15.2   The Challenge of the Coulomb Potential   *137*

15.3   A Primitive Model   *138*

15.4   Schrödinger's Equation for Hydrogen   *139*

15.4.1   Spherical Harmonics – *merci* Monsieur Laplace   *139*

15.4.2   The Angular Equation   *141*

15.4.3   The Shape of the Atomic Orbitals   *142*

15.4.4   Radial Kinetic Energy   *143*

15.4.5   The Radial Equation   *144*

15.5   Discussion   *146*

16   The Periodic Table   *149*

16.1   Introduction   *149*

16.2   Adding More Protons   *150*

16.3   The Periodic Table   *150*

16.4   Molecular Bonds   *152*

16.4.1   Ionic Bonds   *152*

16.4.2   Covalent Bonds   *153*

16.5   Bonds in the Nucleus   *154*

16.6   Virtual Particles   *154*

16.7   Fusion and Fission   *155*

16.8   Module Summary   *156*

16.9   Module Memory Jogger   *157*

Module IV   Relativistic Quantum Mechanics   *159*

17   Spin   *161*

17.1   Intrinsic Angular Momentum: Spin   *161*

17.2   Spin-half Particles and the Pauli Exclusion Principle   *162*

17.2.1   The Stern-Gerlach Experiment   *162*

17.2.2   Spin-half and Spinors   *163*

17.2.3   The Pauli Exclusion Principle   *164*

17.2.4   The Pauli Matrices   *165*

17.3   Integer-spin: The Photon   *168*
17.3.1 Photon Polarisation   *169*
17.4   Bell's Inequality and the Aspect Experiment   *170*
17.5   Summary   *172*

18   The Dirac Equation   *173*
18.1   Yet Another Equation?   *173*
18.2   Bi-spinors and Four-component Wave Functions   *174*
18.3   The Dirac Equation   *175*
18.3.1 The Ingredients   *175*
18.3.2 Dirac's Crazy Insight   *176*
18.3.3 Dirac's Matrices   *177*
18.3.4 We Are Finally There: Dirac's Equation   *179*
18.4   Spin-half Is Built in   *180*
18.5   Interpreting the Dirac Equation   *182*
18.5.1 Zero Momentum: Distinct Spin and Antiparticles   *182*
18.5.2 The Dirac Equation and Minkowski Spacetime   *182*
18.5.3 Particle and Antiparticle States   *183*
18.5.4 Moving Frame   *184*
18.6   The Dirac Equation and Hydrogen   *185*
18.7   Dirac Equation: Modern Formulation   *186*
18.8   The Aftermath: Physics Falls Apart Again   *186*

19   Quantum Field Theory   *189*
19.1   Changing the Question   *190*
19.2   Quantum Fields Win the Day   *190*
19.2.1 The Quantum Field Structure   *191*
19.2.2 Quantum Fields and Spin   *192*
19.2.3 Creation and Annihilation   *192*
19.2.4 Bosons Like to Party   *193*
19.2.5 Conservation of Energy and Momentum   *194*
19.3   Non-relativistic Path Integrals and Action   *195*
19.4   QFT Path Integrals: A Relativistic Twist   *197*
19.5   Energy and Time   *197*
19.6   QFT Field Development Pathways   *198*
19.7   The Klein-Gordon Lagrangian as a Model   *199*
19.8   Global Gauge Invariance to Phase   *200*
19.9   Summary   *201*

20   Local Gauge Invariance   *202*
20.1   Introduction to Local Gauge Invariance   *202*
20.2   The Infinity Swimming Pool – an Analogy   *204*
20.3   Refresher in Electromagnetics (EM)   *205*
20.3.1 EM Refresher (1): The Basics   *205*
20.3.2 EM Refresher (2): The Vector Potential   *206*
20.4   The EM Quantum Field and Lagrangian   *208*
20.5   EM Gauge Invariance   *210*
20.6   U(1) Local Gauge Invariance: Putting Together the Pieces   *210*
20.6.1 The Swimming Pool: The Electron Field   *210*
20.6.2 The Balancing Tank: The EM Field   *211*
20.6.3 The Connection   *211*
20.6.4 The Interaction   *211*

20.6.5   The Infinity Pool: Combined Electron and EM Fields   *211*
20.7   The Dirac Lagrangian   *212*
20.8   Interaction and the Pathway of Stationary Action   *213*
20.9   The Photon Must Be Massless   *214*
20.10   Summary   *214*

21   QED and Feynman Diagrams   *216*
21.1   Feynman Diagrams   *216*
21.2   Example: Electron-positron Annihilation   *218*
21.3   Off-shell Drift and the QED Interaction   *219*
21.4   Feynman Rules   *221*
21.4.1   The Vertex and the Coupling Constant   *221*
21.4.2   The Propagator   *222*
21.4.3   Illustrative QED Calculation (Simplified)   *223*
21.4.4   From Amplitude to Cross Section   *224*
21.5   Resonance and the Search for New Particles   *225*
21.6   Do Virtual Particles Exist?   *225*

22   Renormalisation and EFT   *227*
22.1   Troublesome Loops   *227*
22.2   The Dressed Electron   *228*
22.3   Using Feynman Diagrams   *229*
22.4   Renormalisation   *230*
22.5   Ken Wilson's Effective Field Theory (EFT)   *232*
22.6   Summary   *232*

23   The Strong Force   *234*
23.1   The Elementary Particles   *234*
23.2   The Strong Force: An Overview   *235*
23.2.1   Colour Charge   *236*
23.2.2   QCD, Gluons and Confinement   *236*
23.2.3   Strong Force Coupling Constant   *237*
23.3   QCD Local Gauge Invariance   *238*
23.3.1   SU(3) Symmetry and Colour   *238*
23.3.2   A Short Detour into Group Theory   *240*
23.3.3   The QCD Lagrangian   *241*
23.3.4   Gluons and the Generators   *242*
23.3.5   Summary: QCD As an Infinity Swimming Pool   *243*
23.4   The Residual Strong Force   *244*
23.5   Oh No! Here Comes Jill Again!   *245*

24   The Weak Force and Higgs Field (1)   *246*
24.1   Idealised Weak Force and SU(2) Symmetry   *246*
24.2   The Real Weak Force   *248*
24.2.1   Weak Isospin   *248*
24.2.2   Weak Interactions   *249*
24.2.3   Massive Weak Bosons   *250*
24.2.4   Wu and the Weak Left-handed Bias   *250*
24.3   What About SU(2) Gauge Symmetry?   *251*
24.4   Mass, Chirality and the Higgs Field   *252*
24.4.1   Mass as an Interaction   *252*
24.4.2   Chirality Versus Helicity   *253*

24.4.3   Chiral Dirac Equation   *254*
24.5   The Story So Far   *255*

25   **The Weak Force and Higgs Field (2)**   *257*
25.1   The Higgs Interaction   *257*
25.2   The Higgs Field and Mechanism   *258*
25.3   The Maths of the Higgs Field   *259*
25.4   Visualising the Higgs Field   *259*
25.5   Spontaneous Symmetry Breaking   *260*
25.6   The Maths of the Higgs Mechanism   *260*
25.6.1   The Starting Point   *261*
25.6.2   The Potential of the Higgs Field   *261*
25.6.3   Rotational Fluctuations of the Higgs Field   *262*
25.6.4   Putting It All Together   *262*
25.7   The Discovery of the Higgs Boson   *264*
25.8   Electroweak Unification   *264*
25.8.1   The Z Boson   *265*
25.8.2   The Photon   *266*
25.9   Summary   *266*

26   **The Standard Model and Beyond**   *269*
26.1   The Standard Model Lagrangian   *269*
26.2   From Einstein and de Broglie to Higgs   *271*
26.3   Questions and Problems   *271*
26.4   General Relativity and Quantum Mechanics   *272*
26.5   Supersymmetry (SUSY)   *273*
26.6   String Theory   *274*
26.6.1   Gravity in String Theory   *274*
26.6.2   Difficulties with String Theory   *275*
26.7   Loop Quantum Gravity (LQG)   *276*
26.7.1   LQG Space as a Quantum Entity   *277*
26.7.2   LQG Background Independence: Spin Networks   *278*
26.7.3   Difficulties with LQG   *279*
26.8   That's All Folks!   *280*
26.9   Module Memory Jogger   *280*

Index   *282*

Introduction

Why Write This Book?

Most popular science books can offer only a light introduction to quantum mechanics because they tend to feature little, if any, maths. On the other hand, books that include the maths tend to be complicated and hard to digest. There seemed to be something missing between the two: a book that offers students a detailed explanation of quantum mechanics in an intuitive, engaging, and even fun way.

Quantum Untangling is my attempt to fill that gap. It is designed to be accessible to non-experts, but at the same time includes the maths necessary for a full understanding. The text is structured to build knowledge piece by piece, introducing ideas gradually with mathematical detail as required. As the book progresses, the material becomes steadily more complicated. By the time readers reach the later chapters, they will, I hope, be prepared with the bedrock of knowledge needed to handle the more advanced topics.

Why Read This Book?

Is this the right book for you? If you abominate mathematics and loathe equations it may not be your first choice for holiday reading, but for most of it, although some familiarity with complex numbers will help, basic calculus is all you need. Knowledge equivalent to A-level maths (UK) or an AP calculus course (USA) should be enough, and science undergraduates past and present will find the maths very familiar.

The first two modules of the book, *Special Relativity* and *Essential Quantum Mechanics*, introduce the essential elements of the subject while keeping the maths to an absolute minimum. You will need Pythagoras' theorem and simple differentiation, but no more.

The next module, *Complex Quantum Mechanics*, builds on the Schrödinger equation and covers material found in most mainstream undergraduate courses. The maths is stepped up and complex numbers introduced, but slowly and steadily, in readily comprehensible stages. Clear examples and simple illustrations help to pave the way.

The last module, *Relativistic Quantum Mechanics*, covers advanced material such as the Dirac equation, quantum field theory, the Standard Model and the Higgs Mechanism. These are complicated subjects so I have been careful to keep the text accessible by referring to, and building on, the material set out in earlier chapters.

My challenge as author is to explain this highly technical subject as simply and clearly as possible. If there are sections that you don't understand, then that is a failure on my part. If however there is something that enhances your interest or sheds light on some aspect of quantum mechanics that you didn't previously understand, I shall consider my efforts worthwhile and be a happy man.

Let me end with a couple of thoughts that inspired me in writing this book ...

If you can't explain it simply, you don't understand it well enough.

- attributed to Albert Einstein

The real problem in speech is not precise language. The problem is clear language.

- Richard Feynman

Acknowledgements

I am greatly in debt to Dr James Millen at King's College, London, and Matthew von Hippel of the Niels Bohr Institute in Copenhagen. James scoured through my draft manuscript with an unerring eye for sloppy thinking and errors. If he ever tires of science, he definitely has a future as a book editor. Matt is a wonderful communicator and showed great patience with my efforts to explain advanced relativistic topics in an approachable intuitive way. Any errors that remain are my fault and mine alone.

My thanks to Diane Pengelly who steered me through the unfamiliar process of developing a book proposal and contacting publishers. On the literary side, I am grateful to Martin Preuss, Monica Chandrasekar, Christy Michael, and Angela Cohen at Wiley. Without Martin's enthusiasm and support for the book, I fear *Quantum Untangling* would have remained a perhaps/sometime/maybe project.

Finally, I must thank my family, especially my long-suffering wife Rachel. Her brilliant mind keeps me on my toes, but physics is not her thing. She, and my two daughters, have been tortured with endless hours of lectures, to say nothing of a kitchen full of maths and science books. And there is no comfort now this book is finished... sorry darling, I have already started on the next!

Module I

Special Relativity

Chapter 1

Special Relativity

CHAPTER MENU

1.1 Special Relativity: Simple, Yet Baffling
1.2 The Speed of Light Is Constant: So What?
1.3 The Invariant Interval Equation
1.4 Time Distortion Quantified
1.5 Length Distortion
1.6 Leading Clocks Lag
1.7 Lorentz Transformations and Invariance
1.8 Summary: Are You Joking Mr Einstein?

1.1 Special Relativity: Simple, Yet Baffling

I will make the bold (hopefully correct) assumption that you have some notion of special relativity and are familiar with Einstein's two postulates that for all observers, however fast they are moving relative to each other:

- the speed of light is constant for all observers
- the laws of physics are the same for all observers

These two postulates are driven by a common logic. Let me explain. It was well known that a moving electric field creates a magnetic field and, conversely, a moving magnetic field creates an electric field. James Maxwell combined these effects to show that, if the conditions were just right, a moving electric field would generate a moving magnetic field that would generate a moving electric field. Each field generates the other, creating that is, a self-sustaining wave that fitted perfectly with light so it all tied together neatly. There was one niggle. Maxwell's equations required that light move at a specific speed ($c \approx 300,000$ kilometres per second). But this specific speed is a speed relative to what?

Most scientists made the rather sensible assumption that light waves move through some underlying substance (that they called *ether*) in the same way as water waves move through water. If true, Maxwell's speed of light would only be constant relative to the ether. It would vary for different observers depending on the observer's own motion relative to the ether. Simple! Searches were made in the late 1800s, the best known being the Michelson-Morley experiment, but no evidence was found for any background ether. The implication is that there is no special *something* that light moves through so c is its speed through a vacuum, that is, its speed through nothing.

This fitted with Einstein's view. His *theory of relativity* requires no special substance or ether. It assumes there is no preferred vantage point. If two observers are moving relative to each other you can never say one is moving and one is stationary. If this is true, there can be no background ether for an observer to measure against, nor any experimental result that would distinguish a particular observer as stationary. If indeed there is no preferred vantage point, both of Einstein's postulates must apply. The speed of light must be constant for all observers *and* the laws of physics must be the same for all observers.

We now know that these postulates underpin how the universe works. They sound simple. Indeed the mathematics, as you will see, is extremely simple involving nothing more than Pythagoras' theorem for the sides of a triangle ($a^2 + b^2 = c^2$).

Quantum Untangling: An Intuitive Approach to Quantum Mechanics from Einstein to Higgs, First Edition. Simon Sherwood.
© 2023 John Wiley & Sons Ltd. Published 2023 by John Wiley & Sons Ltd.

But the logical consequence of all this is quite bizarre. We will step through all of this in detail, but in summary, if you observe an object moving relative to you, you will see:

1) Its time change (aging more slowly)
2) Its length contract in the direction of motion
3) Its clocks fall out of sync (leading clocks lag)

This book covers special relativity in two chapters: one on the basics and another on the paradoxes. That worries me. The maths of special relativity is so simple that it is easy to find shortcuts, but I am not sure that speed is your friend if you want a deep understanding of the subject. For those who want more, there is an excellent free online lecture series by Dr Larry Lagerstrom at Stanford University.[1] In contrast to Lagerstrom's detailed course, we will focus only on what we need for quantum mechanics: the invariant interval equation and a passing knowledge of the three distortions (listed earlier) that are consequences of special relativity. The maths is simple, but the concepts are mentally challenging. In short, you need the maths of an 11 year old to do the calculations, but the brain of an Einstein to take the idea seriously.

Special relativity throws up paradox after paradox. For example, if I observe a rocket pass me at high speed, then I see the rocket's clock running more slowly than mine. But from the perspective of an observer on the rocket, it is me that is moving and he/she sees my clock running more slowly. I mean *what*? And don't get me started on the Twin Paradox. Chapter 2 covers a few of the better known paradoxes, but you can skip it if you want to push ahead quickly. In our discussion on quantum mechanics, you will not need an in-depth understanding of special relativity as long as you have a grasp of the invariant interval equation.

1.2 The Speed of Light Is Constant: So What?

So what? So, a lot! To get to grips with special relativity it is essential you understand why the constant speed of light screws with physics. Let's start with a simple example. Imagine that you are in the park throwing a ball for your dog Pooch to chase. You want to know how fast Pooch and the ball move so you mark out evenly spaced stripes on the ground. You throw the ball which Pooch chases. One second later, you take a photo which allows you to see exactly how far the ball and Pooch have moved.

Suppose that your photo shows that after one second, the distance between you and Pooch is 5 metres and the distance between Pooch and the ball is 5 metres. This is illustrated at the top of Figure 1.1. From your perspective, you conclude that the speed of Pooch relative to you is 5 metres per second and that the speed of the ball relative to Pooch is also 5 metres per second. Classically you would conclude that Pooch also sees the ball moving away from him (for he is a male dog) at 5 metres per second. I am sorry to tell you that you would be wrong.

The good news is that you would only be very, very, very slightly wrong at these speeds. However, at higher speeds the discrepancy becomes significant. Let me illustrate by introducing you to turbo-Pooch, an imaginary dog that can travel at half the speed of light. You repeat your experiment, but this time you flash a pulse of light for turbo-Pooch to chase. Turbo-Pooch runs at half the speed of light ($0.5c$). Again, you take a photo after one second. At that moment, there are 150,000 kilometres between you and turbo-Pooch and 150,000 kilometres between turbo-Pooch and the light pulse. This is illustrated at the bottom of Figure 1.1.

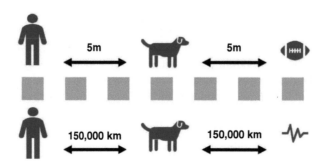

Figure 1.1 The problem with a constant speed of light.

You conclude that turbo-Pooch moved 150,000 kilometres in one second which is $\frac{c}{2}$, and that the light pulse moved 300,000 kilometres in one second which is c. That all makes sense, but what about the speed of the light pulse relative to turbo-Pooch? The distance between them is 150,000 kilometres so it must be $\frac{c}{2}$ *from your perspective*. All of this is true, but the final three words in the last sentence are crucial so I will repeat them: *from your perspective*.

Let's switch now and examine what happens from turbo-Pooch's point of view. In physics this is called *changing reference frame*. If turbo-Pooch measures how fast the light moves away from him, it *must always* be c. That is Einstein's law. Therefore, turbo-Pooch, making the same measurement as you, sees the light pulse 300,000 kilometres away after one second. This is very different from the 150,000 kilometres after one second that you measure.

I want to be clear that this conundrum is unavoidable if the speed of light is constant. Let me repeat it. If, from your perspective, turbo-Pooch is moving in the direction of the light pulse at $\frac{c}{2}$, you will measure the distance between dog and light pulse as 150,000 kilometres after one second on your clock, but turbo-Pooch must measure it as 300,000 kilometres after one second on his clock because that is the speed of light.

What can possibly explain these two very different views of the same thing? The reason is that your and turbo-Pooch's measurements of time and space differ. In terms of time, your two clocks run at different speeds. The two of you disagree on how long a second is. In terms of space, you will measure distances differently. The two of you will disagree on the distance between the stripes on the ground. Let's take a look at the mathematics of these distortions.

1.3 The Invariant Interval Equation

The speed of light is constant for all observers. As a result, time and space are distorted when viewed by a moving observer. We can derive the underlying mathematics of this distortion by considering how a model clock looks if viewed stationary versus in motion. Our clock is going to be a light clock. This simplifies the mathematics because we know the speed of light is the same for all observers. So the light in a stationary clock travels at c and, if you observe that clock moving relative to you at velocity v, you will still see the light in it travel at c. Simple!

So, let's consider clocks that work by emitting a flash of light up and down. The clocks tick every time the light flash arrives. We measure the time between each tick for a stationary clock. I will label this with a bold capital **T**, so a short moment on the stationary clock is $\Delta\mathbf{T}$. This is shown as (a) in Figure 1.2[2].

If we take an identical clock and observe it fly by at high speed, then from our perspective the clock physically moves along between each tick. We track the path of the *light* between ticks. It is slightly longer than for the stationary clock. This is shown as (b) in Figure 1.2. As the light travels a longer path in the moving clock and the speed of light is constant, the time between ticks for the moving clock must be longer. To us, the moving clock ticks more slowly. We will call the time between ticks of the moving clock Δt.

The relationship between the tick of a stationary clock ($\Delta\mathbf{T}$) and a moving clock (Δt) can be determined from the triangle in the bottom right hand corner of Figure 1.2. The distance travelled by the light in the stationary clock (a) is

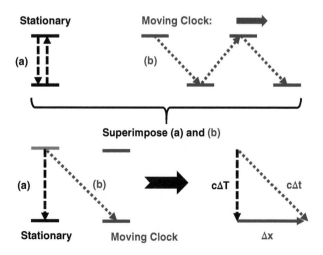

Figure 1.2 Determining the distortion of time and space.

the time taken multiplied by the velocity of light: $c\Delta \mathbf{T}$. By a similar calculation, the light in the moving clock (*b*) travels $c\Delta t$. The clock itself moves Δx distance over the same time. Simple Pythagoras gives us: $(c\Delta \mathbf{T})^2 + (\Delta x)^2 = (c\Delta t)^2$. This is called the invariant interval equation and, now that we have derived it, we have the maths of special relativity at our mercy. It is more typically written as Equation 1.1 (we use notation such as dx to indicate an infinitesimally small change).

$$c^2\,d\mathbf{T}^2 \ = \ c^2\,dt^2 - dx^2 \tag{1.1}$$

This represents an enormous shift in the maths of space and time. In Newtonian physics time was assumed to be *absolute*. We learn now that the tick of a clock (i.e. the passage of time) changes for an object that is observed moving. We have a new understanding of space and time. Physicists refer to this as *Minkoswki spacetime*. Equation 1.1 is shown using only one spatial dimension. The motion of the clock can be extended simply to three spatial dimensions. The important thing is the total distance that the clock travels during the interval, so the full invariant interval equation becomes:

$$c^2\,d\mathbf{T}^2 \ = \ c^2\,dt^2 - dx^2 - dy^2 - dz^2 \tag{1.2}$$

The quantity on the left, $c^2\,d\mathbf{T}^2$, and therefore also $d\mathbf{T}$, is invariant for all observers. Consider a certain light clock that is assessed by multiple observers travelling past it at different speeds. Each will measure it to have a different length of tick. But they can all calculate and agree on the invariant interval $d\mathbf{T}$ which is the length of the clock tick as measured by an observer who is stationary relative to the clock. This is called *proper time*.

The invariant interval, $d\mathbf{T}$, can be expressed in units of space rather than time, often shown as ds. And the equation can be shown with dx^2, dy^2, and dz^2 positive and the time components negative (see Box 1.1). It is not that the square of time or space is actually *negative*. That can never be. It is that time and space *offset* each other in terms of the invariant interval.

Box 1.1   The invariant interval equation and the metric

The invariant interval equation quantifies the relationship between space and time so it defines Minkowski spacetime. It has several common forms such as:

1) $c^2d\mathbf{T}^2 \ = \ c^2dt^2 - dx^2 - dy^2 - dz^2$
2) $ds^2 \ = \ -c^2dt^2 + dx^2 + dy^2 + dz^2$

The crucial thing is that the left hand term in the equation is invariant for all observers. In the first version, this invariant interval is expressed in units of time, in the second in units of space. Physicists refer to the relationship between time and space as the *metric*. They use the notation $< + - - - >$ and $< - + + + >$ to indicate if time or space is positive. While they are mathematically equivalent, you must be careful to remember which you are using and to be consistent in order to avoid $+/-$ errors appearing in your work.

1.4   Time Distortion Quantified

We have seen that time dilates for a moving object in the sense that a second of time that we observe on a moving clock will be longer than a second on the same clock when it is observed stationary. If we observe a moving clock, or man, or woman, or anything, we will see each second of its/his/her life last a little longer than a second of our life. Moving things age more slowly.

How does the speed of motion change the observed time, dt versus the proper time, $d\mathbf{T}$? We can derive this from the invariant interval Equation 1.1, shown again below for simplicity in one spatial dimension (i.e. the clock is moving in the x direction). With some simple maths, we can derive the relationship between dt and $d\mathbf{T}$ as shown in Equation 1.4. Remember that $\frac{dx}{dt}$ is velocity v.

$$c^2\, d\mathbf{T}^2 \;=\; c^2\, dt^2 - dx^2 \tag{1.3}$$

$$\frac{c^2\, d\mathbf{T}^2}{c^2\, dt^2} \;=\; 1 - \frac{dx^2}{c^2\, dt^2}$$

$$\left(\frac{d\mathbf{T}}{dt}\right)^2 \;=\; 1 - \frac{v^2}{c^2}$$

$$\left(\frac{dt}{d\mathbf{T}}\right)^2 \;=\; \frac{1}{1 - \frac{v^2}{c^2}}$$

$$dt \;=\; \sqrt{\frac{1}{1 - \frac{v^2}{c^2}}}\; d\mathbf{T}$$

$$dt_{\text{(moving)}} \;=\; \gamma \; d\mathbf{T}_{\text{(static)}} \qquad \textit{Where time dilation is:} \quad \gamma \;=\; \sqrt{\frac{1}{1 - \frac{v^2}{c^2}}} \tag{1.4}$$

From now on we will use the generally accepted physics notation γ (pronounced gamma) for the time dilation effect as shown as Equation 1.4. γ varies from 1 when $v = 0$ up to infinity when $v = c$ (i.e. if the clock moved at light speed). It is helpful to remember that γ is always greater than 1. The second/hour/day we observe on a moving clock is always longer than the second/hour/day on a stationary clock. It is worth reminding ourselves why, by quickly referring back to Figure 1.2 and remembering that we see the flash of light in our moving light clock travel further than that of an identical stationary clock.

How does this affect elapsed time? It is important to get your head around this. It means that the elapsed time shown on a moving clock is *shorter* than the elapsed time on the stationary clock. For a moving clock, we see time dilate. Each second/day/year is longer so if something is moving, it ages more slowly. I find it helps to think of time dilation as the moving clock *slowing down*. Consider a scenario where $\gamma = 1.5$ (about $0.75c$ or 220,000 kilometres per second). We see each second on the moving clock take 1.5 seconds on our stationary clock. By the time four years have passed for the moving clock, our stationary clock reads 6 years. Elapsed time is shorter for something moving. For elapsed time the relationship is:

$$dt_{\text{(moving)}} \;=\; \gamma \, d\mathbf{T}_{\text{(static)}} \qquad\Longrightarrow\qquad t_{\text{(elapsed)}} \;=\; \frac{\mathbf{T}_{\text{(elapsed)}}}{\gamma} \tag{1.5}$$

This can be confusing. Try to remember that *proper time* (**T**) has the shortest seconds so elapsed *proper time* is always longest. How big is this distortion? We can quantify the effect at different velocities using γ to calculate the discrepancy in elapsed time at various speeds. This is shown in Table 1.1.

Table 1.1 The effect of time dilation.

Speed	Time Discrepancy	Scenario
0	None	Stationary
360 km/h	1 second per 600,000 years	Formula 1 racing car
1800 km/h	1 second per 20,000 years	Supersonic jet
40,000 km/h	1 second per 50 years	Spacecraft (Apollo 11)
30,000 km/s ($0.1c$)	1 second per 3 days	Almost round the world in 1 second
$0.5c$	1 second per 6 seconds	Half the speed of light
$0.998c$	Time x 15	Muon created by cosmic ray
$0.99999999c$	Time x 7,000	Proton in the Large Hadron Collider

The effect is small even for objects that we consider to be moving fast, so it is not surprising that the distortion was missed pre-Einstein. However it becomes very significant at high velocities. The muon example in Table 1.1 is interesting. Muons are created when cosmic rays hit the earth's upper atmosphere. They are very short lived (about 2 microseconds) and can be detected at the earth's surface only because their high speed increases their lifespan 15-fold from our perspective.

For non-relativistic velocities we can use the approximation for time dilation shown in Equation 1.6 which is $\gamma \approx 1 + \frac{1}{2}\frac{v^2}{c^2}$. This is based on a Taylor expansion (see Box 1.2). This approximation is accurate to within 1 in 100 (1% error) up to 30,000 kilometres per second ($0.1c$). At slower speeds it is even more accurate. Up to speeds of 2,000 to 3,000 kilometres per second, the approximate speed of electrons in a hydrogen atom, it is good to better than one part in 10,000 (0.01% error).

$$\frac{dt}{d\mathbf{T}} = \gamma = \sqrt{\frac{1}{1 - \frac{v^2}{c^2}}} \approx 1 + \frac{1}{2}\frac{v^2}{c^2} \qquad (1.6)$$

Box 1.2   Taylor expansions are useful for making approximations

We can use a Taylor expansion to approximate γ when v is small compared to c. If you have not come across Taylor expansions, it is worth reading about them although we will use only a couple of simple ones in this book. They express a function in terms of an infinite series that can be used to create useful approximations. For γ, we use the following Taylor series:

$$\sqrt{\frac{1}{1-x}} = 1 + \frac{1}{2}x + \frac{3}{8}x^2 + \frac{5}{16}x^3 \ldots \quad \Rightarrow \quad \sqrt{\frac{1}{1-\frac{v^2}{c^2}}} = 1 + \frac{1}{2}\frac{v^2}{c^2} + \frac{3}{8}\frac{v^4}{c^4} + \frac{5}{16}\frac{v^6}{c^6} \ldots$$

At non-relativistic speeds, terms of $\frac{v^4}{c^4}$ and higher can be ignored so: $\gamma \approx 1 + \frac{1}{2}\frac{v^2}{c^2}$.

This raises an obvious question. What happens if we were able to accelerate an object up to the speed of light? In this case, γ would increase to ∞ (the symbol for infinity) so we would see time for the object slow down and finally stop. Clearly our concept of time does not apply to light in any normal way.

1.5   Length Distortion

In addition to the distortion in time, there is a distortion in length. As an object moves faster, a stationary observer will see its length shorten in the direction of motion. Imagine somebody flying by at very high speed holding a metre ruler upwards (i.e. not in the direction of motion). The metre would appear the same length to both of you. However, if our high speed friend shifts the metre ruler so it is pointing forwards (i.e. in the direction of motion) then you will see it shrink in length. This is called length contraction. Why does it happen?

Box 1.3   Just for laughs: Einstein and Newton are in a bar …

Einstein says to Newton, *I've discovered that time and space are distorted if you travel at high speed, because the speed of light is constant for all observers however fast they are moving.*

Newton replies, *Do go on.*

Einstein explains, *Imagine two people on a train moving quickly and a flash of light …*

Newton interrupts, *What the hell is a train?*

I wonder what Newton would have thought of all this (see Box 1.3 for a laugh). Let me give you a flavour of what is going on (the paradoxes in Chapter 2 give more detail if needed). Two observers always agree on the relative velocity between them. This is at the heart of relativity because both observers have an equivalent right to consider themselves stationary and the other as moving. Suppose our two observers both measure *v* as they pass. Each makes a measurement of distance in the direction of motion and elapsed time, then calculates the velocity *v*. One of our observers looks at his measurements and compares them with those of his passing friend. As per Equation 1.5 elapsed time will be distorted. For the two observers to agree on *v*, distance in the direction of motion (shown as *d* below) must be distorted in the same way as *elapsed* time:

$$t_{(elapsed)} = \frac{T_{(elapsed)}}{\gamma} \quad d_{(moving)} = \frac{d_{(static)}}{\gamma} \quad \Longrightarrow \quad \frac{d_{(moving)}}{t_{(elapsed)}} = \frac{d_{(static)}}{T_{(elapsed)}} = v \tag{1.7}$$

The distortion of length is small for everyday objects. The supersonic jet listed in Table 1.1 shrinks in length about one part in a billion billionth, but it is significant at higher speeds. For example, a 100 metre long rocket passing at 0.5*c* will be shrunk to about 87 metres due to this length contraction. Similarly, if you are stationary relative to a planet 10 light years away and then suddenly speed up towards it at 0.5*c* (or you can consider it moving towards you), the distance from you to the planet shrinks to 8.7 light years. If you are unfamiliar with a light year as a measure, check out Box 1.4.

Box 1.4 What is a light year?

A light year is a measurement of *distance*, not time. It is the distance that light travels in a year and is equivalent to about 9,500 billion kilometres. For context, that is over 60,000 times further than the distance to the sun which is about 8 light minutes away.

1.6 Leading Clocks Lag

Another important effect of special relativity is that clocks placed apart on a moving object appear out of sync. Consider the following scenario (see Figure 1.3). Two clocks (A and B) are synchronised using a flash from a light exactly between them. When the flash of light hits the clocks, they are programmed to show the same time, say 12.00. Somebody stationary relative to the clocks (as in the top of Figure 1.3) sees them trigger at the same time – perfectly in sync.

How does this appear to an observer who sees the clocks moving at high speed to the right (as in the lower part of Figure 1.3)? Things look very different. Clock B moves to the right as the flash of light travels towards it. The flash of light arrives at clock B before the flash of light arrives at clock A. To that observer, clocks A and B are *not* in sync ... clock A is slightly behind clock B.

In order to quantify how much moving clocks are out of sync, let's focus on clock B (see Figure 1.4). For the stationary observer, the distance from the light to clock B is *d* and the flash of light takes time $\frac{d}{c}$ to travel to clock B. Now let's consider a rocket moving at high speed that passes at exactly the same moment that the light flashes (if you worry about making this simultaneous, you can imagine the rocket hitting something that triggers the light to flash just as the rocket passes). A spaceman on the rocket has the right to consider that he is stationary and that clock B is rushing towards him. In Figure 1.4 the speed of the rocket is shown as 0.4*c* so there is a 40% discrepancy between the distance that the spaceman sees the flash of light travel versus that seen by the stationary observer.

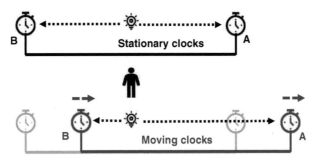

Figure 1.3 Leading clocks lag: if the clocks are moving, the synchronising light flash must travel further to reach the leading clock, so the time on that clock lags.

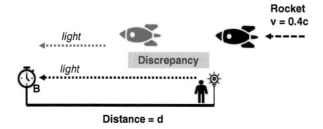

Distance = d

Figure 1.4   For the stationary observer the light takes time $\frac{d}{c}$ to reach B. From the perspective of the rocket observer, clock B is moving towards him. At 0.4c the light travels 40% less distance. The discrepancy in timing is: $\frac{d}{c} \times \frac{v}{c} = \frac{dv}{c^2}$.

If the rocket travels at 0.4c the discrepancy is 40% of the path, so 0.4d. This means that the spaceman sees the light hit clock B at a time $0.4\,\frac{d}{c}$ earlier than the stationary observer sees. If the rocket was travelling at 0.5c the discrepancy would be 50% of the path. The discrepancy in the length of the path d is the velocity of the rocket expressed as a fraction of the speed of light. From that, it is simple to calculate the time discrepancy:

$$Length\ discrepancy = \frac{v}{c}\,d \quad \Longrightarrow \quad Time\ discrepancy = \frac{v}{c}\frac{d}{c} = \frac{dv}{c^2} \tag{1.8}$$

If two clocks are synchronised when stationary, then if those clocks are viewed moving, the leading clock will lag the rear clock by $\frac{dv}{c^2}$ where v is the speed of motion and d is the *proper distance* between the clocks in the direction of the motion, that is, the length of the distance between the two clocks when stationary (see Box 1.5).

Box 1.5   To calculate how leading clocks lag, always use *proper distance*

When calculating synchronisation, always use the distance as calculated by an observer who is stationary relative to the clocks so your result is not distorted by length contraction.

1.7   Lorentz Transformations and Invariance

To express the coordinates of something in a different inertial frame (i.e. for an observer moving at a different speed) you need to adjust for these distortions. This process is called *Lorentz transformation*.

Consider two observers, 1 and 2, with Observer 2 moving at velocity v in the x-direction relative to Observer 1. If the observers record the movement of an object such as a rocket through spacetime, they are going to measure its track through space and time differently. How can we quantify this? Let's say that Observer 1 measures coordinates (t, x, y, z) while Observer 2 measures coordinates (t', x', y', z'). Providing we set the origins of both $(0, 0, 0, 0)$ at the same point in spacetime, we can transform between the two as follows:

$$x' = \gamma\,(x - vt) \qquad y' = y \qquad z' = z \qquad t' = \gamma\left(t - \frac{vx}{c^2}\right) \tag{1.9}$$

The y' and z' coordinates are the same as y and z because the motion is only in the x direction. The x' and t' coordinate values are affected by the classical effect of the motion along the x axis *and* by the distortions that are reflected in the value of γ. If you have understood these distortions that were discussed in Section 1.4, then the adjustments to the x and t coordinates should make some sense:

$$x' = \overset{b}{\overbrace{\gamma}}\ (x - \overset{a}{\overbrace{vt}}\ ) \qquad\qquad t' = \overset{d}{\overbrace{\gamma}}\ (t - \overset{c}{\overbrace{\frac{vx}{c^2}}}\ )$$

a: Adjust x to account for movement in x direction of vt over time t
b: Adjust scale for length distortion with velocity
c: Adjust for change in sync of clocks (leading clocks lag)
d: Adjust scale for time distortion with velocity

To transform the other way round (that is from Observer 2 to Observer 1) is easy. Simply change the relative velocity between the observers in the equations to $-v$ instead of $+v$. This has no effect on γ because its value depends on v^2 which is unaffected by a change in the sign of v. Just in case you need it, a derivation of the Lorentz transformation is shown in Box 1.6.

Things that are the *same* for all observers, such as the speed of light c, are not changed when you do a Lorentz transformation. They are called *Lorentz invariant* or sometimes *Lorentz scalars*. This is an important concept and you will see the term *Lorentz invariant* used frequently in this book when something is the same for all observers.

Box 1.6   **Derivation of Lorentz transformation**   **Optional**

For this exercise we ignore the y and z coordinates that are unchanged by movement in the x direction. First, we describe the Lorentz transformation as shown below on the left with K, L, M and N undefined. We can quickly determine L and N. Consider the two observers viewing a moving object (at less than light speed). If we choose Observer 1 to be stationary relative to the object then $x = 0$ for all values of t. In this scenario, the object is moving at v from the perspective of Observer 2. We can deduce two things from this. The time dilation from Equation 1.5 shows that $t' = \gamma t$ so $L = \gamma$. From this we can deduce that $x' = vt' = v\gamma t$ so $N = v\gamma$:

$$t' = Kx + Lt \quad x' = Mx + Nt \quad \implies \quad t' = Kx + \gamma t \quad x' = Mx + v\gamma t$$

We then use the invariant interval relationship to derive expressions relating K and M:

$$c^2t^2 - x^2 = c^2(t')^2 - (x')^2 = c^2(Kx + \gamma t)^2 - (Mx + v\gamma t)^2$$
$$= c^2(K^2x^2 + 2K\gamma xt + \gamma^2 t^2) - (M^2x^2 + 2v\gamma Mxt + v^2\gamma^2 t^2)$$
$$= t^2(c^2\gamma^2 - v^2\gamma^2) + x^2(c^2K^2 - M^2) + 2\gamma xt(c^2K - vM)$$

Equating the t^2 term gives: $c^2\gamma^2 - v^2\gamma^2 = c^2$. This is true as you can check with the definition of γ in Equation 1.4. The x^2 and xt terms give two equations for K and M:

$$-1 = c^2K^2 - M^2 \quad \implies \quad K^2c^2 = M^2 - 1$$

$$0 = c^2K - vM \quad \implies \quad K = \frac{Mv}{c^2} \quad \implies \quad K^2c^2 = \frac{M^2v^2}{c^2}$$

This allows us to calculate the values of M and K:

$$\frac{M^2v^2}{c^2} = M^2 - 1 \implies M = \sqrt{\frac{1}{1 - \frac{v^2}{c^2}}} \quad \implies \quad M = \gamma \quad K = \frac{\gamma v}{c^2}$$

... finally revealing the full Lorentz transformation:

$$t' = \gamma t - \frac{\gamma vx}{c^2} \quad x' = \gamma x - \gamma vt \quad y' = y \quad z' = z$$

1.8   Summary: Are You Joking Mr Einstein?

The last few sections are quite heavy going but we have arrived. We have covered all three of the distortions. Let's summarise them. For an object observed in motion:

1) Time dilation so each second is longer and elapsed time is shorter: $dt_{(moving)} = \gamma \, dT_{(static)}$

2) Lengths appear shorter in the direction of motion: $\text{distance}_{(moving)} = \dfrac{\text{distance}_{(static)}}{\gamma}$

3) Clocks appear out of sync as leading clocks lag by: $\dfrac{\text{distance}_{(static)} \, v}{c^2}$

These distortions are forced on us if we want a consistent framework of space and time with the speed of light constant. What a mess this makes of things. Are you joking Mr Einstein? But it is no joke. This is how things work.

This chapter started with two simple and sensible sounding postulates: *the speed of light is constant for all observers* and *the laws of physics are the same for all observers*. But for these postulates to hold, we must radically change our understanding of time and space.

Our analysis used scenarios with clocks that work on flashes of light. This simplifies things because the speed of light is constant for all observers so the law of addition of velocities is simple when light is involved ($c + v = c$), that is, add any speed to c and you still get c. Comparing stationary and moving light clocks gave us the invariant interval equation and led us to three distortions that occur to objects that are moving: time dilation, length contraction and moving clocks appearing out of sync. Einstein's two simple postulates wreak havoc on the old common sense view that space and time are independent. Newtonian space is replaced by Minkoswki spacetime, a new, more intricate picture of the fabric of the universe (see Box 1.7).

Box 1.7   Hermann Minkowski (1864 – 1909)

Minkowski, one of Einstein's professors, famously introduced his lecture on spacetime with:

Henceforth space by itself, and time by itself, are doomed to fade away into mere shadows, and only a kind of union of the two will preserve an independent reality.

He died suddenly of appendicitis at only 44 years old.

To follow the coming modules on quantum mechanics, you do *not* need to remember the precise details of these distortions. You just need to be aware that they exist. However, we will be using the invariant interval equation (as shown earlier in Equation 1.1). It shows the relationship between an interval of time, such as a second, for an object observed moving (dt) versus the *proper time* of the same object when observed stationary ($d\mathbf{T}$):

$$c^2\,d\mathbf{T}^2 \;=\; c^2 dt^2 - dx^2 - dy^2 - dz^2 \tag{1.10}$$

And keep in mind the time dilation relationship that we derived from the invariant interval in Equation 1.6.

$$\frac{dt}{d\mathbf{T}} \;=\; \gamma \;=\; \sqrt{\frac{1}{1 - \frac{v^2}{c^2}}} \;\approx\; 1 + \frac{1}{2}\frac{v^2}{c^2} \tag{1.11}$$

Chapter 2 covers some of the strange paradoxes that surface when you start to think more deeply about the distortions of Minkowski spacetime. This material is not essential in any study of quantum mechanics, but it is fun. At least, I think so. Even Punch magazine thought so (see Box 1.8).

Box 1.8   There was a young lady …   *Limerick published in Punch magazine, 1923*

There was a young lady named Bright
Whose speed was far faster than light;
She set out one day
In a relative way
And returned on the previous night.

Notes

1 https://online.stanford.edu/courses/som-y0009-understanding-einstein-special-theory-relativity (as at December 2022).

2 Technical note: you may see τ used for this in other books, but I am saving the symbol τ for a slightly different role when we discuss energy, momentum, and then get into quantum mechanics.

Chapter 2

Paradoxes of Special Relativity

CHAPTER MENU
2.1 Journey to a Distant Planet (1)
2.2 Journey to a Distant Planet (2)
2.3 The Twin Paradox
2.4 Experimental Proof

Consider the following paradox – in this chapter, we will show how the three distortions of special relativity discussed in Chapter 1 combine and then we will address the infamous Twin Paradox. For the examples in this chapter, I will use an Earth man compared with a Rocket woman passing at high speed. No sexism is intended. It allows me to distinguish using *he* and *she*. That helps the writing flow a little more smoothly.

2.1 Journey to a Distant Planet (1)

Length contraction in the direction of motion is essential for special relativity to be consistent. Consider the following that is illustrated in Figure 2.1. Earth man is stationary relative to a planet orbiting a star that is 8 light years distance away (don't forget that a light year is a measure of distance, not time). A rocket passes him travelling towards the planet at $0.8c$. By Earth man's clock, the rocket takes 10 years to reach the planet (8 light years $\div$ $0.8c$ = 10 years). But the rocket clock is moving relative to him so it runs more slowly due to time dilation. At $0.8c$, γ is about 1.67 so elapsed time for Rocket woman on her clock upon arrival at the planet is only 6 years versus the 10 years shown on Earth man's clock (10 years $\div$ 1.67 = 6 years). This is our first paradox. Has Rocket woman travelled to a planet 8 light years away in 6 years? Isn't that faster than the speed of light?

The solution is to compare the perspective of Earth man versus Rocket woman in terms not only of time but also of distance. Earth man is stationary relative to the planet so we do not need to worry about length contraction. His clock shows 10 years for the rocket to travel the 8 light years to the planet. This is consistent as the rocket travels at $0.8c$ relative to him and the planet.

Rocket woman's perspective of distance is very different. She sees the planet rushing towards her at $0.8c$ so the distance to the planet is contracted by γ (1.67). For Rocket woman, the planet is only 4.8 light years away, not 8 light years away. Her rocket clock shows 6 years on arrival consistent with her view that she has travelled 4.8 light years at $0.8c$. So, both parties agree that the rocket's velocity is $0.8c$ relative to the planet and that the rocket does not travel faster than the speed of light. Taking into account length contraction resolves the paradox.

But something emerges that is surprising to many. Suppose you want to travel in your lifetime to a planet 500 light years away. Is this allowed by special relativity? The answer is *yes*. For example, if you could accelerate up to $0.998c$ (similar to the muon in Table 1.1) then the distance to the planet would contract from 500 light years to about 30 light years. At $0.998c$ you in your rocket would arrive in just over 30 years on your clock. But for me watching your progress from here on earth,

Quantum Untangling: An Intuitive Approach to Quantum Mechanics from Einstein to Higgs, First Edition. Simon Sherwood.
© 2023 John Wiley & Sons Ltd. Published 2023 by John Wiley & Sons Ltd.

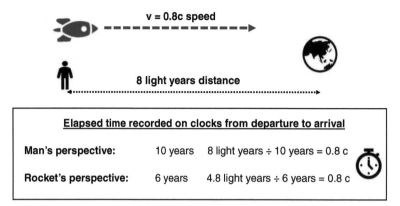

v = 0.8c speed

8 light years distance

Elapsed time recorded on clocks from departure to arrival		
Man's perspective:	10 years	8 light years ÷ 10 years = 0.8 c
Rocket's perspective:	6 years	4.8 light years ÷ 6 years = 0.8 c

Figure 2.1   Does the rocket move faster than the speed of light? Only 6 years passes on the rocket clock while, from the man's perspective, the rocket travels 8 light years distance. The solution to the paradox is that length contracts by γ from the perspective of the rocket, so for Rocket woman the planet is only 4.8 light years away.

there is no shortcut so I would see your journey take just over 500 years and be worm food by the time you arrive unless you let me come along for the ride (please!). Of course this is all just theory. The energy involved in accelerating to such a speed is beyond fantastic (for more details on this, see Box 2.3 at the end of this chapter).

2.2   Journey to a Distant Planet (2)

At some stage, most students wonder at the symmetry of special relativity. If a rocket passes at great speed, the Earth man sees the rocket's clock run slowly. But from the perspective of Rocket woman, it is the earth that is rushing past so she will see the Earth man's clock run slowly. How can we reconcile this? This is our second paradox. Let's get to work.

Figure 2.2 gives more detail on the scenario we just studied. In addition to looking at time and distance for our two protagonists, we are going to look at clock synchronisation. This will prove to be very important.

Let's start with the Earth man's view of what happens (see the clock times in the left box of Figure 2.2). The Earth man is stationary with respect to the distant planet and has earlier synchronised his clock with a clock on the planet (we will call this the planet clock). As the rocket passes earth, both Earth man and Rocket woman mark the time on their clocks and set it as zero. So upon departure, the Earth man's view is that the earth, rocket and planet clocks all read zero. The Earth man sees the rocket travel a distance of 8 light years at 0.8c so it takes 10 years. However, the Earth man sees time dilate for the rocket clock. The time dilation factor γ is 1.67 so the elapsed journey time on the rocket clock is 6 years.

I want to pause for a moment to clarify an important issue. The box on the left of Figure 2.2 shows the *Earth man's view* of the clocks. Let's be clear what this means. The Earth man will not actually *see* the rocket arrive at the planet in 10 years time. No, no, no! If Earth man uses a telescope to watch, he will see the rocket arrive at the planet after 18 years on his clock because it will take 10 years journey time on his clock *plus* 8 years more for the light showing the rocket's arrival to get back from the distant planet to him on earth.

However, there are a couple of ways that Earth man can know at what time on his clock the rocket arrives at the planet. The simplest is that Earth man *calculates* that the rocket arrives after 10 years. He can do this by taking the 18 years and subtracting the 8 years light travel time. Alternatively, Rocket woman on arrival at the planet can take a photo of the planet clock reading 10 years and her rocket clock simultaneously reading 6 years, then send the photo back to Earth man. The earth and distant planet are stationary with respect to each other and their clocks are synchronised. Either way, Earth man can be categorical that the earth clock reads 10 years when the rocket arrives, albeit it will take him another 8 years to get the data.

I should add that it is fairly straightforward to come up with scenarios to synchronise clocks that are stationary relative to each other. For example, Earth man can bounce a light pulse off the distant planet and see how long it takes to get back.

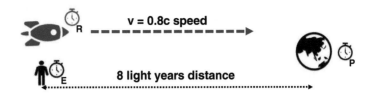

Earth man's Clock View			
(years)	**Depart**	**Arrive**	**Change**
Earth clock:	0	10	+10
Rocket clock:	0	6	+6
Planet clock:	0	10	+10

Rocket woman's Clock View			
(years)	**Depart**	**Arrive**	**Change**
Earth clock:	0	3.6	+3.6
Rocket clock:	0	6	+6
Planet clock:	6.4	10	+3.6

Figure 2.2 Different observers, different views. To make sense of the scenario, remember that the Rocket woman sees the planet clock advanced at departure i.e. ahead of the earth clock (leading clocks lag).

He can then use this information to beam a clock time to the planet noting the time adjustment needed so that the earth and planet clocks are in sync. And, if you are worried that the earth and planet may not be stationary relative to each other, Earth man can check for any Doppler shift in the original light pulse.

And how can Rocket woman know the clock times on the earth and planet (see the right box in Figure 2.2)? This is a bit more complex but still theoretically possible (and that is all we need). She could trail a synchronised clock with a camera the right distance behind her to take a photo of the earth clock upon her arrival at the planet. The trailing clock is stationary relative to the rocket so will stay in sync with the rocket clock. When she arrives at the planet, she sees her rocket clock reading and then hunts out the photo taken by the trailing clock at the same time. This photo will show what the earth clock was reading from her perspective. All of this is highly theoretical, but the only important thing is that this information is real and defined.

Let's get back to our paradox and turn to the Rocket woman (see the clock times in the right box of Figure 2.2). She sees the distant planet rushing towards her at 0.8c. This creates length contraction so the distance to the planet from her perspective is not 8 light years but 4.8 light years ($\gamma = 1.67$). At 0.8c the journey takes her 6 years (4.8 ÷ 0.8 = 6 years) so she agrees with the Earth man that her rocket clock reads 6 years upon arrival. That all makes sense (we went through this when we did the scenario earlier).

From Rocket woman's perspective the earth and planet are moving at 0.8c so she sees the earth and planet clocks run slowly compared to hers. During the journey, she sees only 3.6 years pass on the earth and planet clocks compared to the 6 years that pass on her clock ($\gamma = 1.67$).

How can the Earth man see 10 years elapse on the planet clock while the Rocket woman sees only 3.6 years elapse on the same clock? For the Rocket woman, the earth and planet are moving. It may help you to imagine a huge solid ruler connecting the earth and the planet. For the Rocket woman this ruler moves past her with the earth leading at the front and the planet at the rear. So she sees the earth and planet clocks out of sync (leading clocks lag). On departure she sees the earth clock set at zero, but for her, the planet clock already reads 6.4 years ($\frac{dv}{c^2} = 8 \times 0.8$). For our Rocket woman the planet clock reads 6.4 years on departure plus 3.6 years that elapse during the journey so the planet clock does indeed read 10 years when she arrives.

Upon arrival at the planet, from the Rocket woman's perspective the earth clock reads 3.6 years while the planet clock reads 10 years. Why? Well, she is still moving at 0.8c and leading clocks lag so, to her, the earth clock is still 6.4 years behind the planet clock. Everything is neat and consistent. This can be confusing, but please try to persevere. We will now use this same scenario to look at what happens if the Rocket woman turns round and flies back to earth: the Twin Paradox. One word of warning if you are required to do problems and calculations: while two observers will agree on their mutual relative velocity, do not forget that you cannot add velocities in the usual way (see Box 2.1).

Box 2.1   Addition of velocities in special relativity

This is a useful formula although we will not need it for our scenarios. Imagine Rocket A is travelling past us at velocity v. Rocket A fires out from its nose Rocket B at velocity u relative to Rocket A. How fast does Rocket B move relative to us? Classically the velocity relative to us (that we will call v') is: $v' = u+v$. This simple addition of velocities breaks down in special relativity and becomes:

$$v' = \frac{u+v}{1+\frac{uv}{c^2}} \qquad \textit{note that if } v = c, \textit{ then...} \qquad v' = \frac{u+c}{1+\frac{uc}{c^2}} = c\left(\frac{u+c}{c+u}\right) = c$$

As shown here, this is consistent with a constant speed of light for all observers. If something moves at the speed of light for one observer then it will move at c for all observers because if $v = c$ then $v' = c$. Even if u and v are both equal to c, the combined velocity is still c. You cannot travel faster than the speed of light.

2.3   The Twin Paradox

If you are still with me, then congratulations on working your way through. Your reward is the *big one*, the most famous of all, the Twin Paradox. For all the mysteries of special relativity, there is an enduring fascination with this paradox. Talk about special relativity at the dinner table and I can bet the conversation will soon veer towards the Twin Paradox. Why? Because it seems to fly in the face of logic and experience. I am sure you all know the basic paradox. Two twins live together on earth. One buys a round trip ticket on a rocket to a distant planet orbiting a star that is far far away. He returns to earth to find his earth-bound twin much older than him. It is a very cute paradox and it is a true requirement of special relativity.

Instead of twins, let's return to our Earth man and Rocket woman (as in Figure 2.2) and look at what happens if our Rocket woman turns around and returns to earth. The Twin Paradox says she will have aged less than the Earth man? Why? You might argue there is a symmetry between the Earth man and the Rocket woman. Relative to each other, they move apart and then back together. How can it matter which one moves? There is a crucial difference. In changing her direction of motion to return to earth, the Rocket woman changes which clocks are *leading*. This means that the synchronisation of the clocks changes. But we are getting ahead of ourselves.

Figure 2.3 shows the clock times for the outbound and return trips from the perspective of the Earth man and Rocket woman. From the Earth man's perspective (the left box in Figure 2.3), the return trip is exactly the same as the outbound. To him, the rocket takes 10 years each way so his clock reads 20 years when the Rocket woman arrives back. The Earth man sees the rocket clock equally slow on the return trip as the outbound (time dilation depends only on speed, not direction) so the rocket clock reads 12 years when Rocket woman returns. He has aged 20 years while she has aged only 12 years.

For the Rocket woman (the right box in Figure 2.3), the outbound and return trips also are similar in many ways. On the return trip, Rocket woman sees another 6 years pass on her clock while from her perspective 3.6 years pass on the earth and planet clocks just like the outbound leg. Her change in direction does not alter these elapsed times. But her switch in direction of travel does result in one massive change: on the return leg, the earth clock is no longer leading. It is this that drives the Twin Paradox.

On the outbound journey the rocket is travelling to the right (Figure 2.3). For the Rocket woman, the earth and planet move to the left with the earth leading. Leading clocks lag. On arrival at the planet, she is still going at 0.8c so the planet clock reads 10 years versus the lagging earth clock at 3.6 years. Now the *big* difference. Rocket woman turns around and accelerates to 0.8c in the *other* direction. The earth clock goes from being the leading clock to being the rear clock. For Rocket woman the earth and planet clocks are now out of sync the other way. The earth clock is no longer 6.4 years *behind* the planet clock but is now 6.4 years *ahead* of it, a change of 12.8 years. When Rocket woman arrives at the planet travelling to the right, for her the earth clock reads 3.6 years and the planet clock 10 years. Once she has turned and is going in the opposite direction, for her the earth clock reads 12.8 years more, that is, 16.4 years in total (see the underlined numbers in the right box of Figure 2.3).

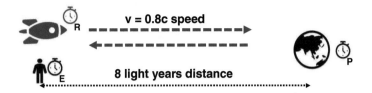

Earth man's Clock View *(years)*			
OUTBOUND	Depart	Arrive	Change
Earth clock:	0	10	+10
Rocket clock:	0	6	+6
Planet clock:	0	10	+10
RETURN	Depart	Arrive	Change
Earth clock:	10	20	+10
Rocket clock:	6	12	+6
Planet clock:	10	20	+10

Rocket woman's Clock View *(years)*			
OUTBOUND	Depart	Arrive	Change
Earth clock:	0	3.6	+3.6
Rocket clock:	0	6	+6
Planet clock:	6.4	10	+3.6
RETURN	Depart	Arrive	Change
Earth clock:	16.4	20	+3.6
Rocket clock:	6	12	+6
Planet clock:	10	13.6	+3.6

Figure 2.3 The Twin Paradox. The return trip is symmetrical with the outbound except the Rocket woman's change in direction means the earth clock is now the rear clock, adding 12.8 years.

In summary, Rocket woman returns younger than Earth man because it is she that turned around. The switch in direction does not affect time dilation or length contraction. That is why the elapsed times are the same for the outbound and return legs. But Rocket woman's switch in direction changes which clock (earth versus planet) is the leading clock to her. That changes the synchronisation of the clocks for her (see Box 2.2). It is this asymmetry between her and the Earth man that drives the Twin Paradox.

Box 2.2 Twin Paradox: the effect of Rocket woman changing direction

Rocket woman changes her velocity from $+0.8c$ to $-0.8c$. The change in direction has no impact on time dilation or length contraction because γ is affected by v^2 and that is unchanged if $+v$ becomes $-v$:

$$t_{(moving)} = \frac{T_{(static)}}{\gamma} \qquad d_{(moving)} = \frac{d_{(static)}}{\gamma} \qquad \gamma = \sqrt{\frac{1}{1-\frac{v^2}{c^2}}}$$

The change in direction *does* change the synchronisation of clocks seen in motion (leading clocks lag) as this depends on v, not on v^2:

$$Sync\ change\ =\ \frac{d_{(static)}\,v}{c^2} \qquad so\ if\ \ +v \Rightarrow -v \qquad \frac{d\,v}{c^2} \Rightarrow -\frac{d\,v}{c^2}$$

For Rocket woman the earth clock changes from 3.6 years to 16.4 years because of the change in her direction of travel. This raises a question. What actually happens? Theoretically, if she instantaneously jumps from velocity $0.8c$ to $-0.8c$, then for her the earth clock jump forwards. But that would require infinite acceleration so let's dig a bit deeper. In order to change her direction of travel, Rocket woman must accelerate towards the earth. This acceleration will take some time. As Rocket woman accelerates, her velocity will change $0.8c$ to $0.6c$ to $0.4c$, to zero, to $-0.6c$ and finally to $-0.8c$. Let's suppose that she takes a day for each $0.2c$ change of velocity. This is ridiculously fast and would squash her to a bloody pulp ... but bear with me. We can calculate what each clock reads for Rocket woman using our rule of *leading clocks lag*.

In Table 2.1, the rocket clock shows the days of acceleration measured by Rocket woman. The change in velocity does not affect her clock's synchronisation with the planet clock because she is right beside the planet ($d = 0$). But for her, the acceleration has a major impact on the earth clock. From her perspective the earth clock speeds up. For her, 1.6 years pass on the earth clock for every day on her rocket clock during this period of acceleration.

Table 2.1 Rocket woman's perspective of clock readings while accelerating beside the planet.

Velocity	+0.8c	+0.6c	+0.4c	+0.2c	Zero	−0.2c	−0.4c	−0.6c	−0.8c
Rocket clock *days*	0	1	2	3	4	5	6	7	8
Earth clock *years*	3.6	5.2	6.8	8.4	10.0	11.6	13.2	14.8	16.4
Planet clock *years*	10	10	10	10	10	10	10	10	10

Table 2.2 Rocket woman's perspective of clock readings while decelerating beside the earth.

Velocity	−0.8c	−0.6c	−0.4c	−0.2c	Zero
Rocket clock *days*	0	1	2	3	4
Earth clock *years*	20	20	20	20	20
Planet clock *years*	13.6	15.2	16.8	18.4	20

Let me emphasise again that Rocket woman does not actually *see* the time on the earth clock. She is light years away from it. But she can calculate (or in theory use an attached line of trailing clocks) to determine what the earth clock reads from her perspective.

At the end of her acceleration, Rocket woman is heading back to earth at 0.8*c*. Her rocket clock is still at 6 years, the planet clock at 10 years and the earth clock at 16.4 years (see Table 2.1). When she arrives back to earth, the rocket clock reads 12 years, the planet clock 13.6 years and the earth clock 20 years (see Figure 2.3). So, arriving back at earth, the earth clock and planet clock are out of sync from her perspective. But her twin on earth sees them in sync (see the left box of Figure 2.3). Oops! How can this be? What we have forgotten is that Rocket woman arrives at earth travelling at 0.8*c*. If she wants to stop and shake hands with her twin, she must first decelerate.

To complete the story we must consider the effect on synchronisation of this deceleration. Let us assume Rocket woman reaches earth and decelerates alongside it from −0.8*c* (shown negative to distinguish from the outbound leg) to 0 in 4 days. This is the same rate that we just modelled for her turn at the planet. During the deceleration process the distant clock (this time the planet clock) speeds up rapidly from her perspective. At the end of the deceleration, Rocket woman is stationary relative to the earth and planet so the *leading clocks lag* effect will have disappeared and she sees the earth and planet clocks back in sync (see Table 2.2).

At the end of our tale, with the rocket, earth and planet all stationary relative to each other, everybody experiences the same clock readings: 20 years for the earth and planet clocks versus 12 years for the rocket clock. The Twin Paradox is complete.

2.4 Experimental Proof

The more cynical among you may ask for proof. How can we possibly know? The most direct way is to stick a clock on a jumbo jet and fly it round the world. And yes, this has been done in the Hafele-Keating experiment. In 1971, four caesium atomic clocks were flown twice round the world with results that matched the time dilation from special relativity and the gravitational effect of general relativity. Rather impressively, this test cost under $8000 for the plane tickets (caesium clocks not included). Now that is science on a budget! Equally impressive evidence is the increase in the lifespan of particles at colliders such at the Large Hadron Collider in Cern. The particles move at close to light speed and have much longer half-lives in line with the prediction of special relativity. That experiment is a teensy weensy bit more expensive. The cost of the LHC was about $4.75 billion.

It is well worth spending time to fully understand the Twin Paradox. The significance of the change in synchronisation is the first hint of something very important. Acceleration has a profound effect on time. Technically this lies somewhat outside the scope of special relativity, but it gives a flavour of the issues that must have surfaced in Einstein's mind when he started to think about the interplay of time, acceleration and gravity. Perhaps this was one of the factors that started him in the search for his theory of general relativity, but that will have to be another book.

 In the next chapter we will focus on the implications of special relativity for energy and mass. Implications that, taken together with the distortions inherent in Minkowski spacetime, help to shape the weird world of quantum mechanics.

Box 2.3   Is interstellar travel feasible?

First, let's give a huge cheer for what we have achieved. With the Voyager missions, two spacecraft are on escape trajectories from the solar system (i.e. they're not coming back). Off they head into the void – not bad for a species that invented powered flight only 120 years ago. *But* – they are travelling at about 35,000 miles per hour. At that speed it would take 80,000 years to cover the 4 light years distance to our nearest neighbouring star, Alpha Centauri. We need to go much faster and reach speeds of about $0.1c$ for a meaningful mission (40–50 years). You might think that time dilation and length contraction would help a space traveller, as mentioned in Section 2.1, but both effects do not really play a role until speeds of over $0.5c$.

 Massive acceleration with an on-board power source faces the *rocket equation*: energy means fuel, fuel means more on-board mass and that requires more energy to accelerate. And the more efficient engine designs tend to offer lower acceleration. It took the ion engine of NASA's Dawn spacecraft 4 days to accelerate from 0 to 60 miles per hour. What about an off-board fuel source? Solar energy sounds good, but that is only effective near the sun unless you use a huge solar sail and that means more weight and we are back to the problem of the rocket equation ... Doh!

 One option with theoretical promise is to use an off-board laser with a long enough range to steadily accelerate a miniature probe during its journey to the edge of our solar system, launching it beyond at high speed. Challenging? Definitely. No laser of this power exists and you would need to avoid the laser burning up the probe itself, but still an idea with potential. So perhaps some day an unmanned probe will send back passing photos of a nearby star as it streaks by (we would have no way of slowing it). That would be a huge achievement and seems the very best we can hope for, short of a massive leap in technology; perhaps based around nuclear fusion, perhaps something completely new; given the achievements of the past, who dares discount the possibility?

Chapter 3

Einstein's Famous Equation

CHAPTER MENU

3.1 Mass, Energy, Momentum – and Particle Time
3.2 How Did Albert Figure It Out?
3.3 Three Beautiful Equations
3.4 How Wrong Were We?
3.5 One Further Equation
3.6 Summary

3.1 Mass, Energy, Momentum – and Particle Time

Since Einstein's original paper in 1905, over 100 years ago, the predictions of special relativity have been tested and validated. Atomic clocks are now accurate enough to show the effect of time dilation. As noted in Chapter 2, atomic clocks have been flown on jet airliners (including an around-the-world trip) and results are in line with those expected from time dilation (the Hafele-Keating experiment). To this we can add copious evidence of particle half-lives increasing when moving at speed in particle accelerators such as the Large Hadron Collider at CERN.

In this chapter, we will investigate how special relativity leads to a redefinition of the relationship between mass, energy and momentum. I hope I will be able to convince you that Einstein's famous equation $E = mc^2$, that relates energy E with mass m, falls out from the framework of special relativity in an obvious and intuitive way.

Before we start, it is helpful to introduce some new nomenclature for time. In the last chapter we compared the perspective of two observers, Earth man and Rocket woman. We considered how Rocket woman's clock (*her* proper time **T**) appears to Earth man. We then switched perspective to consider how Earth man's clock (*his* proper time **T**) appears to Rocket woman. Each observer has his own proper time which is the time on a clock stationary relative to him or her, so proper time **T** can be confusing unless we are explicit about whose proper time we are discussing.

In the rest of this book we discuss how moving particles and objects appear. To keep things clear, I replace **T** with a new symbol τ that I call *particle time*. It is the time measured on a clock that is stationary relative to the particle or object. So, from now on, remember that τ is the proper time of the particle or object, whereas the symbol t is the time measured by observers (see Box 3.1).

Box 3.1 Particle time τ

Particle time τ is the *proper time* of the particle or object under discussion.

Quantum Untangling: An Intuitive Approach to Quantum Mechanics from Einstein to Higgs, First Edition. Simon Sherwood.
© 2023 John Wiley & Sons Ltd. Published 2023 by John Wiley & Sons Ltd.

3.2 How Did Albert Figure It Out?

How did Albert Einstein jump from special relativity to $E = mc^2$? Incidentally, Einstein did win a Nobel prize but not for this equation (see Box 3.2). It seems a leap, but all the clues are there. Einstein knew that kinetic energy and momentum are not conserved quantities from the perspective of special relativity. A stationary object has no kinetic energy and no momentum. But viewed by a moving observer, the same object has both kinetic energy and momentum. How can energy and momentum be important underlying variables if they do not appear the same to all observers? Are they related? Could kinetic energy ($\frac{1}{2}mv^2$) and momentum (mv) be two different sides of the same thing? Einstein must have known that something was missing.

3.2.1 The Ingredients

I suspect that he figured things out very early on, for reasons I hope will become obvious to you too. This analysis requires only a couple of simple ingredients from the last chapter. We need to use the invariant interval equation for Minkoswki spacetime (see Equation 1.1) to compare the proper time τ of a moving object with the time t as measured by an observer. For simplicity this is shown in one dimension:

$$c^2\, d\tau^2 = c^2\, dt^2 - dx^2 \tag{3.1}$$

We also need γ, the relationship between dt and $d\tau$ (i.e. time dilation) and the approximation for γ at low velocities will also prove useful (see Equation 1.6 for the derivation if you need a reminder):

$$\frac{dt}{d\tau} = \gamma = \sqrt{\frac{1}{1 - \frac{v^2}{c^2}}} \approx 1 + \frac{1}{2}\frac{v^2}{c^2} \tag{3.2}$$

It might be more accurate to say we need one *ingredient*, rather than *ingredients*, because the formula for γ falls straight out of the invariant interval equation as we showed in Equation 1.4, so they are really one and the same thing. This is all we need to derive $E = mc^2$.

Box 3.2 Einstein's Nobel Prize (1921)

Einstein published his special theory of relativity in 1905 and was nominated for the Nobel Prize a total of 66 times (yes – 66!). The Nobel committee succumbed to pressure and awarded him the prize in 1921, but did so primarily for his work on the photoelectric effect rather than for his work on relativity. Oh, and they delayed the award until 1922 on the grounds that none of the nominations in 1921 really met the criteria.

Why? Doubts about relativity? Growing antisemitism in Germany? Jealousy? We may never know, but reportedly Einstein considered it a bit of a slap in the face.

3.2.2 The Calculation

If we want to consider how special relativity affects momentum, then an obvious thought is to look at the rate of change of the invariant interval equation. We must be careful that the left side of the equation remains invariant for all observers. For the first step, we divide by invariant *particle time* $d\tau$. Then, if we want a momentum term (mv), we need to develop a velocity term $\frac{dx}{dt}$ and introduce mass. We define mass m as *rest mass* to be sure it is invariant for all observers. This all can be done in short order and there is no way Einstein would have missed it:

$$c^2\, d\tau^2 = c^2 dt^2 - dx^2$$

$$c^2\, \frac{d\tau^2}{d\tau^2} = c^2\, \frac{dt^2}{d\tau^2} - \frac{dx^2}{d\tau^2} \tag{3.3}$$

$$c^2 = c^2\, \frac{dt^2}{d\tau^2} - \frac{dx^2}{dt^2}\frac{dt^2}{d\tau^2} \qquad \textit{Remember } \frac{dt}{d\tau} = \gamma$$

$$c^2 = c^2 \gamma^2 - v^2 \gamma^2 \tag{3.4}$$

$$m^2 c^2 = (mc\gamma)^2 - (mv\gamma)^2 \tag{3.5}$$

$$m^2 c^2 = (mc\gamma)^2 - (p\gamma)^2 \qquad \textit{Where p is classical momentum (mv)} \tag{3.6}$$

Equation 3.6 is still Lorentz invariant on the left side: $m^2 c^2$. On the right side there is a term $p\gamma$ that includes momentum and another term has appeared: $mc\gamma$. What might that be?

3.2.3 The Intuition

(1) What is $p\gamma$?

Look back at where it came from in Equation 3.3. Momentum p is mass multiplied by the rate of change of x with respect to time $\frac{dx}{dt}$. The term $p\gamma$ is mass multiplied by the rate of change of x with respect to invariant *particle time* $\frac{dx}{d\tau}$. To be clear, the term $p\gamma$ is a slightly different definition of momentum that is based on τ rather than t. The difference between p and $p\gamma$ is only significant at relativistic speeds when the value of γ diverges noticeably from 1. It is not a major leap to define $p\gamma$ as relativistic momentum (which we will label **p** in bold to distinguish it from the classical quantity).

Where does this take us? Look back again at Equation 3.6. On the left side we have $m^2 c^2$. That is two constants multiplied together and is Lorentz invariant, that is, the same for all observers. Therefore the right side of the equation must also, in total, be the same for all observers. The term $p\gamma$ looks exactly like classical momentum (mv) except at huge speeds. Clearly, this momentum term is *not* the same for all observers. The faster an observer is moving relative to an object, the higher the momentum he or she will measure. This increase must be offset according to Equation 3.6 by an increase in the other term that involves $mc\gamma$ so that the equation balances to the constant $m^2 c^2$. An observer passing at speed sees momentum increase *and* something else increase. Put yourself inside Einstein's head. Can you imagine the excitement mounting?

(2) What is $mc\gamma$?

Einstein must have wondered if $mc\gamma$ is related to relativistic energy. After all, energy and momentum both increase as an observer's speed increases. He would have searched for that telltale $\frac{1}{2}mv^2$ of kinetic energy. He knew that for non-relativistic velocities γ is approximately $\left(1 + \frac{1}{2}\frac{v^2}{c^2}\right)$ so:

$$mc\gamma \approx mc\left(1 + \frac{1}{2}\frac{v^2}{c^2}\right) = mc + \frac{1}{2}m\frac{v^2}{c} = \frac{1}{c}\left(mc^2 + \frac{1}{2}mv^2\right) \tag{3.7}$$

As velocity increases, the last term $(mc^2 + \frac{1}{2}mv^2)$, increases with $\frac{1}{2}mv^2$ so exactly as classical energy increases due to kinetic energy (see Box 3.3). Let's neaten things up and multiply both sides by the speed of light:

$$mc^2\gamma \approx mc^2 + \frac{1}{2}mv^2 \tag{3.8}$$

The obvious leap is that relativistic energy **E** is $mc^2\gamma \approx (mc^2 + \frac{1}{2}mv^2)$. At anything but enormous speed this increases in line with kinetic energy. By now, Einstein must have been breaking out the champagne (actually, he was not a big drinker and preferred cognac). Based on this we can substitute with $mc\gamma = \frac{E}{c}$. Using this and our definition of **p**, Equation 3.6 becomes:

$$m^2 c^2 = (mc\gamma)^2 - (p\gamma)^2$$

$$= \left(\frac{\mathbf{E}}{c}\right)^2 - \mathbf{p}^2$$

$$m^2 c^4 = \mathbf{E}^2 - \mathbf{p}^2 c^2 \tag{3.9}$$

Equation 3.9 gives the relationship between relativistic energy **E** and relativistic momentum **p** under the rules of special relativity. If you are not familiar with this relationship, then look long and hard. It will be a key factor when we move on to discuss quantum mechanics.

Einstein's famous equation for the energy of a stationary mass, $E = mc^2$, is easily derived from Equation 3.9 because at zero velocity, $\mathbf{p} = 0$. It can also be derived from the approximation for relativistic energy ($mc^2 + \frac{1}{2}mv^2$) setting v to zero.

This is not a mathematical proof, but it is intuitive and logical. Remember that all we have done is look at the rate of change of the invariant interval equation. Nothing more.

I suspect that Einstein figured this out quite quickly. He attempted to construct a pure mathematical proof of $E = mc^2$ based on light emission. This avoids the need for intuition, but his argument is seen by most as circular and, to me, is much less convincing than the simple approach seen earlier. As is usual in physics, the final proof of $E = mc^2$ came in the form of experimental results, and both it and the other tenets of special relativity have been confirmed by countless particle experiments.

Box 3.3   Taylor expansion for energy: $mc^2\gamma$

In Equation 1.6 we used a Taylor expansion to approximate γ when v is small compared to c:

$$\sqrt{\frac{1}{1-x}} = 1 + \frac{1}{2}x + \frac{3}{8}x^2 + \frac{5}{16}x^3 \dots \quad \Rightarrow \quad \sqrt{\frac{1}{1-\frac{v^2}{c^2}}} = 1 + \frac{1}{2}\frac{v^2}{c^2} + \frac{3}{8}\frac{v^4}{c^4} + \frac{5}{16}\frac{v^6}{c^6}\dots$$

so.. $$\mathbf{E} = mc^2\gamma = mc^2\sqrt{\frac{1}{1-\frac{v^2}{c^2}}} = mc^2\left(1 + \frac{1}{2}\frac{v^2}{c^2} + \frac{3}{8}\frac{v^4}{c^4} + \frac{5}{16}\frac{v^6}{c^6}\dots\right) = mc^2 + \frac{1}{2}mv^2 + \frac{3}{8}m\frac{v^4}{c^2}\dots$$

Terms containing $\frac{1}{c^2}$ and higher can be ignored at low speeds as they are insignificant, leaving us with a (very) accurate approximation of $\mathbf{E} = mc^2 + \frac{1}{2}mv^2$ at non-relativistic speeds.

3.3   Three Beautiful Equations

We have generated three equations that are at the heart of special relativity. I want to convey the beauty of these equations and how neatly they tie together. Remember that we use bold $\mathbf{E}$ and $\mathbf{p}$ for *relativistic* energy and *relativistic* momentum to distinguish from the classical quantities. The approximations that are shown work for non-relativistic speeds.

$$\mathbf{E} = mc^2\gamma \approx mc^2 + \frac{1}{2}mv^2 \tag{3.10}$$

$$\mathbf{p} = mv\gamma \approx mv \tag{3.11}$$

$$m^2c^4 = \mathbf{E}^2 - \mathbf{p}^2c^2 \tag{3.12}$$

What do our three equations tell us? They tell us to completely rethink the relationship between energy and momentum. Observers viewing at different velocities (or changing velocity) calculate different values for the energy and momentum of a system, but $\mathbf{E}^2$ and $\mathbf{p}^2c^2$ always increase exactly in tandem. Kinetic energy and momentum are simplified and linked through mass. They tie together in a way that nobody could have imagined. Yet, for an observer staying in one frame of reference (i.e. not accelerating), we still have conservation of energy and momentum, but with the twist that the definition of energy has been expanded to include mass at an exchange rate of $\mathbf{E} = mc^2$ with major consequences (see Box 3.4). As soon as Einstein saw all of this, he must have known for sure that his special relativity was not theory but fact.

These equations become even more compelling if switched to *natural units*. For natural units we scale the units so the speed of light c equals 1. In simple terms this means that velocity is shown as a fraction of the speed of light and if we measure time in seconds, we must measure distance in light seconds. It gets rid of all those pesky c factors in the equations. Equation 3.12 becomes the more elegant: $m^2 = \mathbf{E}^2 - \mathbf{p}^2$. And the relationship between momentum and energy reduces to velocity: $\frac{\mathbf{p}}{\mathbf{E}} = v$ (you can calculate this easily from Equations 3.10 and 3.11). At the speed of light in natural units $v = 1$, so $\mathbf{E} = \mathbf{p}$ and m must be zero. Only objects with zero rest mass can travel at the speed of light. So much information is contained in Equations 3.10, 3.11 and 3.12. They are indeed beautiful.

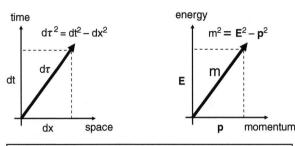

Figure 3.1   Spacetime versus energy-momentum space.

Equation 3.12 shows the relationship between mass, energy and the *total* momentum of an object. For completeness, I will show how to expand this to all three spatial dimensions. It is simple to do:

$$\mathbf{p_{total}}^2 \;=\; \mathbf{p}_x^2 + \mathbf{p}_y^2 + \mathbf{p}_z^2 \qquad so$$

$$m^2 c^4 \;=\; \mathbf{E}^2 - \mathbf{p}^2 c^2 \quad \Longrightarrow \quad m^2 c^4 \;=\; \mathbf{E}^2 - (\mathbf{p}_x^2 + \mathbf{p}_y^2 + \mathbf{p}_z^2)\,c^2 \tag{3.13}$$

Box 3.4 $\mathrm{E} = \mathrm{mc}^2$ and the atomic bomb

Progress in science is often controversial, but never more so than at the end of the Second World War. The atomic bomb that was dropped on Hiroshima in August 1945 unleashed on the city an explosive power equivalent to 15,000 tons of TNT. The mass that was converted to energy by the bomb is estimated at 0.7 grams, about the same as that of a paper clip.

It is worth pausing for a moment to reflect on the symmetry between the invariant interval equation and the equation we have developed for energy-momentum. This is shown in Figure 3.1 using natural units ($c = 1$). The left side of the figure shows spacetime coordinates. The right side of Figure 3.1 shows a similar framework, but using energy and momentum as coordinates and based on the relationship established in Equation 3.12.

The symmetry should not surprise you because we developed the energy-momentum relationship by considering the rate of change of the invariant interval equation. We derived one equation from the other. Later in this book you will learn that advanced (very advanced) quantum mechanics works predominantly with energy-momentum calculations. This is one way that quantum mechanics drifts away from the mathematics in Einstein's theory of gravity, and one of the challenges in drawing them back together. By the way, it is worth noting that what we call rest mass is not actually resting (see Box 3.5).

3.4   How Wrong Were We?

Let's focus back on Equation 3.10. Special relativity has shown us that our understanding of basic physics was wrong. Our definitions were wrong. Our ideas about the relationship between energy and momentum were wrong. Pre-Einstein, we happily defined kinetic energy as $\frac{1}{2}mv^2$. But how wrong were we? This is important because the classical definition of energy is baked into a lot of physics, even quantum mechanics (Schrödinger's equation is built on it). Thankfully, our non-relativistic approximations are very accurate at anything but extremely high speeds.

As you can see in Table 3.1, the classical approximation is extremely accurate with less than one in a hundred discrepancy even at 30,000 kilometres per second (which is about $0.1c$). The discrepancy is less than one part in 10,000 at $2,000$ to $3,000$ kilometres per second, the approximate speed of electrons in a hydrogen atom. However, at very high velocities such as

Table 3.1 How accurate is our classical definition of kinetic energy?

Speed	Kinetic Energy Discrepancy	Scenario
0	None	Stationary
360 km/h	0.00000000000008:1	Formula 1 racing car
1,800 km/h	0.000000000002:1	Supersonic jet
40,000 km/h	0.000000001:1	Spacecraft (Apollo 11)
2,200 km/s (0.007c)	0.00004:1	Electron in hydrogen atom
30,000 km/s (0.1c)	0.008:1	Almost round the world in 1 second
0.5c	0.2:1	Half the speed of light
0.99c	12.4:1	High speed accelerated particle

the example shown at 0.99c, the discrepancy becomes huge with the correct relativistic calculation being over 12 times the classical approximation.

Box 3.5   Rest mass: a massive misconception

Rest mass is the mass measured by an observer who is stationary relative to the centre of mass of the object. But in a composite object, the internal components are moving and interacting, so what most people think of as rest mass actually isn't resting at all.

The matter we interact with day to day is made of atoms that are in turn made of protons, neutrons and electrons. The mass of the electron is about 0.5 MeV (equivalent to about 10^{-30} kg) compared to over 900 MeV for each proton and neutron.

Each proton and neutron consists of three quarks that together have a rest mass of only about 9 MeV. How do we get from 9 MeV to over 900 MeV? The difference comes from the rapid movement of the quarks plus their interaction with other quarks.

Only a few percent of what we regard as atomic mass is due to the rest mass of the elementary constituents of the atom. The vast majority of atomic mass is the result of a hive of activity within the protons and neutrons.

3.5   One Further Equation

At the risk of blinding you with equations there is another important relationship that ties together energy-momentum and spacetime coordinates in Minkowski spacetime. Imagine that we have several observers track the path of an object, such as a particle or rocket, as it moves through spacetime. We assume the object moves at sub-light speed and is not accelerating. If we set the origin of the coordinates of all the observers at $(0, 0, 0, 0)$ at some point in the track of the object, the following relationship is true of the track as measured by all observers (shown below for two observers a and b with the object moving in the x direction):

$$\mathbf{E_a}\, t_a - \mathbf{p_a}\, x_a = \mathbf{E_b}\, t_b - \mathbf{p_b}\, x_b = mc^2 \tau \tag{3.14}$$

This relationship will prove very important when we discuss quantum mechanics, so let me go through it step-by-step. Take any two points on the object's track through spacetime and compare the perspective of different observers. In the coordinates of Observer a the distance between the two points is x_a of space and t_a of time. He or she can use this information and the object's rest mass to calculate its energy $\mathbf{E_a}$ and momentum $\mathbf{p_a}$. Observer b is moving relative to Observer a and so observes a different distance and time between the points because of the relative motion and the effects of that on time dilation, length contraction and clock synchronisation. Therefore Observer b will calculate the object to have different energy and momentum ($\mathbf{E_b} \neq \mathbf{E_a}$ and $\mathbf{p_b} \neq \mathbf{p_a}$). However, Observers a and b will agree on the quantity $\mathbf{E}t - \mathbf{p}x$. This quantity is Lorentz invariant.

And what of an observer who is *stationary* relative to the object? This perspective is shown on the far right of Equation 3.14. The observer sees the object move through time but not through space. For him, the object has no momentum, has rest mass energy mc^2 and time is the proper time τ of the object because his clock is stationary relative to it.

The distortions caused by motion are intricate, but we can derive the relationship easily by using the formula from Equation 1.9 for Lorentz transforms shown again below for motion in the x direction.

$$t' = \gamma \left(t - \frac{vx}{c^2} \right) \qquad x' = \gamma \left(x - vt \right) \tag{3.15}$$

In the case of an observer stationary relative to the object, the object is not moving so $x' = 0$ for all time t'. His clock is stationary relative to the object so $t' = \tau$. This gives Equation 3.16. For Equation 3.17, multiply both sides by mc^2 (all observers agree on rest mass) and use the definitions of relativistic $\mathbf{E}$ and $\mathbf{p}$ from Equations 3.10 and 3.11.

$$\tau = \gamma \left(t - \frac{vx}{c^2} \right) \quad \Longrightarrow \quad mc^2\tau = mc^2\gamma\, t - mv\gamma\, x \tag{3.16}$$

$$= \mathbf{E}\, t - \mathbf{p}\, x \tag{3.17}$$

Equation 3.17 shows the values of $\mathbf{E}t$ and $\mathbf{p}x$ partially offset each other in this relationship. This makes sense. For an observer seeing the object move very quickly, it has a higher value of $\mathbf{E}t$ *and* a higher value of $\mathbf{p}x$ than if it were moving slowly. Conventions vary, but the norm is to show the energy term negative and the momentum term positive as is shown in Equation 3.18. This does not imply the object has negative energy. It merely reflects the offsetting relationship of $\mathbf{E}t$ and $\mathbf{p}x$. Equation 3.19 shows the same relationship including all three spatial dimensions.

$$\mathbf{p}\, x - \mathbf{E}\, t = -mc^2\tau \qquad \textit{motion in the x direction} \tag{3.18}$$

$$\mathbf{p}_x\, x + \mathbf{p}_y\, y + \mathbf{p}_z\, z - \mathbf{E}\, t = -mc^2\tau \qquad \textit{motion in any direction} \tag{3.19}$$

3.6 Summary

In this chapter, we used the invariant interval and the equation for time dilation of special relativity, and applied them to a moving object. To do this, we looked at the rate of change of the invariant interval equation with particular reference to the time τ of a clock that is stationary relative to the object. We developed three equations that define relativistic energy, relativistic momentum and their relationship to each other and mass. We derived a fourth from the formula for Lorentz transformations (all shown below for motion in the x direction):

$$\mathbf{E} = mc^2 \frac{dt}{d\tau} = mc^2\gamma \tag{3.20}$$

$$\mathbf{p} = m \frac{dx}{d\tau} = mv\gamma \tag{3.21}$$

$$m^2c^4 = \mathbf{E}^2 - \mathbf{p}^2 c^2 \tag{3.22}$$

$$-mc^2\tau = \mathbf{p}\, x - \mathbf{E}\, t \tag{3.23}$$

In the next chapter, we start the module on *Essential Quantum Mechanics*. For this, you will need to understand sinusoidal waves and basic calculus. I will do my best to keep things simple and add the occasional joke (Box 3.6).

Box 3.6   Just for laughs: An atom is sitting at a bar …
The atom says to the bartender, *I think I've lost an electron.* The bartender asks, *Are you sure?* The atom replies, *I'm positive.* A customer whispers to the bartender, *You can't trust atoms. They make up everything.*

Module II

Essential Quantum Mechanics

Chapter 4

Wave-particle Duality

CHAPTER MENU
4.1 Classical Physics Cannot Explain...
4.2 Quanta of Light and the Photoelectric Effect
4.3 De Broglie's Crazy Idea
4.4 The Double-slit Experiment
4.5 Schrödinger's Mistreated Cat
4.6 Summary

We will now dig into the mysterious world of quantum mechanics. I intentionally delay introducing complex numbers in these early chapters. Some students are confused by complex notation for waves ($e^{i\theta}$) so I cover the basics using more conventional mathematics albeit a little more cumbersome. I also delay introducing Schrödinger's equation. This is intentional to emphasise the relativistic nature and origin of quantum mechanics (Schrödinger's equation is built on a non-relativistic approximation). But don't worry, we will cover all these things in good time.

Some of you will have a passing knowledge of the subject from popular science books. All the better if you do. There are many excellent texts such as John Gribbin's *In search of Schrödinger's cat*. I am a fan and have three copies. All of you will have heard something about wave-particle duality and will know that particles have some sort of wave-like nature. But what does that mean and why is it important?

4.1   Classical Physics Cannot Explain...

Around the early 1900s when Einstein wrote his paper on special relativity, serious cracks had begun to appear in classical (i.e. pre-quantum) physics. Something was very wrong.

One major problem was explaining the stability of atoms. It is clear that the atom is composed of a small dense positive nucleus surrounded by electrons. This was shown beautifully by *Rutherford scattering*. Alpha particles were fired at gold foil. Nobody was sure what alpha particles were, but they knew they were high energy and positively charged (actually they are helium nuclei so composed of two protons and two neutrons). The gold foil caused only a slight deviation in the path of most of the alpha particles that flew straight through. However, 1 in 8, 000 of the alpha particles bounced back. This was clear evidence that the gold atoms in the foil have tiny dense positively charged cores and raised a seemingly impossible question. Why don't the negatively charged electrons spiral down into the positive atomic core?

One suggestion was that the atomic structure is like our solar system and the electrons orbit the atomic nucleus in the way that the planets orbit the sun. But this cannot work. A planet in orbit is accelerating because it is pulled by the attraction of the sun and this force accelerates it away from the natural straight line path ($F = ma$). We cannot use a planetary model for electrons because accelerating electrons radiate energy and would not hold a stable orbit. So classical physics cannot explain the existence of stable atoms. That is a big problem.

Another problem for classical physics is that electrons in atoms emit radiation at very specific and precise wavelengths that are distinct for each element. For example, for hydrogen there are several grouped wavelengths of emission called the

Quantum Untangling: An Intuitive Approach to Quantum Mechanics from Einstein to Higgs, First Edition. Simon Sherwood.

Lyman, Balmer and Paschen series. The wavelengths are so precise that physicists nowadays can observe a distant star and judge its motion from the slight deviation in the wavelength of these emissions due to the star's velocity relative to us. This tells us that there is a clear and defined pattern to the way electrons change when excited by heat or some other energy source. Classical physics sheds no light on why.

A further mystery emerged. Calculations in classical physics suggested that there should be much larger amounts of high-frequency (short-wavelength) ultraviolet radiation emitted by hot objects than actually occurs in nature. Theoretically, for a perfect radiating body (called a *black body radiator*), it should increase to an infinite level as the frequency of the radiation grows, but experiments showed a rapid fall off in radiation at higher frequencies. This conundrum was called the *ultraviolet catastrophe*. What is it that limits the emission of high-frequency radiation?

In 1900, Max Planck discovered that setting the energy E of each unit of radiation to $E = hf$, where f is the frequency of the radiation and h is about 6.6×10^{-34} joules-seconds, gave an answer that matched experimentally determined emissions. The reason this equation works is that the higher frequency radiation requires more energy than lower frequency radiation. The amount of radiation falls at higher frequencies because there simply is not enough energy to produce it. Planck looked on this as a mathematical work-around rather than a physical reality. It was Einstein (*him again!*) five years later who was audacious enough to propose that light actually comes in quanta of energy. In Einstein's words, light *consists of a finite number of energy quanta, which ... can only be produced and absorbed as complete units.*

Einstein was able to point to several pieces of experimental evidence to support his proposal, especially the photoelectric effect. If you shine a beam of light on some metals, the energy of the light beam can knock electrons away from the metal atoms. This can be encouraged or discouraged by negatively or positively charging a collector nearby so any electrons that shake loose are attracted or repelled.

4.2   Quanta of Light and the Photoelectric Effect

The set-up is shown on the left of Figure 4.1. Monochromatic (single frequency) light, shown as wavy lines in the figure, is shone on to the metal emitter (E). This light excites electrons in the emitter and, given enough energy, the electrons can escape and reach a nearby collector (C) creating a current (I) across the gap. The voltage (V) between the emitter and collector can be varied to measure the energy of the emitted electrons. The voltage can be set to repel the electrons (so only energetic electrons get across the gap) or to attract them (encouraging even the weaker laggards to arrive). The results on the right of the figure show that there is a specific voltage at which current starts to flow. This means that the electrons all receive a similar amount of energy from the light.

Experiments show that the energy kick given to each electron by the light beam depends on its frequency, not its intensity. Furthermore, the monochromatic light delivers quanta of energy, now called *photons*, in the relationship $E = hf$. In the words of Einstein again: *the simplest possibility is that a light quantum transfers its entire energy to a single electron.*

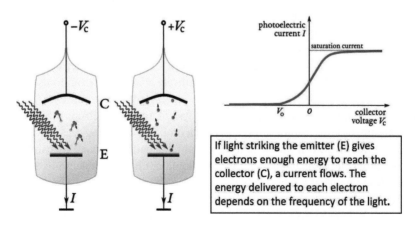

Figure 4.1   The photoelectric effect.

Box 4.1   Max Planck (1858–1947)

Planck's most famous contribution to physics was to propose the equation $E = hf$ for which he received the 1918 Nobel Prize and his name was immortalised in Planck's constant. By his own admission he did not recognise the enormity of his breakthrough admitting it was *an act of despair* and *a purely formal assumption ... actually, I did not think much about it ...* At 16 years old, Max was advised by his physics professor Philipp von Jolly that the subject was: *nearly fully matured ...* and required only *putting in order a jot here and a tittle there, but the system as a whole is secured.* How wrong Jolly was and how fortunate that Planck ignored him.

Planck's life was beset by family tragedy. He had four children with his first wife Marie. She died of tuberculosis, his two daughters both died giving birth, his eldest son died in the First World War and the younger son was hanged for involvement in a plot to kill Hitler in 1945. Stoic to the end, Planck's favourite slogan was: *Persevere and continue working.* Having married again, had a further son and survived both World Wars, Max Planck died at the ripe old age of 89 from a heart attack. His reputation survives as one of the first physicists to eschew dogma and follow experimental evidence. In his words,

The laws of physics have no consideration for the human senses; they depend on the facts, and not upon the obviousness of the facts.

Initially even Max Planck, who had first proposed $E = hf$, rejected Einstein's ideas. Why did Planck react so negatively to the idea of light quanta? Because the idea flew in the face of what had been known about light for a century. Everybody knew that light was a wave so how could it be composed of particles? It was hard for physicists to get their heads around the idea of light having the characteristics of both waves and particles, but the evidence for photons was overwhelming. In the end Planck accepted it and won the Nobel Prize for his work (see Box 4.1).

4.3   De Broglie's Crazy Idea

Louis de Broglie (later to become the 7th Duc de Broglie) was a young French PhD student with a disarmingly simple idea that he was crazy enough to use as the topic for his thesis. He wondered how things would appear if the relationship $E = hf$ applied to *all matter*, meaning that all matter (not just light) has wave-like characteristics. His hypothesis turned out to be correct and earned him the 1929 Nobel Prize.

De Broglie's starting point was to consider how $E = hf$ would manifest itself for an object or particle viewed by an observer stationary relative to it. The object would not be moving, but would have a frequency – some sort of expression with a value that is increasing and decreasing periodically. De Broglie called this rhythmic pulsing the *internal periodic phenomenon* of the particle or object. Mathematically, a stationary periodic function of frequency f is expressed as the value of a sinusoidal wave such as $\sin(2\pi f t)$ or $\cos(2\pi f t)$. This is because the value of the sinusoid repeats itself with every increase of 2π (for example the cosine wave has crests at $\cos 2\pi, \cos 4\pi, \cos 6\pi ... = 1$). If you need a reminder take a look at the refresher Box 4.2.

Box 4.2   Refresher on sinusoidal waves

$A \sin\left(\frac{2\pi}{\lambda}\right)x$ is a stationary sinusoidal wave of amplitude A and wavelength λ

$A \sin\frac{2\pi}{\lambda}(x - vt)$ is the same wave travelling to the right at velocity v

$A \sin\left(\frac{2\pi}{\lambda}x - 2\pi f t\right)$ shows this wave in terms of wavelength and frequency as $f = \frac{v}{\lambda}$

$A \sin(kx - \omega t)$ is the same travelling wave written in a form often used in physics.
k is called the *wave number* and ω is called the *angular frequency*: $k = \frac{2\pi}{\lambda}$ $\omega = 2\pi f$

For an observer stationary relative to the particle, we know that the energy is $\mathbf{E} = mc^2$. This means that the frequency is $f = \frac{E}{h} = \frac{mc^2}{h}$. In the stationary frame the observer's time and particle time are the same so we use the label τ for particle time. This gives the following expression for a particle's periodical function in the stationary frame (I am using a sine wave, but a cosine wave is equally valid):

$$\sin \frac{2\pi}{h} mc^2 \tau \qquad \textit{viewed by an observer stationary relative to the object} \qquad (4.1)$$

De Broglie's next step was to consider how this appears to an observer who is moving relative to the particle or object. Such an observer will see the periodic phenomenon moving through space as well as oscillating with time. The observer's measurements of time and space are altered by the distortions of special relativity: time dilation, length contraction and a change in clock synchronisation. Fortunately, we can short-cut to the answer using the Lorentz transformation shown below (this is from Equation 3.17):

$$-mc^2 \tau = \mathbf{p}x - \mathbf{E}t \qquad (4.2)$$

Applying this to Equation 4.1 gives Equation 4.3. This is how the periodic oscillation appears when an observer sees the particle or object moving. The convention is to switch the sign so energy is negative and momentum positive in the equation. This makes no difference to the frequency as it simply changes the oscillation from $+1,-1,+1,-1...$ to $-1,+1,-1,+1...$

$$\sin \frac{2\pi}{h} (\mathbf{E}t - \mathbf{p}x) \qquad \textit{shown by convention as:} \qquad \sin \frac{2\pi}{h} (\mathbf{p}x - \mathbf{E}t) \qquad (4.3)$$

If you compare this with the standard expression for a sinusoidal wave (see the refresher in Box 4.2), you can see that its wavelength is $\lambda = \frac{h}{\mathbf{p}}$ (which is de Broglie's famous result) and frequency is $f = \frac{\mathbf{E}}{h}$ which is in line with the energy quantisation seen in light waves.

Let's summarise what we have done. We started from de Broglie's idea to extend the rule that applies for light, $E = hf$, to all particles. We used $E = mc^2$ to derive 4.1 for a stationary particle. By substituting with Equation 4.2, we have performed a Lorentz transform that shows how this periodic oscillation appears when it is observed moving. The transform takes into account all the distortions of special relativity. The result is that, when observed moving, de Broglie's internal periodic phenomenon has the spatial wavelength $\lambda = \frac{h}{\mathbf{p}}$.

Not everybody reacted with enthusiasm to de Broglie's hypothesis. For example Max von Laue, who won the Nobel Prize in 1914 for his work on x-ray diffraction, is reputed to have quipped: *if that turns out to be true, I will quit physics.* However, Einstein warmed to the notion. I suspect he realised that this might help explain why electrons do not spiral down into the atomic nucleus (see Box 4.3). We now know that de Broglie was right and that all particles and objects exhibit wave-particle duality. Not bad for a young student's PhD thesis!

Box 4.3 Wave-particle duality and the stability of the atom

Unlike classical physics, quantum physics gives a stable atomic model. If an electron is wave-like and is constrained in a tight space such as the core of an atom, then only certain wavelengths will be stable. This is analogous to sound waves in a musical instrument. If only specific wavelengths are stable, then only specific electron energies are allowable because $\mathbf{p} = \frac{h}{\lambda}$. The electron can jump only between allowable energy levels. The atomic emission spectrum is explained.

And what of an electron emitting radiation and spiralling down into the nucleus? An electron constrained in a space as small as the nucleus, would have a very small wavelength. Therefore it would have to have exceedingly high momentum and kinetic energy; the tighter the constraint, the higher the energy required. Don't worry if you are a bit confused. We will address this in much more detail later.

4.4 The Double-slit Experiment

Some of you may look at Equation 4.3 and think: *Huh ... so there is some weird periodic phenomenon associated with matter ... so what?* Let's discuss why this is so significant and why physicists initially were resistant to the implications of wave-particle duality.

We discussed the photoelectric effect that shows light is composed of photons. However, physicists could not ignore over one hundred years of unequivocal evidence that light is also a wave. Thomas Young had proved this in 1801 with his Double-slit experiment (see Figure 4.2).

Young's experiment is simple. Monochromatic light (i.e. one wavelength) is shone from a point source, shown in the figure as a hole in a screen (S1), onto another barrier with two slits (S2). Behind these two slits is a screen (F). The result is an interference pattern on the screen as shown on the right of Figure 4.2. The same would happen with water waves because the waves going through the top slit interfere with the waves going through the bottom slit. For example, at certain locations on the screen, the crest of the waves from the top slit will cancel out with troughs of waves that have travelled a slightly different distance because they have gone through the bottom slit. The interference pattern in this experiment shows that coherent waves emerge from both of the slits and interfere with each other.

For those interested in the maths, you can see how the interference works in Figure 4.3. There are paths for the light to travel to the screen through slit 1 and through slit 2. The two paths to the screen differ in length by $d \sin \theta$ where d is the gap between the slits and θ is the angle to the screen. For small angles (measured in radians) $d \sin \theta \approx d\theta$. This approximation is accurate to better than 1% for angles up to 0.2 radians (about 10 degrees). For constructive interference the difference in path length needs to be a full wavelength (or multiple of the wavelength) so that the peaks and troughs match up. At each point on the screen where θ has increased by $\frac{\lambda}{d}$ there is another peak of constructive interference. This creates the interference pattern shown on the right of Figure 4.2.

The original Double-slit experiment was done with light, but the same effect has been shown with electrons. For the experiment with electrons the slits are about 60 nanometres wide which is one thousandth the width of a human hair. This interference effect has even been shown for oligo-tetraphenylporphyrins enriched with fluoroalkylsulfanyl chains. These particles are about 25,000 times the mass of a hydrogen atom. Dr James Millen of Kings College London, to whom I am eternally grateful for extensive help with this book, was part of the research team and described it as an *amazing experiment, truly beautiful.*

This behaviour makes *absolutely no sense* if you consider the light in the original experiment, or the electron beam (if that is used), as a stream of particles. Classically it is nonsense (see Box 4.4 for some of the history surrounding this). You would expect each particle to go through one slit or the other and create two points on the screen. But no, there is interference.

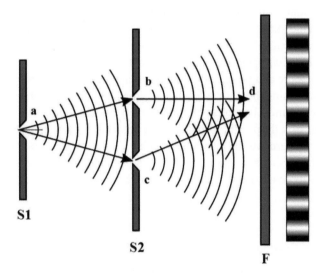

Figure 4.2 Young's Double-slit experiment.
(Lacatosias / Wikimedia commons / CC BY SA 3.0).

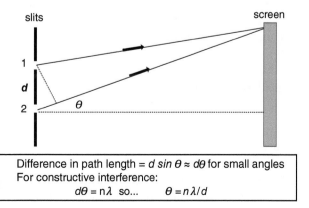

Figure 4.3 Double-slit path difference.

Open only the top slit and the interference disappears as the beam passes in a straight line through the slit to the screen. The same happens if you open only the bottom slit. But open them both and the interference appears again. Yet, the particles arrive locally at the screen as little dots. How can a particle going through the top slit affect a separate particle going through the bottom?

Later experiments show that this pattern appears even if you send the photons or electrons through the apparatus one at a time. This means that each individual particle is interfering with itself. The only possible explanation is that each individual particle is affected by the presence of both available pathways. We describe this as the particle being in a *superposition* of travelling via both available pathways.

This sort of experiment has been replicated countless times and experimenters have come up with all sorts of cunning ways to measure whether each particle goes through the top slit or the bottom slit. For example, you can put a polarising filter in front of the top slit that lets through only vertically polarised photons and a horizontal filter in front of the bottom slit. By measuring the photon polarisation at the screen you would be able to identify which slit each photon has gone through. Do this and the interference disappears.

The result is always the same. If you use an experimental set-up that does not allow you to know which slit the particle goes through, you get interference. But if you use an experimental set-up that allows you in principle to tell whether a particle goes through the top slit rather than the bottom slit, then the interference pattern disappears. This is *obvious*, you might think. After all, if the particle definitely went through the top slit, then it cannot possibly be in a superposition of going through the bottom slit. And you would be right. The logic is clear, but it is difficult to explain *how* this can happen.

You have to abandon the concept of a particle moving along one particular pathway. Instead you must embrace the idea of particles existing in a superposition of states. You can see why physicists were hesitant to accept de Broglie's proposal until it was backed by rigorous experimental proof. The famous quantum physicist Richard Feynman said the Double-slit experiment *contains all of the mystery of quantum mechanics.*

Feynman pointed out that the Double-slit experiment leads us down the following inescapable logical rabbit hole. Suppose, for example, we do the Double-slit experiment with electrons. If there are two slits, the path of the electron to the screen is a superposition of the two available paths. So if there are three slits, the path to the screen would be a superposition of three available paths. What if we open a fourth slit, fifth slit, sixth slit ...? Logically the path of the electron to the screen must be a superposition of more and more available paths that will interfere with each other to produce the result on the screen. What if we keep increasing the number of slits in our now perforated barrier until there is no barrier left at all? The resulting path of the electron must be a superposition of every possible path to the screen!

It turns out mathematically that all of the possible paths interfere and most of them cancel out with surrounding pathways except for the shortest path. So we perceive the electrons as going directly from source to screen, but what is actually happening is that the electrons travel in a superposition of all possible paths between the source and the screen. This is called the Feynman path integral approach. Don't worry if you find this confusing. We will cover it in much more detail in Chapter 6.

Box 4.4   The Thomsons and the electron: like father, like son

Joseph John (JJ) Thomson (1856–1940) is credited with discovering the electron. He studied cathode ray emissions and concluded small particles that he called corpuscles were being knocked out from the gas atoms in the tube. He won the 1906 Nobel Prize and in his lecture outlined his *proof that cathode rays are negatively charged particles.*

George Paget (GP) Thomson (1892–1975) was one of the pioneers who managed to conduct an experiment on electrons that showed they diffract in the same way that light diffracts in the Double-slit experiment. George won the 1937 Nobel Prize and in his lecture outlined how his *experiments in diffraction confirm so beautifully the de Broglie-Schrödinger wave theory.*

So JJ Thomson won a Nobel Prize for proving the electron is a particle and his son (yes, his son) won one for proving it is a wave. Discussions at family dinners must have been fun!

4.5   Schrödinger's Mistreated Cat

The important thing to understand is how far we have strayed from the comfortable world of classical physics. Schrödinger took things to an extreme with his thought experiment, the infamous *Schrödinger's cat* paradox. He thought about

Figure 4.4 Our cat Stella is not fond of dogs.

the superposition implied by the Double-slit experiment and questioned what happens if the two available paths in a superposition are linked to significantly different outcomes.

I suspect you are familiar with his thought experiment. Stick a cat in a black box with a vial of poison triggered by some quantum event like the decay of an unstable atom in such a way that there is a 50:50 chance that the atom decays, the poison triggers and the cat dies. Does the cat die? If you put the cat in a black box so that you have no way of measuring the outcome then the situation is analogous to the Double-slit experiment. In that experiment, there is no way of knowing whether the particle goes through the top or bottom slit so the result is a superposition of both options. In the same way Schrödinger argued, if there is no way of telling if the atom decays or not, then the result will be a superposition of the atom decaying and not decaying, and therefore of the vial of poison breaking and not breaking. Thus our experimental cat will be in a superposition of being alive and dead! I suspect Einstein would have disapproved. He owned a cat (Box 4.5).

Schrödinger developed this paradox to highlight what he saw as the absurdity of parts of quantum theory. Would the cat be in a superposition of being alive and dead, or would its very presence be a measurement and destroy the superposition? We discuss this further in Chapter 7. Eugene Wigner in the 1960s made the paradox even weirder by asking what would happen if it were a human, rather than a cat, in the black box. This amusingly is called the *Wigner's friend* paradox. I cannot imagine his subject would have stayed a friend for long! This sort of paradox highlights how challenging quantum mechanics is to our normal notion of common sense.

Box 4.5   Tiger

Einstein had a tomcat named Tiger that used to get depressed on rainy days. Reputedly Einstein used to comfort his cat with the words:

I know what's wrong, dear fellow, but I don't know how to turn it off.

To reassure readers that no cats were harmed in the writing of this book, I attach a photograph of my daughter's cat, Stella, whom we love dearly (Figure 4.4). If Stella had invented Schrödinger's paradox, she definitely would have put a dog in the box.

4.6   Summary

In the early 1900s it became clear from experiments that light comes in quanta of energy (photons) according to the relationship $E = hf$ where E is energy, f is the light frequency and h is Planck's constant (check out Box 4.6). The fact that light is composed of photons is hard to reconcile with its wave-like behaviour in interference experiments such as the Double-slit experiment. If each photon were a classical particle, you would expect it to travel through only one of the slits. However, the interference pattern shows it is affected by both slits. This is wave-particle duality.

Box 4.6   Fixing the value of Planck's constant in 2019

Planck's constant has units of kg m² s⁻¹ and was measured in the early 1900s from both the pattern of black body radiation and the photoelectric effect (see Figure 4.2). Over time very much more accurate measurement techniques were developed such as the Kibble balance, Josephson junctions and magnetic resonance.

   The definition of the metre and the second are set using the hyperfine transition time of the caesium atom and the speed of light, but the value of h became more accurate than the value of the kilogram, so in 2019 the process was reversed. The value of h was *fixed* at $6.626070150 \times 10^{-34}$ and this now is used to define the value of the kilogram.

De Broglie proposed that $E = hf$ apply not just to light, but to all matter. We can calculate the wave function for an observer stationary relative to the particle, because the relativistic energy is $\mathbf{E} = mc^2$ so $\frac{mc^2}{h} = f$. We use the rules of special relativity (Lorentz transform) to show Equation 4.4 is how the particle wave appears to observers moving at different speeds relative to the particle. Note that typically in quantum mechanics we need $\frac{h}{2\pi}$ so there is a special symbol for this, $\hbar$ (pronounced h-bar) that we will use from now on.

$$\sin \frac{2\pi}{h} (\mathbf{p}x - \mathbf{E}t) = \sin \frac{1}{\hbar} (\mathbf{p}x - \mathbf{E}t) \qquad where \ \hbar = \frac{h}{2\pi} \tag{4.4}$$

In the next chapter we will look more closely at this wave function and what it tells us about the nature of particles at the quantum level. And check out Box 4.7 for the sad story of the man who first named the photon.

Box 4.7   Gilbert Lewis: the Nobel Prize that never was

Gilbert Newton Lewis (1875–1946) enters our story as the man who coined the name *photon*. He made huge contributions in the field of chemistry including discovering the covalent bond and inventing the Lewis-dot notation for molecular bonds that is still used today. An obvious candidate for the Nobel Chemistry prize, he was nominated 41 times with no success. One possible explanation is that he had made a powerful enemy by drawing attention to errors in the work of his old professor, Walther Nernst, who never forgave him. Nernst had a friend on the Nobel committee and may repeatedly have blocked any award. In 1946, Lewis returned morose from a lunch with his long-time rival, the Nobel laureate Irving Langmuir. That afternoon Lewis was found in his laboratory dead from fumes of hydrogen cyanide. It may have been an accidental leak, but some suspect he took his own life.

Chapter 5

Superpositions and Uncertainty

CHAPTER MENU
5.1 The Free Particle Wave Function
5.2 From Sinusoid to Uncertainty
5.3 Superposition
5.4 Heisenberg's Uncertainty Principle
5.5 In Praise of Fuzziness
5.6 God Plays Dice: The Role of Probability
5.7 Summary
5.8 What Is This Wave Function?
5.9 The Role of Rest Mass

5.1 The Free Particle Wave Function

Equation 4.3 that we derived by generalising $E = hf$ is shown again in Box 5.1, including all three spatial dimensions. If you want to refer back to the Lorentz transform that sits behind this, you can find that in Section 3.5. This is called the *free particle wave function*. You will frequently see the symbol ϕ and it referred to as the particle *state* of a free particle.

Box 5.1 The free particle wave function: $\quad \phi = \sin\frac{1}{\hbar}(\mathbf{p}x - \mathbf{E}t)$

$$\phi = \sin\frac{1}{\hbar}(\mathbf{p}_x x + \mathbf{p}_y y + \mathbf{p}_z z - \mathbf{E}t) = \sin\frac{1}{\hbar}(-mc^2\tau) \quad \textit{in all three spatial dimensions}$$

The equation above shows a sine wave. The formula could be a sine wave, cosine wave or any phase in between. We will generalise this using complex numbers later. The left hand part of the longer equation shows the perspective of an observer relative to whom the particle is moving with momentum $\mathbf{p}_x$ in the x direction, $\mathbf{p}_y$ in the y direction, $\mathbf{p}_z$ in the z direction and has relativistic energy $\mathbf{E}$. Time t is measured on the observer's clock. The far right of the equation shows the perspective of an observer moving at the same speed as the particle. Time τ is particle time, measured on a clock stationary relative to the particle. In this frame, momentum $\mathbf{p}$ is zero and energy $\mathbf{E}$ reduces to the rest mass energy mc^2.

What do we mean by *free particle*? Quite simply this is a particle that is not constrained in any way and is not interacting with any other particles. An example of the opposite, a constrained particle would be an electron in an atom that is constrained by the electromagnetic attraction from the nucleus. A free particle is a theoretical entity because particles are always interacting, but it is a useful and simple starting point for our studies.

Quantum Untangling: An Intuitive Approach to Quantum Mechanics from Einstein to Higgs, First Edition. Simon Sherwood.
© 2023 John Wiley & Sons Ltd. Published 2023 by John Wiley & Sons Ltd.

5.1.1 The Phase of the Wave

I am using a sine wave for the free particle wave function, but I might equally use a cosine wave. The waves $\sin(kx - \omega t)$ and $\cos(kx - \omega t)$ have the same frequency (see refresher in Box 4.2). The only difference between them is a difference in *phase*. This is illustrated in Figure 5.1. If you add $\frac{\pi}{2}$ radians to the phase of a sine wave you get a cosine wave (2π radians is 360 degrees). Add another $\frac{\pi}{2}$ radians to get a negative sine wave and so on.

$$\sin\left(\theta + \frac{\pi}{2}\right) = \cos\theta \qquad\qquad \sin(\theta + \pi) = -\sin\theta \qquad\qquad \sin\left(\theta + \frac{3\pi}{2}\right) = -\cos\theta \quad ... \qquad (5.1)$$

In quantum mechanics, you will find that the starting phase is unimportant (indeed, you will learn that the phase must be undefined to comply with special relativity). However, while we can be quite relaxed about whether we choose a sine wave or cosine wave as the *starting phase* in our calculations, we must account vigilantly for changes in phase because these lead to interference effects such as those in the Double-slit experiment.

5.1.2 Derivatives of the Free Particle Wave Function

In this section we will examine how the rate of change of the free particle wave function ϕ over *time* is linked to energy, and the rate of change over *space* is linked to momentum.

Box 5.2 Derivatives of sinusoidal functions: a refresher

Consider the sinusoidal function: $y = \sin Ax$ where A is a constant (i.e. not a function of x)

$$y = \sin Ax \qquad \frac{dy}{dx} = A(\cos Ax) \qquad \frac{d^2y}{dx^2} = A^2(-\sin Ax) \qquad \frac{d^3y}{dx^3} = A^3(-\cos Ax)$$

Each successive differentiation multiplies by a factor of A and adds $\frac{\pi}{2}$ radians to the phase of the sinusoidal component. For example, when a sine wave is differentiated, the phase of the resulting wave is a cosine wave because $\sin(\theta + \frac{\pi}{2}) = \cos\theta$. This pattern continues with each successive differentiation: $\sin \rightarrow \cos \rightarrow -\sin \rightarrow -\cos \rightarrow \sin...$

The partial derivatives of the free particle wave function with respect to t and x are shown below. When you differentiate a sinusoidal wave, the phase of the sinusoidal component shifts by $\frac{\pi}{2}$ radians (see Box 5.2). To make things more general, I am introducing new notation such that $\phi_{(\frac{\pi}{2})}$ means that it is out of phase from ϕ by $\frac{\pi}{2}$ radians. In the right hand column below, the derivatives with respect to t and x are shown with this new notation. The phase change in the derivative expressions in the right hand column is correct whether you differentiate a sine wave or a cosine wave or any phase in between. Check out Box 5.3 on partial derivatives if you need a refresher.

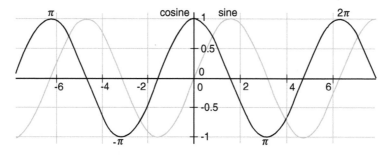

Figure 5.1 Sine and cosine waves are the same but shifted by $\frac{\pi}{2}$ radians. This shift is called a difference in the phase of the wave. Cosine waves are described as symmetric as they are symmetrical about zero. Sine waves are described as asymmetric.

$$\phi = \sin \frac{1}{\hbar} (\mathbf{p}_x x + \mathbf{p}_y y + \mathbf{p}_z z - \mathbf{E}t) \tag{5.2}$$

$$\frac{\partial \phi}{\partial t} = -\frac{\mathbf{E}}{\hbar} \cos \frac{1}{\hbar} (\mathbf{p}_x x + \mathbf{p}_y y + \mathbf{p}_z z - \mathbf{E}t) \qquad\qquad \frac{\partial \phi}{\partial t} = -\frac{\mathbf{E}}{\hbar} \phi_{\langle \frac{\pi}{2} \rangle} \tag{5.3}$$

$$\frac{\partial \phi}{\partial x} = \frac{\mathbf{p}_x}{\hbar} \cos \frac{1}{\hbar} (\mathbf{p}_x x + \mathbf{p}_y y + \mathbf{p}_z z - \mathbf{E}t) \qquad\qquad \frac{\partial \phi}{\partial x} = \frac{\mathbf{p}_x}{\hbar} \phi_{\langle \frac{\pi}{2} \rangle} \tag{5.4}$$

Box 5.3   Total derivative or partial derivative?

This is an important difference when something is affected by two or more variables. As a simple example, consider the function: $k = 2t + 4x$. The *partial* derivative is shown by ∂ and is the change with respect to *one* variable (say, t) while keeping the other (x) constant. In this case, increasing t by 1, increases k by 2 so $\frac{\partial k}{\partial t} = 2$. Imagine that we track k from (t, x) of $(0, 0)$ to $(4, 3)$. In this simple example the partial derivatives are constant values so we can calculate the change in k as:

$$\Delta k = \frac{\partial k}{\partial t} \Delta t + \frac{\partial k}{\partial x} \Delta x = (2)4 + (4)3 = 20$$

The *total* derivative is shown using the symbol d and is calculated when all variables are allowed to vary. On the track above, k has changed by 20 over $t = 4$ seconds so for that track: $\frac{dk}{dt} = 5$

It is simple to repeat the process and differentiate again to generate the second derivatives as shown in Equation 5.5 (we will need these derivatives later when we discuss the Schrödinger equation). Differentiating twice shifts the phase by π which is 180 degrees. If you refer back to Figure 5.1 you will see this gives the negative value i.e. $\phi_{\langle \pi \rangle} = -\phi$.

$$\frac{\partial^2 \phi}{\partial t^2} = \frac{\mathbf{E}^2}{\hbar^2} \phi_{\langle \pi \rangle} = -\frac{\mathbf{E}^2}{\hbar^2} \phi \qquad\qquad \frac{\partial^2 \phi}{\partial x^2} = \frac{\mathbf{p}_x^2}{\hbar^2} \phi_{\langle \pi \rangle} = -\frac{\mathbf{p}_x^2}{\hbar^2} \phi \tag{5.5}$$

5.1.3 Linking Back to Special Relativity

The relationship of energy and momentum with the derivatives of the wave function is of fundamental importance in analysing how quantum systems develop. These relationships, as shown in Equations 5.3 and 5.4, may not jump out as obvious so it may help you to see how they are grounded in the symmetries of special relativity. Let me explain.

The value of the free particle wave function ϕ is Lorentz invariant. As shown in Box 5.1, the free particle wave function can be expressed in terms of $\hbar$, m (rest mass), c and τ (particle time). All of these quantities are Lorentz invariant. They are not affected by the speed of the observer. If we had chosen to focus on a measure that is not Lorentz invariant, such as the wavelength of the wave function (which varies with the particle's momentum), we would always have to specify the speed of the observer. This would complicate things unnecessarily. This is the same logic that we use when we discuss other fundamental properties of particles. For example, we prefer to use as a key measure the *rest mass energy* of an electron which is Lorentz invariant rather than its *total energy* which is observer-dependent.

So, all observers agree on the value of the wave function ϕ and on particle time τ because both are Lorentz invariant. Therefore the quantity $\frac{d\phi}{d\tau}$ must also be Lorentz invariant. We can use this to reveal why the derivatives of ϕ take the form they do by using the multivariable chain rule for derivatives as follows:

$$\frac{d\phi}{d\tau} = \frac{\partial \phi}{\partial t} \frac{dt}{d\tau} + \frac{\partial \phi}{\partial x} \frac{dx}{d\tau} + \frac{\partial \phi}{\partial y} \frac{dy}{d\tau} + \frac{\partial \phi}{\partial z} \frac{dz}{d\tau} \tag{5.6}$$

The left side of Equation 5.6 shows the change in the particle state ϕ from the perspective of an observer who is at rest relative to it. The right side shows the change in ϕ from the perspective of an observer who is moving relative to the object or particle. Time τ is the proper time of the particle. We already know a lot about $\frac{dt}{d\tau}$, $\frac{dx}{d\tau}$ (and so forth) from special relativity and that they are related respectively to relativistic energy and relativistic momentum. We derived these relationships from the invariant interval equation (see Subsection 3.2.2):

$$\mathbf{E} = mc^2 \frac{dt}{d\tau} \qquad \Longrightarrow \qquad \frac{\mathbf{E}}{mc^2} = \frac{dt}{d\tau} \tag{5.7}$$

$$\mathbf{p}_x = m \frac{dx}{x\tau} \qquad \Longrightarrow \qquad \frac{\mathbf{p}_x}{m} = \frac{dx}{d\tau} \tag{5.8}$$

$$m^2 c^4 = \mathbf{E}^2 - \mathbf{p}_x^2 c^2 - \mathbf{p}_y^2 c^2 - \mathbf{p}_z^2 c^2 \tag{5.9}$$

The next step is obvious. We insert the special relativity relationships from Equation 5.7 and 5.8 into Equation 5.6. Starting again with Equation 5.6:

$$\frac{d\phi}{d\tau} = \frac{\partial \phi}{\partial t} \frac{dt}{d\tau} + \frac{\partial \phi}{\partial x} \frac{dx}{d\tau} + \frac{\partial \phi}{\partial y} \frac{dy}{d\tau} + \frac{\partial \phi}{\partial z} \frac{dz}{d\tau}$$

$$= \frac{\partial \phi}{\partial t} \frac{\mathbf{E}}{mc^2} + \frac{\partial \phi}{\partial x} \frac{\mathbf{p}_x}{m} + \frac{\partial \phi}{\partial y} \frac{\mathbf{p}_y}{m} + \frac{\partial \phi}{\partial z} \frac{\mathbf{p}_z}{m} \tag{5.10}$$

If the left side of Equation 5.10 is Lorentz invariant, then the total of the right side must also be Lorentz invariant. $\mathbf{E}$, $\mathbf{p}_x$, $\mathbf{p}_y$ and $\mathbf{p}_z$ are *not* Lorentz invariant. Observers moving at different velocities relative to the object see different values for energy and momentum. However, once we substitute in the derivatives of ϕ from Equations 5.3 and 5.4, the energy and momentum balance, so the total is Lorentz invariant:

$$\frac{d\phi}{d\tau} = \frac{\partial \phi}{\partial t} \frac{\mathbf{E}}{mc^2} + \frac{\partial \phi}{\partial x} \frac{\mathbf{p}_x}{m} + \frac{\partial \phi}{\partial y} \frac{\mathbf{p}_y}{m} + \frac{\partial \phi}{\partial z} \frac{\mathbf{p}_z}{m} \tag{5.11}$$

$$= -\frac{\mathbf{E}}{\hbar} \phi_{\langle \frac{\pi}{2} \rangle} \frac{\mathbf{E}}{mc^2} + \frac{\mathbf{p}_x}{\hbar} \phi_{\langle \frac{\pi}{2} \rangle} \frac{\mathbf{p}_x}{m} + \frac{\mathbf{p}_y}{\hbar} \phi_{\langle \frac{\pi}{2} \rangle} \frac{\mathbf{p}_y}{m} + \frac{\mathbf{p}_z}{\hbar} \phi_{\langle \frac{\pi}{2} \rangle} \frac{\mathbf{p}_z}{m}$$

$$= -\frac{1}{\hbar mc^2} \left(\mathbf{E}^2 - \mathbf{p}_x^2 c^2 - \mathbf{p}_y^2 c^2 - \mathbf{p}_z^2 c^2 \right) \phi_{\langle \frac{\pi}{2} \rangle}$$

$$= -\frac{m^2 c^4}{\hbar mc^2} \phi_{\langle \frac{\pi}{2} \rangle} = -\frac{mc^2}{\hbar} \phi_{\langle \frac{\pi}{2} \rangle} \tag{5.12}$$

The important point is that the structure of the free particle wave function and the relationship of its partial derivatives $\frac{\partial \phi}{\partial t}$ and $\frac{\partial \phi}{\partial x}$ with energy and momentum is not a coincidence. This relationship exists because ϕ is Lorentz invariant.

5.1.4 Consider a Rocket...

The multivariable chain rule used in Equation 5.11 allows you to analyse the overall distortion of special relativity, but it does not give you any flavour of how the individual components add up. So, my final attempt to provide an intuitive feel of what is going on is an analogy using a space rocket.

Let me lay out a highly schematic scenario. Our space rocket has three windows. Behind each window a passenger holds up a number card. From the perspective of the rocket (and any observer stationary relative to the rocket) the passengers all hold up the same number at the same moment. Over time they hold up successive numbers: 1, 2, 3... as shown on the left in Figure 5.2 which shows the numbers seen in the rocket windows increasing over time.

This is (very) loosely analogous with an observer who is stationary relative to a particle observing the value of the wave function change. To assess the change over space you take a horizontal sliver. Let's take the moment when it is (2,2,2). There is no difference in the numbers at each window so the partial derivative $\frac{\partial}{\partial x} = 0$. This is true at any given moment in the stationary frame.

Compare this with an observer who sees the rocket moving at high speed. This is shown on the right of Figure 5.2. The three distortions (see Section 1.8) are time dilation, length contraction and, often forgotten, a change in clock synchronisation (leading clocks lag). The last of these distortions means that the observer in the moving frame does not see the same number at all the windows. For the moving observer the leading clock lags so the number in the front window will be lower than the rear window. If we analyse the change over space with a horizontal sliver, then instead of seeing (2,2,2), the

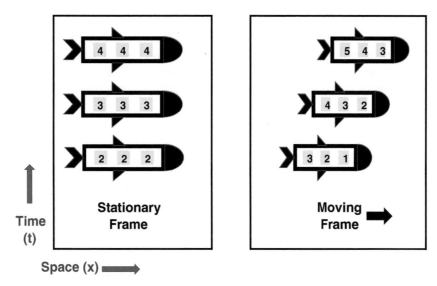

Figure 5.2 Schematic illustration of the effect of special relativity on how the derivatives of a mathematical function change when observed from a moving reference frame (see text for details).

observer might see (4,3,2). The derivative $\frac{\partial}{\partial x}$ has a value. The faster the rocket is moving, the higher its momentum and the higher the absolute value of $\frac{\partial}{\partial x}$.

We can assess the partial derivative with respect to time by taking vertical slivers up the left side (stationary) and right side (moving) of Figure 5.2. The stationary frame shows (1,2,3) while a vertical sliver up the right frame shows (1,3,5). The effect of different clock synchronisation and the motion of the rocket combine to increase the change. The faster the rocket is moving, the higher its energy and the higher the absolute value of $\frac{\partial}{\partial t}$. Let's compare this effect on $\frac{\partial}{\partial t}$ with the effect on $\frac{\partial}{\partial x}$ that we discussed in the previous paragraph:

$$\frac{\partial}{\partial t} \qquad\qquad (1,2,3) \implies (1,3,5) \qquad\qquad stationary \implies moving \qquad\qquad (5.13)$$

$$\frac{\partial}{\partial x} \qquad\qquad (2,2,2) \implies (4,3,2) \qquad\qquad stationary \implies moving \qquad\qquad$$

I hope this extremely crude model helps you understand the pattern. Even if the partial derivative $\frac{\partial}{\partial x}$ is zero for a stationary observer, it will not be zero when measured moving. And take note that while both $\frac{\partial}{\partial t}$ and $\frac{\partial}{\partial x}$ increase when measured moving, the change is in the opposite direction so we must label one as positive and the other negative. The increase in the rate of change versus the stationary frame is by a factor of γ for $\frac{\partial}{\partial t}$ and $-\frac{\gamma v}{c^2}$ for $\frac{\partial}{\partial x}$. You can spot these factors in the Lorentz transform equations for time (see the last term of Equation 1.9). It is these relativistic distortions that drive the derivative relationships of the free particle wave function:

$$\frac{d\phi}{d\tau} = -\frac{mc^2}{\hbar}\phi_{\langle\frac{\pi}{2}\rangle} \implies \frac{d\phi}{dt} = -\frac{\gamma mc^2}{\hbar}\phi_{\langle\frac{\pi}{2}\rangle} = -\frac{\mathbf{E}}{\hbar}\phi_{\langle\frac{\pi}{2}\rangle} \qquad\qquad (5.14)$$

$$\implies \frac{d\phi}{dx} = \frac{\gamma v}{c^2}\frac{mc^2}{\hbar}\phi_{\langle\frac{\pi}{2}\rangle} = \frac{\gamma mv}{\hbar}\phi_{\langle\frac{\pi}{2}\rangle} = \frac{\mathbf{p}}{\hbar}\phi_{\langle\frac{\pi}{2}\rangle} \qquad\qquad (5.15)$$

5.2 From Sinusoid to Uncertainty

The free particle wave function contains the periodicity of $\mathbf{E} = hf$ built-in and, because it is Lorentz invariant, has derivative relationships with energy and momentum. However, on a standalone basis, it contains a mind-bending flaw. You will have heard of Heisenberg's uncertainty principle: if you know the exact momentum of an object then you cannot know its exact location. It sounds crazy, but believe it or not, it is staring you in the face if you take a close look at the free particle wave function. The fact that $\mathbf{p}$ sits right beside x in the wave function manifests uncertainty.

The equation contains four variables: $\mathbf{p}$, x, $\mathbf{E}$ and t. In classical physics (pre-quantum) it was always assumed that objects have a definite value for all these variables. However, what does the free particle wave function reveal about a particle with definite (i.e. exact values of) momentum and energy? In order to keep things clear we will label the momentum as $\mathbf{p} = p_0$ and energy as $\mathbf{E} = E_0$ to show that $\mathbf{p}$ and $\mathbf{E}$ are each a definite precise value (not a variable). Let's simplify further by considering one specific moment in time and set it at $t = 0$. This gives us Equation 5.16 that is a sinusoidal wave for x:

$$\phi = \sin \frac{1}{\hbar} (p_0 \, x - E_0 \, t)$$

$$= \sin \frac{p_0}{\hbar} x \qquad \text{at moment in time } t = 0 \tag{5.16}$$

Setting an exact value for $\mathbf{p}$ creates a sinusoidal wave in variable x with wavelength $\lambda = \frac{2\pi\hbar}{p_0}$

The interesting fact is not the wavelength, but that this is a sinusoidal wave stretching from minus infinity to plus infinity over space in the x direction. Furthermore this sinusoidal wave can be any phase (note Box 5.4). This means that, if the particle has an *exact* value for momentum, then the value of the particle state $\phi(x)$ must be the same for any and all values of x.

We know from the Double-slit experiment that a particle's wave-like nature, embodied in the wave function, is important in determining where it arrives at the screen. This is why we see an interference pattern. So we have a paradox. If the momentum of an object has an exact value, there is no information on its location. To be clear, it is not just that we *lack information* on where the object may be. The equation tells us that, for an object with exact momentum, such information *cannot and does not exist!*

By a similar argument, the wave function for a particle with an exact value for position (which I will label x_0) is a sinusoidal wave in the variable $\mathbf{p}$ as shown in Equation 5.17. If the location of an object has an exact value, there is no information on its momentum.

$$\phi = \sin \frac{x_0}{\hbar} \mathbf{p} \qquad \text{choosing one moment in time at } t = 0 \tag{5.17}$$

Setting an exact value for x creates a sinusoidal wave in variable $\mathbf{p}$ with wavelength $\lambda = \frac{2\pi\hbar}{x_0}$

Box 5.4   A niggling problem with the free particle wave function

Specifying the phase in Equation 5.16 creates a problem. For example, as a sine wave it is value zero at $x = 0$. That makes $x = 0$ a special point which conflicts with relativity because the particle is free so there are no constraints to justify any special points. After all, the difference between a sine wave and a cosine wave is simply where you set the origin $x = 0$. We need a wave function that is more general and I will introduce complex numbers later to handle this, but to keep things simple, I will stick with a simple sinusoidal wave for a little longer. Just remember that there is a problem.

5.3 Superposition

The last section has most likely left you with an uncomfortable feeling. Particles and objects with exact momentum but no defined position. What? It is time to tie things back to the reality we observe in our day-to-day lives. For this section, I will use cosine waves rather than sine waves for the free particle wave function for no other reason than it makes superpositions easier to describe.

5.3.1 Superposition Saves the Day

At first glance we have created a horrific paradox that conflicts with reality. Objects that have momentum are not *everywhere*. In an old-style TV an electron is given momentum and fired towards the screen. It hits the screen at the expected location. The electron certainly does *not* hit the screen *everywhere* or at complete random. It has an approximate position *and* an approximate momentum. The same is true for macro objects. You hit a golf ball and you have some idea of its position *and* momentum. Have you seen on TV how they show a trail following the flight of the ball? The golf ball is not some amorphous object that is everywhere and nowhere. We appear to have invented a weird fantasy world. *Aarrgghh!*

If your head is spinning, take a cold shower. We are not giving up! Our mistake is to assume that a particle is normally in a state with an *exact* defined momentum. The way to solve our paradox is to accept that a particle/object generally is not in a single state of momentum or position, but is in a combination (*superposition*) of states. This should not be too much of a surprise. The result of the Double-slit experiment is the superposition of two coherent waves that exist simultaneously, one associated with each slit. Using superposition, we can reconcile the nature of the structure of the free particle wave function with the reality we experience.

5.3.2 Combining Eigenstates

Later in the book, we will look closely at the mathematics of superposition. In this section, the aim is to convey the flavour of superposition and how it solves our paradox. It is important to understand the concept. We will worry about the maths (Fourier transforms) later.

The classical view of a particle is that it has a *single* definite and precise value for its position and a *single* definite and precise value for its momentum. However, in quantum mechanics we have seen that superpositions exist. This means a particle can simultaneously have multiple values for position and momentum. You are about to learn that the *smaller* the superposition in the value of the particle position, the *greater* the superposition in the value of its momentum and vice versa. This may strike you as strange, but it is a requirement of the free particle wave function.

Box 5.5   Of eigenstates and superpositions...

Things will be clearer if I am careful with notation. A particle in a single momentum state (i.e. it has a definite precise momentum), is said to be in a *momentum eigenstate* that I will label ϕ_p. Similarly, a particle in a single position state (i.e. it has a precise value for x), is said to be in a *position eigenstate* that I will label ϕ_x. More generally, I will label particle states as Ψ to cover all particle states including *superpositions* that are composed of a combination of eigenstates. I will explain the reasoning behind eigenstates later, but we will use the term so get used to it!

As an example, I will now show that the only way to create a wave function with welldefined position is for it to be in a large superposition of momentum states. As noted in Box 5.5, I will describe a particle with a single precise value of momentum as being in a momentum *eigenstate* ϕ_p. We know from 5.16 that a particle in a single momentum eigenstate has no defined position. However, consider a particle in a superposition of values of momentum. This is very different. Such a particle can have spatial definition (i.e. some sort of approximate position in space). As an example, let's build a particle state Ψ that has most of its amplitude localised somewhere close to the point $x = 0$.

Let's start with a particle in a superposition of two momentum eigenstates. Its momentum has two possible values: p_1 and p_2. This is no weirder than the particles in the Double-slit experiment appearing to travel the combination of two paths. The particle is simultaneously in a state with momentum p_1 that we will call eigenstate ϕ_{p_1} *and* momentum p_2 that we will call ϕ_{p_2}. Viewed over space the wave function is a superposition (addition) of two waves. To demonstrate the point, I am going to use cosine waves. Both have a peak at $x = 0$ but they have different wavelengths. So at $x = 0$ the peaks of the two cosine waves add up, but there will be interference at other values of x.

We can take this further and assume the particle is in a superposition of even more momentum states: p_3, p_4, p_5, etc. This means adding more and more waves of different wavelengths. This is shown in Equation 5.18. Remember that ϕ refers to a single state (an *eigenstate*) and Ψ refers to a more general total particle state that may be a superposition of eigenstates. I have added an amplitude (A_1, A_2, A_3...) for each eigenstate because the amount of each eigenstate in the mix can vary.

$$\Psi = A_1 \phi_{p_1} + A_2 \phi_{p_2} + A_3 \phi_{p_3} + A_4 \phi_{p_4} + A_5 \phi_{p_5} ....$$

$$\Psi(x) = A_1 \cos \frac{p_1}{\hbar} x + A_2 \cos \frac{p_2}{\hbar} x + A_3 \cos \frac{p_3}{\hbar} x + A_4 \cos \frac{p_4}{\hbar} x + A_5 \cos \frac{p_5}{\hbar} x .... \tag{5.18}$$

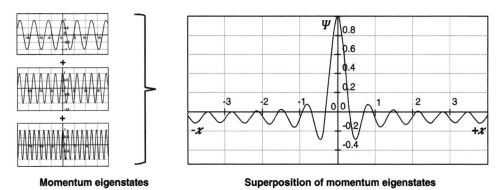

Momentum eigenstates   **Superposition of momentum eigenstates**

Figure 5.3 A particle with exact momentum is in a momentum eigenstate and has a wave function that is spread out across space in a sinusoidal wave so the particle has no defined position. For a particle to have a more narrowly defined position, it must be in a superposition of momentum eigenstates that is, its momentum is not exactly defined.

The peak of the superposition of cosine waves adds up at $x = 0$ but the waves interfere elsewhere. The result of the superposition of these waves is an unambiguous peak around $x = 0$ but the interference elsewhere mostly cancels out (see Figure 5.3). In this way, we can compile a bunch of momentum eigenstates, each of which is a pure sinusoidal wave with no spatial definition, to create a particle state with spatial definition but only at the cost of using multiple momentum states.

It turns out that you can create virtually any shape for Ψ by using the right combination of sinusoidal waves. The process for working out the required combination of sinusoidal waves needed to generate a particular shape is called Fourier analysis. It is named after Joseph Fourier who was a good friend of Napoleon. In spite of war and military expeditions, Fourier still found time to develop his theories (see Box 5.6).

I wrote *virtually* any shape at the start of the last paragraph for a reason. There are some limits. For this to work, function Ψ must be continuous, be single valued (i.e. only one value at any point), and be possible to integrate. In layman language, Ψ cannot be too weird. For example, the function $\tan x$ does not work because it regularly shoots off to $+\infty$ only to reappear at $-\infty$. The rules of Fourier analysis constrain the possible shape of wave functions.[1]

The outcome is the following. To define the position of a particle more narrowly, you need more momentum eigenstates in the superposition, so the less accurate is the particle's momentum. And similarly, to define the momentum of a particle more accurately, you must have fewer momentum eigenstates in the superposition, so the less accurately defined is the particle's position. This manifests the uncertainty relationship that is an intrinsic part of wave-particle duality and plays an important role in quantum mechanics.

Box 5.6   Joseph Fourier (1767–1830)

Fourier was from a poor family and orphaned at the age of ten. At 21 he played an active role in the French Revolution only to have his fellow revolutionaries turn on him when he tried to defend some victims from the slaughter that followed (The Terror). His life was spared but he was imprisoned briefly. He was well liked by Napoleon with whom he travelled to Egypt, mixing military roles with his studies of mathematics and physics. Most famous for the Fourier transform, he also studied heat flow and calculated the temperature of the earth to be higher than expected so, in essence, he discovered *the greenhouse effect*. Fourier is one of 72 French scientists whose names are inscribed on the Eiffel Tower. A bronze bust was erected in his birth place, Auxerre. It was melted down to make armaments in the Second World War.

5.4   Heisenberg's Uncertainty Principle

Heisenberg developed a formula to quantify this uncertainty relationship. His uncertainty principle tells us that $\Delta x\, \Delta \mathbf{p} \geq \frac{\hbar}{2}$. The uncertainty in the position of a particle multiplied by the uncertainty in its momentum (both measured as standard deviations, see Box 5.7) is always equal to or greater than $\frac{\hbar}{2}$. The more defined the particle's position, the less defined its momentum and vice versa.

What is Heisenberg saying? He is saying that particles are *always* in a superposition either of position or momentum (or more usually both). This has the consequence that you cannot know the precise outcome of a measurement of position *and* of momentum, but only calculate the probability of each outcome. The particle is in a superposition so there is some uncertainty about one measure or the other. For now, I ask that you accept Heisenberg's result. In a later chapter, we will discuss it further and I will show you how to quantify it.

Box 5.7   Standard deviation and variance: a refresher

Heisenberg's uncertainty is the *standard deviation* of the range of possible values, weighted according to how likely each one is. The square of the standard deviation is called the *variance* and is calculated by taking how much each result differs from the average value, squaring that difference to make it positive, and finally dividing the sum of all these squares by the number of values. It can be summarised mathematically as follows, where μ is the average value for x, and n is the number of values in the sum:

$$Variance = \frac{\sum(x - \mu)^2}{n}$$

One point to note is the uncertainty relationships in three spatial dimensions. Below is the free particle wave function in all three spatial dimensions (as from Equation 5.1). An uncertainty relationship exists between x and $\mathbf{p}_x$ because they sit side by side in the equation, so if one variable is precisely defined the other variable is in the form of a pure sinusoidal wave. A similar uncertainty relationship exists between y and $\mathbf{p}_y$ and of course between z and $\mathbf{p}_z$ (also between t and $\mathbf{E}$ but we will cover that later in the book). An important thing to remember is that there is no uncertainty relationship between for example x and $\mathbf{p}_y$ or any other combination that mixes spatial dimensions.

$$\phi = \sin\frac{1}{\hbar}(\mathbf{p}_x x + \mathbf{p}_y y + \mathbf{p}_z z - \mathbf{E}t) \tag{5.19}$$

Heisenberg's uncertainty principle does not usually affect us in the macro world (by macro, I mean objects that we can see). Consider something macro but fairly tiny such as a raindrop weighing one twentieth of a gram. Relativistic effects can be ignored so we can use $\mathbf{p} = mv$ and thus Heisenberg's uncertainty principle can be written as: $\Delta x\, \Delta v \geq \frac{\hbar}{2m}$ where v is the velocity of the raindrop. Putting in the values of m and $\hbar$ gives: $\Delta x\, \Delta v \geq 10^{-30}$. If Δx and Δv were the same (and there is no reason why they should be) this would give a minimum value for the uncertainty in position of the order of 10^{-15} metres, which is about the size of a proton so way way way too small for us to notice in every day life. However, this uncertainty becomes significant for small mass particles. For an electron that has a mass of about 10^{-30} kg, the same calculation gives $\Delta x\, \Delta v \geq 5 \times 10^{-3}$ which is over 10^{26} times greater than for the raindrop.

5.5   In Praise of Fuzziness

One of the mental challenges of the quantum world is that things are, for want of a better word, *fuzzy*. While an electron interacts as a point particle, the existence of superpositions and uncertainty means that it does not have a fixed infinitely small location. You may ask: *Why?* This is hard to answer, but I feel somewhat justified in replying: *Why not?* A particle with an infinitely small and precise location may seem simpler, but the concept is actually quite challenging.

Consider a particle with a spatial location (x, y, z). We zoom in with a microscope and magnify and magnify. The location of the particle gets progressively more defined. We zoom in and in. Is there any limit? What if we have an infinite zoom? As we pan across with our infinite-microscope we see nothing, nothing, and then suddenly the particle located in an infinitely small piece of space. This sort of thing is a singularity and is not easy to handle mathematically. So, you may find the fuzziness of quantum mechanics strange and unfamiliar, but a universe with infinitely small exactly located particles would not be simple either. As a wise friend (thanks Matt) once told me: *Nature shouldn't have sharp edges.*

5.6 God Plays Dice: The Role of Probability

We have shown that a superposition of momentum eigenstates can produce a particle state Ψ that has an amplitude peak around a narrowly defined location x. But what does that amplitude peak *mean*? What is the relationship between the amplitude of Ψ at a certain position and the probability of physically finding the particle there?

It turns out that the probability of finding the particle in a particular region is proportional to the square of the amplitude of the wave function[2]. Let's do a crude example. Suppose the amplitude of Ψ is value 3 near $x = 0$ and value 1 near $x = 2$. That would mean it is nine times more likely that we will find the particle near $x = 0$ than $x = 2$. Note that relating the square of the amplitude to the probability of finding the particle deals with the problem of the wave function having negative values of amplitude at some points.

This is called the *Born rule*. While I can offer no theoretical proof, you can get a warm fuzzy intuitive feeling by comparing with the mathematics of classical waves (although let me stress again that the following is in no sense a proof). For the waves of classical physics, the *power* of the wave (energy per second) is proportional to the *square* of the wave *amplitude*. This is true for water waves, sound waves and mechanical waves (see Box 5.8). Compare this with photons. Consider a beam of monochromatic light (such as from a laser) shining on a screen. The photons in the beam have a well defined momentum so there must be some uncertainty in the location of each photon. So when will a photon hit the screen? The spatial uncertainty means that the best we can do is to calculate the probability of a photon hitting the screen in a certain time period.

Box 5.8 Waves: amplitude and intensity

It was well know classically that the power (energy per second) of a wave is proportional to the square of the amplitude and it is fairly easy to understand why. Consider a vibrating violin string. For a given frequency of note, each piece of the string will travel up and back down in a fixed time t. The distance travelled up and back down in time t by each piece of string, and therefore its average velocity v, is proportional to the amplitude of the vibration.

$$\text{Distance travelled in } t \propto \text{Amplitude} \implies \text{Average vibrating velocity} \propto \text{Amplitude}$$

$$E = \frac{1}{2}mv^2 \implies \text{Energy per second} \propto (\text{Amplitude})^2$$

What drives this probability? It depends on how many photons per second are in the light beam, which is the *power* of the beam. The more dense the photon stream, the higher the probability that a photon will hit the screen in a certain time period. Thus, Born's law connecting the square of the amplitude of the wave function with the probability of a particle being detected in a certain region makes intuitive sense, but there was much controversy. Born's rule does not tell you what *will* occur but only the *probability* of it occurring. It follows that you may perform *exactly* the same experiment and get different results simply by chance. This was very hard for some physicists (including Einstein) to swallow. Gone is the certainty that $A \to B \to C$ to be replaced by differing results of varying probability such as $A \to B1(\%) \text{ or } B2(\%) \to C1(\%) \text{ or } C2(\%) \text{ or...}$

Einstein and others disagreed (see Box 5.9 and for more on his philosophy, Box 5.11). They proposed that there be some hidden variable that gives the impression of randomness, but secretly governs the outcome. A number of later experiments have searched for such a hidden variable and the results point firmly to there being no such thing (one of the most famous is the Aspect experiment which is covered in detail later in Section 17.4).

Box 5.9 Einstein and probability

Famously, Einstein did not believe that probability plays a role in the laws of physics. He believed the world to be determinate and that quantum physicists are missing some piece of information, some hidden variable, that explains the probabilities. Einstein argued with Bohr that God doesn't play dice with the universe. Bohr countered that Einstein should not tell God what to do.

5.7 Summary

We started this chapter with a detailed look at the free particle wave function. We discussed how Lorentz invariance explains the relationship of its derivatives with energy and momentum. The maths of the equation appeared at first to lead us down a cul-de-sac because a particle with precisely defined momentum has no defined position. Superposition saves the day. For example, we can combine momentum eigenstates to define the particle position more narrowly, but there is a trade-off: the more narrowly defined the position of the particle state, the less narrowly defined its momentum, and vice versa. Superpositions are essential and key to explaining how the world we see is consistent with quantum mechanics. Yippee! But we have to accept that the universe works in a very strange way:

- We must abandon the classical concept of factual precise values for position and momentum: Heisenberg's uncertainty principle: $\Delta x \, \Delta \mathbf{p} \geq \frac{\hbar}{2}$
- We must abandon the concept of a determinate universe $A \rightarrow B \rightarrow C$ and instead embrace the relationship between the wave function and probability: God plays dice
- Superposition means a particle can do weird things such as appear to go through two different slits at the same time: the Double-slit experiment

If you are still reading, it means that you have made it through this section on superposition. You have survived! Give yourself a pat on the back. It is one of the hardest parts of quantum mechanics to get your head around, but it is worth making the effort because it lies at the heart of every type of quantum weirdness that you will encounter as your studies progress (see Box 5.10).

Box 5.10   A universe without superposition

You may find it odd that particles and objects can be in a superposition of eigenstates, but how much weirder it would be if they could not. If superpositions were not allowed, an object or particle could never have even approximate definition of both momentum *and* position. Forget seeing cars going down the road because a moving car has approximate position and momentum. Forget watching the sun rise and set. Forget seeing your lover rush to your embrace. *Superposition is bewildering, but the alternative would be much weirder.*

5.8   What Is This Wave Function?

Let's drop the maths for a moment and think about an obvious question: *what on earth is this* Ψ *thing?* Does it have any physical existence? Some people describe Ψ as a pilot wave guiding the particle, but if so, then a pilot wave through *what*? Others say it is just a method of calculating and we should not worry about what is happening between interactions. Neither answer is very satisfying.

My approach (and this is my *personal* view) is to try to avoid thinking of the wave function as having any sort of physical existence. As we have discussed, the mathematical behaviour of a periodic function like the wave function is largely constrained if it is to appear consistent to all observers. I think of the wave function as a mathematical description that reflects those constraints.

As an analogy, consider π. It describes the relationship between the circumference and diameter of a circle. It does not *exist* in any physical sense. If you take a perfect circle and drop it onto a flat surface then the circumference will be π times the diameter. The value and characteristics of π are dictated by Cartesian geometry[3] A circle in Cartesian space must be like this. Does this answer what the wave function is? Not really. Perhaps it is just a different way for me to avoid the question. Anyway, if *you* can find some physical form and answer definitively what the particle wave is and how it works, then let me know. But before that call Stockholm and collect your Nobel Prize.

5.9 The Role of Rest Mass

Take a look at how mass appears in the free particle wave function as expressed in Equation 5.20. The rate of change of the phase of the wave function in *particle time* is determined by rest mass (or more specifically rest mass energy), that is, how quickly the wave function vibrates from positive to negative amplitude as measured by an observer stationary relative to the particle. Don't forget that the negative sign in the expression does not mean mass or energy is negative. It is just a convention and does not change the rate at which the amplitude of the wave function swings between positive and negative values.

$$\phi = \sin \frac{1}{\hbar}(\mathbf{p}_x x + \mathbf{p}_y y + \mathbf{p}_z z - \mathbf{E}t)$$

$$= \sin \frac{1}{\hbar}(-mc^2\tau) \qquad \text{\textit{in particle rest frame... in this frame } } \mathbf{p} = 0 \text{ \textit{and} } \mathbf{E} = mc^2 \qquad (5.20)$$

How fast is the particle state changing in the stationary frame? For even something as small as an electron, it is vibrating or pulsing (in a mathematical sense) over 10^{20} times per second!

I want to finish this chapter with a question. Why is mass the key variable in the free particle wave function? In the next chapter you will learn the link that ties the rate of change of the wave function together with Newton's equation $F = ma$. At the same time, you will discover that the workings of quantum mechanics are way wackier than we have explored so far and involve superpositions of every possible path through the universe so don't stop reading.

Box 5.11   Einstein's philosophy

After a long and successful career, Albert Einstein (1879–1955) died at the age of 76 from internal bleeding. Offered surgery that might have extended his life, he refused with the words: *I want to go when I want. It is tasteless to prolong life artificially. I have done my share; it is time to go. I will do it elegantly.* He requested cremation and that his ashes be scattered at a secret location but, without permission, his brain was removed during the autopsy and some pieces can still be seen in museums.

Einstein commented on life: *Strange is our situation here upon earth. Each of us comes for a short visit, not knowing why, yet sometimes seeming to divine a purpose. From the standpoint of daily life, however, there is one thing we do know: that man is here for the sake of other men; above all for those upon whose smile and well-being our own happiness depends, and also for the countless unknown souls with whose fate we are connected by a bond of sympathy.*

Notes

1 Look up Dirichlet conditions if you want to learn more about these constraints.

2 Actually, the square of the *absolute value* of the amplitude which is labelled $|\Psi|^2$. The distinction will become important when we start using complex numbers in later chapters.

3 Cartesian geometry is the geometry of flat space as I hope you know. If not, look it up!

Chapter 6

Everything Happens ... Kind of

CHAPTER MENU

6.1 The Feynman Path Integral
6.2 Change in Phase of the Wave Function
6.3 Simplified Path Integral Model
6.4 The Principle of Stationary Action
6.5 Action and the Lagrangian
6.6 From the Lagrangian to the Equations of Motion
6.7 The Uncertainty Relationship: A Different Perspective
6.8 Feynman Diagrams
6.9 Summary

Let me start this chapter with a declaration. For me, the path integral is the greatest discovery in physics *ever*. In my opinion, nothing else so clearly and completely shatters every illusion we have of how the universe works.

6.1 The Feynman Path Integral

In the discussion of the Double-slit experiment (Section 4.4), I mentioned that Feynman developed the following logic. The interference pattern in the Double-slit experiment exists because each particle is in a superposition of travelling to the screen through both slit 1 and slit 2 in the barrier. If we add a third slit to the barrier, we find interference from all three possible routes to the screen. Let's increase the number of slits in the barrier up to say 10. We now have a superposition of all 10 pathways to the screen. Let's keep adding slits until we are left with no barrier at all. Logically the particle is now in a superposition of travelling every possible path to the screen. Let me repeat this. A particle travelling from one point to another (i.e. without a barrier) must be in a superposition of travelling every possible pathway between the two points!

Feynman took this seriously. He thought about how the wave function would appear when it arrived at the screen. He realised that the phase of the arriving wave function will vary depending on the path taken. Some paths will involve greater change in the phase of the wave function than others. Say for example on one path the wave function arrives with amplitude +1, then for another path which involves more change of phase, it might arrive with an amplitude of −1. If so these two paths would cancel out. Feynman showed that normally all the routes cancel out except those closest to the path with the smallest change in phase. This gives us the impression of the electron travelling on a specific path from A to B whereas the result in reality is an amalgam of all the options.

The maths behind the path integral is complicated, but it is possible to demonstrate the principle with some simple examples. I must give credit to Feynman himself for this analysis. He was an amazing teacher as well as a genius and I thoroughly recommend his book *QED: The strange theory of light and matter* (for a little background on Feynman, see Box 6.1). If you find the level of maths in this chapter a little too much, I would recommend Brian Cox and Jeff Forshaw's book *The Quantum Universe: Everything that can happen does happen* for a less mathematical treatment of this topic.

6.2 Change in Phase of the Wave Function

Electrons (and all other particles and objects) do *not* actually travel directly from A to B. Rather their apparent path is the result of a superposition of all available routes from A to B. The resulting wave functions of these paths interfere and cancel out to create the coherent picture of motion that we take for granted. But how does this happen?

Consider again the Double-slit experiment. The paths through the two slits differ in length (shown in detail back in Figure 4.3) leading to an interference pattern on the screen. The wave function is sinusoidal so its value oscillates. Feynman pictured little phase-gauges, like clocks, to represent this. As you move along each pathway, the phase-gauge rotates. If the phase-gauges from two paths both end up showing 12 o'clock, there will be constructive interference and there is a higher chance that the particle will arrive there. But if we choose a point on the screen where the phase-gauge from one pathway shows 12 o'clock and the other 6 o'clock, they cancel out and there is no chance the particle will arrive there. The neat thing is that you can calculate how the amplitudes of the two paths add up by connecting the phase-gauge hands. The length of the combined phase-gauge hands is the amplitude of the resulting wave function.

Figure 6.1 shows how this works. Box A in the figure shows the combination of two waves that arrive in phase. They combine to a wave of the same phase, but with double the amplitude as shown by the length of the blue arrow. Box C shows the other extreme with the two waves out of phase by half a wavelength (π radians). This is the combination of Ψ and $\Psi_{(+\pi)}$ using the notation I introduced back in Subsection 5.1.2. The two waves cancel each other out. Box B shows the case when two waves that are out of phase by $\frac{\pi}{2}$ meet. The result is shown at the bottom of Box B. The length of the blue arrow shows that the combination produces a wave with an amplitude larger than either of the original waves, but not as large as was the case in Box A when the two waves were fully in phase.

Box D shows this can be extended for superpositions involving more paths. It shows how six different paths to a point might combine. Each path is for the same particle so the arriving wave functions are the same except they have different phase. Put the hands of the phase-gauges end to end and, hey presto, you can see the amplitude and phase of the resulting wave (shown in blue). Square this amplitude to get a measure of the relative probability of the particle arriving at that point that you can compare with results for other points.

As you would expect, the amount that the phase-gauge turns on a given pathway is a function of the rate that the particle state changes and how long the path is through space and time.

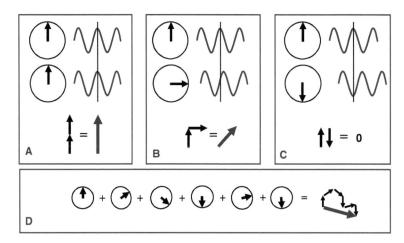

Figure 6.1 Phase change and interference illustrated with phase-gauges.

Box 6.1   Feynman the prankster

Nobody can doubt the intelligence of Richard Feynman (1918–1988), but in spite of an unprecedented perfect score on the Princeton University entrance exams, being acknowledged as having a *practically perfect* record and as *the best undergraduate … for five years at least,* there was hesitation in accepting him. The department head asked:

Is Feynman Jewish? We have no definite rule against Jews, but have to keep their proportion in our department reasonably small because of the difficulty of placing them. In current times, it is hard to believe the formal response was yes, but … *his physiognomy and manner, however, show no trace of this characteristic.*

Feynman did not fit the stereotype of a genius physicist. He was known for bongo-playing and practical jokes, but brilliant he was, and a talented teacher with the ability to simplify rather than complicate. Alongside many other achievements, he was the architect of the theory of quantum electrodynamics (QED) for which he won the 1965 Nobel Prize. Feynman died of cancer, 69 years old. His last words were, *I'd hate to die twice. It's so boring.*

6.3   Simplified Path Integral Model

We are going to use a simplified model to illustrate how and why a coherent picture emerges when you combine all possible pathways that a particle might take. We use the same set-up as the Double-slit experiment but with a very large number of slits. We assume the change in the phase of the wave function depends only on the length of the pathway. This is a bit of a simplification, but very revealing.

The first step is to calculate the difference in length between different pathways. The set-up is shown in Figure 6.2. The particle is emitted from a source on the left and travels through a perforated barrier to the screen on the right. There is a tiny distance Δx between each slit in the barrier. In the figure the gaps are shown magnified so you can see what is going on, but the model assumes Δx is minuscule relative to the distance between source and screen. We will start by comparing the length of the direct shortest route (shown in solid blue) with the shortest route passing through the next closest slit (shown as A in dotted blue). To keep the maths easy we will assume that the source is 1 metre from the barrier and the barrier is 1 metre from the screen so the length of the direct pathway is 2 metres. We calculate the length of pathway A between the source and the slit using Pythagoras and a Taylor expansion (see Box 6.2). We double this to get the full length of pathway A between the source and the screen. It turns out that pathway A is longer than the direct route by $(\Delta x)^2$ as shown by Equation 6.1.

$$\frac{A}{2} = \sqrt{1^2 + (\Delta x)^2} \qquad \textit{Pythagoras' triangle rule} \tag{6.1}$$

$$\approx 1 + \frac{(\Delta x)^2}{2} \qquad \textit{Taylor approximation for small } \Delta x$$

$$A \approx 2 + (\Delta x)^2 \quad \Longrightarrow \quad \textit{Pathway A is longer than direct route by } (\Delta x)^2$$

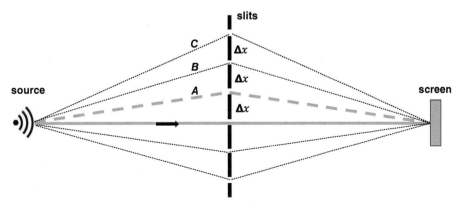

Figure 6.2   Multi-slit simplified model.

The key point is that as the path gets further away from the shortest direct route, its length exceeds the shortest route by the *square* of the distance separating the two paths. Referring back to Figure 6.2 this means that while pathway *A* is only Δx^2 longer than the shortest route, pathway *B* is $(2\Delta x)^2$ longer which is $4\Delta x^2$. Pathway *C* is $9\Delta x^2$ longer. The pathway through the next gap that would be *D* is $16\Delta x^2$ longer, etc.

Box 6.2   Another Taylor expansion:

$$\sqrt{1+x^2} = 1 + \frac{x^2}{2} - \frac{x^4}{8} + \frac{x^6}{16}... \approx 1 + \frac{x^2}{2} \quad \textit{for small } x$$

If we perforate the barrier with a huge number of holes, the difference in length between the shortest route and the closest neighbouring pathway is only Δx^2, but the difference in path length between the 10th and 11th slits is over $20\Delta x^2$, the difference between the length of the 100th and its neighbour is over $200\Delta x^2$ and between, say, the 10000th and its neighbour is over $20000\Delta x^2$. This means a pathway very close to the shortest route has similar phase compared to its neighbouring pathways and so will constructively interfere with them (increasing amplitude). However a pathway further away from the shortest route will have very different phase compared to its neighbouring pathways. They will randomly interfere and cancel each other out.

Now, let's consider a particle that is fired from a source to a screen with no barrier in between. The top of Figure 6.3 shows possible pathways that the particle might take. We would normally assume it flies in a direct line to the screen, but this time we will consider all the options. The shortest path is direct and shown in solid blue. Another 10 paths are shown either side of the shortest path. Each path reaches a point a further Δx away from the shortest route and then swings back towards the screen. This seems crazy, but bear with me. The four closest pathways on either side of the direct path are shown in dotted blue to distinguish them from the ones that are further out.

As calculated earlier, the difference in path length compared to the shortest route is Δx^2 for the closest, $4\Delta x^2$ for the second closest, and $9\Delta x^2$ for the third. The difference grows further: $16\Delta x^2$ for the fourth, $25\Delta x^2$ for the fifth, all the way up to $100\Delta x^2$ for the tenth.

Let's assume that for every extra Δx^2 of pathway length, the phase-gauge moves by the equivalent of a clock minute hand advancing one minute (i.e. $\frac{\pi}{30}$ radians). The arriving phase of each pathway is shown by the line of phase-gauges across the

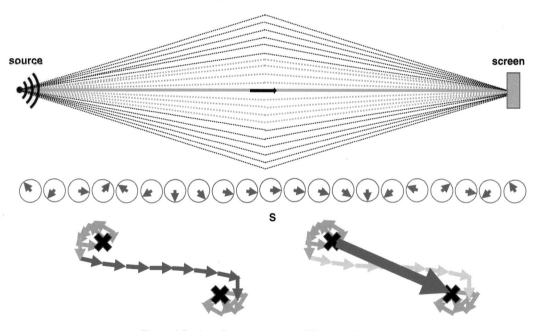

Figure 6.3   Interference between different pathways.

middle of the figure. The one labelled with an S shows that of the direct pathway. I have set it pointing to the right simply so I can fit the results on the page. On either side of that phase-gauge is the next closest path. These differ in phase by only one minute on the phase-gauges. As you move further away from the direct path, the change in phase between neighbouring pathways gets larger because the difference in the length of neighbouring paths grows.

At the bottom left of the figure, I have added all the phase-gauge hands together to combine the arriving wave functions. To the bottom right, the bold blue arrow shows the amplitude and phase of the combined wave functions. The *length* of the arrow shows the *amplitude*. This amplitude squared represents the probability of the particle arriving at the screen. Note that no favouritism has been shown. The little arrows are all the same length so every path is equally weighted. It is the difference in phase between the paths that creates the pattern. We could add up the little arrows in any order and it would give the same result. I have ordered them by position so you can see the contribution of the closest pathways (again shown in blue).

The important point is that the closest pathways to the path with least phase change contribute virtually all of the amplitude. The routes that are slightly farther out add very little. What does this mean? Imagine a scientist experimenting. He takes a barrier, makes a hole in it, and places the hole over the direct route between source and screen. This blocks all the dotted grey routes so the little grey arrows in the sum disappear and only the blue pathways are left. What happens? The amplitude does not change significantly. The particle still arrives at the screen. The scientist concludes that the particle travels through the hole in a straight line from source to screen.

To be sure of his result, he does the reverse. He blocks the direct route between source and screen with a small barrier that cuts off the closer blue pathways while leaving the outer grey pathways clear. What happens? If you remove all the little blue arrows from the mix at the bottom of Figure 6.3, the resulting amplitude becomes vanishingly small. The particle does not arrive at the screen so he concludes again that he has blocked the particle that was travelling in a straight line from source to screen. However, in reality the particle path is not direct. It is in a superposition travelling every possible route. What a mind-blowing illusion!

By the way, the *direction* of the big blue arrow (bottom right of the figure) is the *phase* of the sum. It is slightly different from that of the shortest pathway because the contributing closest routes are slightly longer. It has nothing to do with the direction of motion.

6.4   The Principle of Stationary Action

Allow me to introduce the concept of *Action*. We will get into the maths later. For now, all you need to know is that *Action* is a measure of the change in phase of the wave function along a path[1]. If a path has higher Action, the phase-gauge will turn more. If lower, it will turn less. One good thing is that I no longer need to write *the path along which the phase of the wave function changes the least*. I can write simply *the path of least Action*.

To return to our model, all possible paths contribute with equal weight to the path integral, but they cancel each other out except for those paths nearest to the path of least Action. These nearest paths generate virtually all of the resulting amplitude of the particle state which creates the illusion that the particle travels only along the path of least Action.

Figure 6.4 shows a graph of the Action that the particle experiences compared with the pathway taken. In our simple model, the Action grows with the square of the distance from the shortest pathway. The close-up box shows the pathway of least Action labelled S. If you take the *exact* pathway S and slightly alter it, the Action does not increase because the curve is flat at that point. This flatness means the pathways around S have the same Action and constructively interfere. Physicists describe the Action as *stationary* at S and the overall effect is called the principle of stationary Action. Note carefully that S could also be at a maximum or a saddle point as shown in Boxes B and C in Figure 6.4.

Our model in Figure 6.3 considered only straight pathways close to the direct line between source and screen, but there is no logical justification for stopping there. Curves, loops, twists, a trip around the world – Feynman showed that every pathway however bizarre counts equally and plays its part, but all the weirder ones destructively interfere leaving only those extremely close to the path of stationary Action as the dominant contributors in the combined result. Think what this means. The motion of an electron is the result of the superposition of its wave function along every possible pathway, even to the moon and back. Wow!

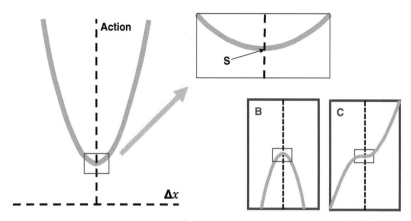

Figure 6.4 Paths close to the point of stationary Action (S) constructively interfere.

6.5 Action and the Lagrangian

How can we measure the Action along a pathway? In our simple model it depended only on the length of the pathway. A glance at the free particle wave function (Equation 6.2) shows that the phase of the wave function also is affected by the momentum and energy of the particle. As each pathway can have twists and turns, the momentum and energy of the particle will change en route so we must calculate the phase change over a very small period of time Δt and then integrate over time to give the result for the whole pathway. This is not as complicated as it sounds.

To keep things simple, we will work in one spatial dimension. The particle moves a small distance Δx along the pathway during the time period Δt. We can compare the phase of the free particle wave function at time (t) with a fraction of a moment later ($t + \Delta t$). The change is shown in Equation 6.3 recognising that $\frac{\Delta x}{\Delta t}$ is velocity v at that moment in time. We can ignore the $\hbar$ factor (it is the same for all pathways) to give us a measure of Action as shown in Equation 6.4.

$$\phi_{(t)} \;=\; \sin \frac{1}{\hbar}(\mathbf{p}x - \mathbf{E}t) \qquad \phi_{(t+\Delta t)} \;=\; \sin \frac{1}{\hbar}(\mathbf{p}(x + \Delta x) - \mathbf{E}(t + \Delta t)) \tag{6.2}$$

$$Phase\ change \;=\; \frac{1}{\hbar}(\mathbf{p}\,\Delta x - \mathbf{E}\,\Delta t) \;=\; \frac{1}{\hbar}\left(\mathbf{p}\frac{\Delta x}{\Delta t} - \mathbf{E}\right)\Delta t$$

$$=\; \frac{1}{\hbar}(\mathbf{p}v - \mathbf{E})\,\Delta t \qquad Phase\ change\ over\ time\ period\ \Delta t \tag{6.3}$$

$$Action \;=\; (\mathbf{p}v - \mathbf{E})\,\Delta t \qquad Action\ over\ time\ period\ \Delta t \tag{6.4}$$

We then can substitute the expressions from Equation 6.5 for $\mathbf{p}$ and $\mathbf{E}$. These are extremely accurate at anything but enormously high speeds (refer back to Section 3.4 if needed). There is one twist. We want the Action for all particle pathways, not just free particles, so we must include the effect of things such as gravitational and electromagnetic fields. To account for this, we add potential energy to the value of $\mathbf{E}$. In quantum mechanics, potential energy is typically represented by a capital V. Please take care not to confuse V (potential energy) with v (velocity):

$$\mathbf{p} \;=\; mv\gamma \approx mv \qquad \mathbf{E} \;=\; mc^2\gamma + V \approx mc^2 + \frac{1}{2}mv^2 + V \tag{6.5}$$

Substituting these into Equation 6.4 gives Equation 6.6 for the Action over the time period Δt:

$$Action \;=\; (mv^2 - mc^2 - \frac{1}{2}mv^2 - V)\,\Delta t \;=\; \left(\frac{1}{2}mv^2 - V - mc^2\right)\Delta t \tag{6.6}$$

To calculate the total Action of the particle along a certain pathway, we integrate the change of the Action from the start (time 0) to the end (time t) as shown in Equation 6.7.

$$Action = \int_0^t \left(\frac{1}{2}mv^2 - V - mc^2\right) dt \tag{6.7}$$

$$= \int_0^t \left(\frac{1}{2}mv^2 - V\right) dt - C \qquad \text{where C is the constant } mc^2t \tag{6.8}$$

This is the Action (which is directly proportional to the change in phase of the wave function) along *one* particular pathway, but we want to compare multiple pathways. Let's say the particle starts from point a at $t = 0$. We want to compare the contributions from all possible pathways that arrive at point b at the same moment t. The pathway of stationary Action is the one with the *lowest* value. We can simplify a bit. The mc^2t factor is a *constant* because mass m and time t are the same for every pathway so we can ignore it. If this puzzles you, it may help to look back at the graph of Action in Figure 6.4. If you add a constant to every point, it does not change the inflection points. It just lifts or drops the curves by the same amount.

In summary, the pathway of least Action gives the smallest possible value of Equation 6.8. It is determined by the *difference* between *kinetic* and *potential* energy over its course. This quantity $\left(\frac{1}{2}mv^2 - V\right)$ is called the *Lagrangian* and its importance was well known in classical physics (see Box 6.3 and for more detail on the photon, Box 6.4).

Box 6.3   Action and the Lagrangian (classical):

Lagrangian: $\frac{1}{2}mv^2 - V$ **Classical Action:** $\int_0^t \left(\frac{1}{2}mv^2 - V\right) dt$

Classical physics uses the Lagrangian to calculate which path a particle or object will take as it passes through a potential. For example, if a ball is thrown up in the air, its path will be the one that minimises the difference between kinetic energy and gravitational potential energy over the course of flight (a parabola). Objects appear to follow the path of stationary Action (sometimes called least Action). The path integral formulation allows us to show for particles how this result can be derived by considering all the available paths and combining the resulting wave functions – yet another neat connection between quantum mechanics and classical physics.

Box 6.4   What about the Action of light photons that are massless?

If Action is proportional to mass (Equation 6.3), then no mass means no phase change. However, photons move at light speed relative to all observers. The longer the pathway, the earlier the photon must have been emitted by the source. This means there is an emission phase difference between pathways even if there is no change in phase en route. As a result the pathway of stationary Action for light is the fastest path (in terms of time).

6.6   From the Lagrangian to the Equations of Motion

The Lagrangian can be used to derive Newton's equations of motion including his formula $F = ma$. We will use a shortcut called the Euler-Lagrange equation (Box 6.5). This equation can be derived directly from the Principle of Stationary Action, as Lagrange discovered back in the 1700s. The derivation is a little taxing so I will not inflict it on you (you can find it online if you are so inclined). The Lagrangian is typically labelled $\mathcal{L}$ and for brevity $\frac{dx}{dt}$ is labelled $\dot{x}$.

Box 6.5   Euler-Lagrange equation:

$$\frac{d}{dt}\left(\frac{\partial \mathcal{L}}{\partial \dot{x}}\right) = \frac{\partial \mathcal{L}}{\partial x}$$

This looks daunting but it is not. It is surprisingly easy to use. The simplest way to show this is to apply it to the classical Lagrangian. First we rewrite the Lagrangian using the $\dot{x}$ symbol and then calculate the derivatives treating x and $\dot{x}$ as separate variables. The potential V is a function of x because potential varies with position, so I have used $V(x)$ to indicate this:

$$\mathscr{L} = \frac{1}{2}mv^2 - V(x) = \frac{1}{2}m\dot{x}^2 - V(x) \tag{6.9}$$

$$\frac{\partial \mathscr{L}}{\partial x} = \frac{dV}{dx} \qquad\qquad \frac{\partial}{\partial t}\left(\frac{\partial \mathscr{L}}{\partial \dot{x}}\right) = \frac{\partial(m\dot{x})}{\partial t} = ma$$

$$\implies \quad \frac{dV}{dx} = ma \quad \textit{(where a is acceleration)} \tag{6.10}$$

Equation 6.10 is more familiar as $F = ma$. Why have I shown you this? It is important because it is the link between the classical Newtonian equations of motion and the path integral formulation of quantum mechanics. It shows why the mass term in Newton's $F = ma$ is the same as the mass term m in the equation for the rate of change of the wave function.

In the absence of a potential, $\frac{\partial V}{\partial x} = 0$ there is no acceleration. The particle's momentum does not change along the pathway of stationary Action as shown in Equation 6.11.

$$\textit{If } \frac{dV}{dx} = 0 \qquad\qquad \frac{\partial}{\partial t}\left(\frac{\partial \mathscr{L}}{\partial \dot{x}}\right) = \frac{\partial(m\dot{x})}{\partial t} = \frac{\partial p}{\partial t} = 0 \tag{6.11}$$

That still leaves a question: why does a particle such as the electron have its particular rest mass? Or to put the same question differently, why does the electron wave function change at that particular rate? I am sorry to report that we do not yet have the answer to this question.

6.7 The Uncertainty Relationship: A Different Perspective

Heisenberg's uncertainty principle provides a lower limit in that $\Delta x\, \Delta \mathbf{p}$ is always greater or equal to $\frac{\hbar}{2}$. However, this does not give us any upper limit. For example, $\Delta x\, \Delta \mathbf{p}$ might be infinite. The result would be a universe of particles with undefined position and undefined momentum – a featureless fuzzy mess. Is that possible? No, it is not. It is impossibly improbable and we can use the path integral to show why.

Let's calculate the change in phase of a free particle along a path from point a to point b. As we have just shown, the combination of all possible pathways very closely resembles that of the shortest path between the two points. If we say the distance between the points is x_0 and the time difference t_0, the change in phase over the path can be derived from the free particle wave function. Remember that there is no potential energy term because we are dealing with a free particle:

$$\sin \frac{1}{\hbar}(\mathbf{p}x - \mathbf{E}t) \quad \implies \quad \textit{Phase change along pathway} = \frac{1}{\hbar}(\mathbf{p}\,x_0 - \mathbf{E}\,t_0) \tag{6.12}$$

What if the particle is not observed to be at b at time t_0 but is at a point c that is a distance Δx further away from a? The combination of all possible pathways to c will also closely resemble that of the shortest path between a and c. That path length is now not x_0 but Δx longer. There is no change to the time t_0 so the phase will differ from that shown in Equation 6.12 by $\frac{\mathbf{p}\Delta x}{\hbar}$.

Now comes the key point. We know from Heisenberg's principle that there is uncertainty. Suppose we design an experiment giving a very high probability of the particle arriving at b. That means going from a to b is the pathway of least phase change. The actual particle state is a superposition of arriving at b, c, d and so on, but pathways that do not lead to b have higher phase change. Those that arrive a significant distance from b cancel out just as we saw with the outer paths in Figure 6.3, but if Δx is very small, the pathway will contribute significantly.

How small? We must assess which pathways we have to include to give a decent approximation of the full path integral and how big is the variation in phase of these pathways. If you refer back to Figure 6.3, adding all the blue pathways gives a fairly accurate result. If you eyeball the line of phase-gauges at the bottom of the figure, you will see that the blue pathways differ in phase by up to about 1 radian ($\approx$ 30 degrees) from path S. For greater accuracy you need to include more pathways, but once the pathway is out of phase by half a turn or a full turn (π to 2π radians) the added precision looks negligible.

In the calculation below I use 1 radian to illustrate, but you could equally argue 2π radians. It is a question of probability so there is no exact answer. This puts a limit on the possible variation in phase before a pathway cancels out. As shown in Equation 6.13, the value of $\mathbf{p}\,\Delta x$ must not exceed $\hbar$ (based on the one radian limit).

$$\Delta\ phase\ <\ 1\ radian \quad \Longrightarrow \quad \frac{\mathbf{p}\,\Delta x}{\hbar} < 1 \quad \Longrightarrow \quad \mathbf{p}\,\Delta x < \hbar \tag{6.13}$$

$$\Longrightarrow \quad \mathbf{p} < \frac{\hbar}{\Delta x} \approx \Delta\mathbf{p} \tag{6.14}$$

$$\Longrightarrow \quad \Delta x\,\Delta\mathbf{p} \approx \hbar \tag{6.15}$$

We now turn to **p**. This is a variable in the equation so we must consider the particle state to be a superposition of possible values for **p**. However, we now know that the variation in **p** can only be up to $\frac{\hbar}{\Delta x}$ before the phase change exceeds the 1 radian limit that I have set (Equation 6.14). This possible range of variation *is* the uncertainty in momentum. It *is* $\Delta\mathbf{p}$. Reordering things gives $\Delta x\,\Delta\mathbf{p} \approx \hbar$ for the 1 radian limit. If you prefer to use a wider limit of 2π radians you get $\Delta x\,\Delta\mathbf{p} \approx 2\pi\,\hbar$ and, of course, $2\pi\hbar$ is h, the original (unreduced) version of Planck's constant.

Let's recap. The variation in the resulting phases of the pathways that combine to give a good approximation of the path integral is not more than about 1 to 2π radians depending on the level of accuracy you use. Quantifying this variation in terms of Δx and $\Delta\mathbf{p}$ gives $\Delta x\,\Delta\mathbf{p} \approx \hbar$ or h. This is an approximate rule of thumb best expressed as *of the order of $\hbar$ or h* (see Box 6.6).

Box 6.6   Position-momentum uncertainty: two guidelines

$\Delta x\,\Delta\mathbf{p} \geq \dfrac{\hbar}{2}$ *Always!* $\qquad\qquad\qquad$ $\Delta x\,\Delta\mathbf{p} \approx \hbar$ *or* h ... *typically... order of...*

6.8   Feynman Diagrams

Returning to our story, Feynman did not stop. He played the bongo drums that he was famous for (really), thought some more, and came to another inescapable logical conclusion. If the particle motion is the result of combining every possible pathway, then surely particle interactions must be the result of combining every possible form of interaction. And, again, he was right.

Suppose a photon is absorbed by an electron in an atom and then re-emitted. This is a common enough scenario. Perhaps you are wearing a red T-shirt. A photon of sunlight hits it and is absorbed by an electron in an atom that is part of a molecule of red dye. It is then emitted as red light (this is how red dye works). Feynman pointed out that based on the path integral the actual interaction will be the sum of contributions from all possible interactions including more intricate ones such as:

- Electron absorbs sun photon coming into shirt
- Electron emits red light photon
- Electron re-absorbs red light photon
- Electron re-emits red light photon that leaves shirt

The possible iterations are infinite. Fortunately, electromagnetic interactions are weak. The probability of any interaction is low so those with multiple steps contribute very little to the final result. In terms of classical physics they are insignificant, but these more complicated interaction pathways do make important contributions to calculations in high energy particle physics. Feynman developed a form of diagram (Feynman diagrams) to keep track of these different interaction pathways. It remains an important tool for particle physicists. Some example Feynman diagrams are shown in Figure 6.5 for two electrons repelling each other. The underlying mechanism is the exchange of photons (called Compton scattering). Each diagram provides a schematic of a possible exchange process. The solid lines represent the electrons and the wavy lines are the photons.

In diagram A two electrons (shown as e^-) exchange a photon (shown as γ). In B the two electrons exchange two photons. Diagram C is even more complicated. The photon emitted by an electron becomes an electron-positron pair that decays back into a photon and then reaches the other electron.

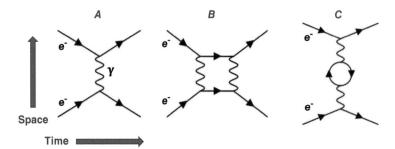

Figure 6.5 Some Feynman diagrams for Compton scattering. The photon is shown with the symbol γ (do not confuse this with the γ of time dilation).

Feynman's theory now is backed by copious evidence from particle physics experiments. To calculate the outcome of high energy particle experiments you must take into account strange possible interactions including the fleeting appearance and disappearance of all sorts of particles. Just as with the path integral, every interaction is the sum of all possibilities and if you don't take that into account, you get the wrong answers. We will delve in greater detail into Feynman diagrams in Chapter 21 which addresses quantum electrodynamics (QED).

6.9 Summary

We started the chapter with a simple model of multiple paths from source to screen. We considered the changing phase of the particle wave function along different pathways. The combined result is that the sum of all the pathways closely follows the Path of Stationary Action. We calculated how Action appears to an observer who sees the particle moving along the pathway. This gave us what is called the classical Lagrangian: $\frac{1}{2}mv^2 - V$ (where V is potential energy).

With the Lagrangian, plus a bit of help from Lagrange (check out Box 6.7) in the form of the Euler-Lagrange equation, we generated the equations of motion and showed why the m that determines the rate of change of the wave function in the rest frame is the m in Newton's formula $F = ma$. We also used the path integral to delve deeper into the position-momentum uncertainty relationship.

Box 6.7 Lagrange, Lavoisier and the French Revolution

Joseph-Louis Lagrange (1736–1813) developed the Principle of Stationary Action in the 1750s while working at the Royal Military Academy of the Theory and Practice of Artillery. He was born in Sardinia, but moved to Paris. In 1793, following the French Revolution, a law called for all foreigners to be arrested and their possessions confiscated. However, the noted chemist Antoine Lavoisier, spoke up and exempted Lagrange from persecution.

Lavoisier himself fared less well. On May 8th 1794, he was tried, convicted and executed, all in a day. Legend has it that the judge answered his wife's pleas for mercy with: *The Republic needs neither scholars nor chemists.* Horrified by the news that Lavoisier had been guillotined, Lagrange famously said it took: *only a moment to cut off that head and a hundred years may not give us another like it.*

Lagrange spent the rest of his life in Paris, enjoying a comfortable second marriage with the daughter of a friend. She was 32 years his junior. In a twist of fate, the judge who presided at Lavoisier's trial, Jean-Baptiste Coffinhal, was sent to the guillotine only three months later.

The good news is that you do not have to worry too much about path integrals until we get to the module on *Relativistic Quantum Mechanics* towards the end of the book. Until then you can use the classical approximation that a free particle moves in a straight line unless it encounters a potential, but don't forget Feynman showed that reality is a lot stranger. His rigorous analysis showed the motion of particles is the result of a superposition of all available pathways, however crazy the shape, however crazy the length, around the world, or even to the moon and back; but as we have illustrated in this chapter, most of these pathways cancel out to produce the coherent pattern of motion that is familiar to us all. This fills me with excitement and awe. I hope I have been able to convey a little of that excitement to you.

As a final flourish on this topic, let's return to the title of this chapter. No, everything that can happen does not actually happen (or at least it is extremely unlikely to) but everything that can happen does *contribute* to the final result so yes, perhaps it is fair to say *everything that can happen does – kind of*.

Note

1 Some of you may wonder if this is the same as *Action* in classical physics. It is, as will be revealed.

Chapter 7

Measurement and Interaction

CHAPTER MENU

7.1 What Can You Know about a Quantum System?
7.2 Collapse of the Wave Function
7.3 When a Body Meets a Body …
7.4 An Electron in a Box
7.5 Collapse of the Wave Function – a Twist
7.6 Decoherence and the Measurement Problem
7.7 When a Body Leaves a Body – Entanglement at a Distance
7.8 Summary

In Chapter 5, we showed that particles exist in superpositions. This means that they have simultaneous distinct possible values for variables such as position or momentum. The probability of measuring a certain value of a variable is proportional to the square of the amplitude of the wave function. For example if we measure the position of the particle, then the probability of finding it at x depends on the square of the amplitude: $\Psi(x)^2$. Actually, we will need to use $|\Psi(x)|^2$ which is the square of the modulus, but that only becomes relevant when we introduce the maths of complex numbers.

But what is it that forces a particle to abandon its superposition and take on a specific value (or at least a smaller range of values) for a particular variable? This process is generally called *collapse of the wave function*. Along with it comes a whole lot more quantum weirdness. Hold on to your hats – the roller coaster ride is going to get wilder with particles that move at high speed but go nowhere, spooky action at a distance and the possibility of multiple quantum worlds all coexisting in one massive superposition.

But first let's review the state of a particle before it interacts. What do we know? What don't we know? What can we know? What can't we know?

7.1   What Can You Know about a Quantum System?

If a particle is in a superposition of locations, then we cannot know where the particle is until there is some sort of measurement. Why? Because the particle does *not have* a precise location until we measure it. After all, if the particle had a precise position beforehand, then it would not have been in a superposition! The superposition, by definition, means that there are multiple possible outcomes when measured. Do not interpret this uncertainty of outcome as some sort of mathematical sloppiness. No, the maths is very precise.

A good comparison is the mathematics of playing dice. Each has six sides. If you roll one, you do not know what number from one to six will come up,[1] but you do know there is precisely a one in six chance for each particular number. You know everything there is to know about each of the dice. The roll comes down to probability, but it does not mean that you lack some knowledge of dice.

Quantum Untangling: An Intuitive Approach to Quantum Mechanics from Einstein to Higgs, First Edition. Simon Sherwood.
© 2023 John Wiley & Sons Ltd. Published 2023 by John Wiley & Sons Ltd.

Quantum mechanics is the same. If you measure where a particle is by interacting with it, the result comes down to probability just as it does with the dice. This does not imply that you lacked knowledge about the particle state. You just have to alter your concept of what perfect knowledge is (see Box 7.1).

Box 7.1   Quantum knowledge

I use the term *quantum knowledge* to mean knowing exactly the probability of every possible outcome of a measurement. To know this, you need to know the shape/values of Ψ for the variable. This is exact knowledge of the quantum state. There is no more that you can know. It is different from the classical concept of knowing the exact outcome of a measurement.

Perfect quantum knowledge (i.e. knowing the probability of each outcome) does not conflict with Heisenberg. The *uncertainty* in Heisenberg's uncertainty principle is uncertainty about the result of measurement. If you compare with dice, then Heisenberg's uncertainty is analogous to the uncertainty in the outcome of rolling the dice. This uncertainty exists even if you have perfect knowledge of how the dice work and know the exact probability for each outcome.

There is another twist to the story. If you know the relative probability of finding a particle in each possible location/position, you also know the relative probability of it having each possible value of momentum. Let me repeat this. If you know the pattern of probability for where the particle is, you can calculate the exact pattern of probability for its momentum. Position and momentum are *not* separate variables.

What does all of this mean for measurement? Position and momentum are dependent variables. If you know one then you know the other. They are Fourier transforms. This has an important consequence: if you change one then it will change the other which brings us to the impact that measurement has and how measurement changes the wave function.

7.2   Collapse of the Wave Function

It is simple to show that measurement can wreak great change on the wave function. Consider a free particle that is in a momentum eigenstate and let's randomly choose its momentum to be exactly 135 (in whatever units you want as it makes no difference). Taking time at $t = 0$ to simplify things, we can express Ψ in terms of momentum, that I will label $\Psi(p)$ *or* in terms of position that I will label $\Psi(x)$. The example below uses a sine wave to illustrate the effect.

$$\text{If } \mathbf{p} = 135 : \qquad \Psi(p) = \delta\,(\mathbf{p} - 135) \qquad \Longleftrightarrow \qquad \Psi(x) = \mathcal{N} sin \frac{135}{\hbar} x \qquad (7.1)$$

The expression for the wave function in terms of momentum $\Psi(p)$ in Equation 7.1 uses the Dirac delta function δ and indicates that the result of a measurement of momentum will definitely be 135 (see Box 7.2). The expression for the wave function in terms of position $\Psi(x)$ on the right side of Equation 7.1 shows that the position of the particle is completely undefined as you would expect if it is in a momentum eigenstate (check back to Equation 5.16 if needed). A factor $\mathcal{N}$ appears in the equation. This is called a *normalisation* factor and acknowledges that the equation needs adjusting so that the probability of all measurement outcomes adds up to one (i.e. 100%). Clearly there must be 100% chance of finding the particle if you include all possible outcomes. The two expressions $\Psi(p)$ and $\Psi(x)$ in Equation 7.1 are the *same* wave function, but expressed with different coordinates.

Box 7.2   The Dirac delta function (δ)

The δ in Equation 7.1 is the Dirac delta function. $\Psi(p) = \delta\,(\mathbf{p} - 135)$ means (by definition) that the value of $\Psi(p)$ is 0 except where $\mathbf{p} = 135$. We cannot write $\Psi(p) = 135$ because $\Psi^2(p)$ gives the *probability* of finding that the particle has momentum $\mathbf{p}$ and it cannot be 135 squared! In this example, the Dirac delta function is zero except at $\mathbf{p} = 135$ where it jumps to infinity. The maths behind it is a bit complicated, but the function is defined so that if we measure $\mathbf{p}$, there is 100% chance that the result will be $\mathbf{p} = 135$.

Now, what happens when we take a measurement of the particle's position? The particle state *changes*. Let's suppose we measure the position of the particle carefully and the result of the measurement is by chance exactly $x = 20$ (this again is purely illustrative as it is impossible to perfectly measure anything). What happens to the wave function? We know $x = 20$. Therefore the particle is now in a position eigenstate that I will call Ψ'. The value of x is fixed so it is now the value of momentum **p** that is undefined and has an equal probability of being any value:

$$If \; x \; = \; 20 : \qquad \Psi'(x) \; = \; \delta(x - 20) \qquad \Longleftrightarrow \qquad \Psi'(p) \; = \; \mathcal{N} sin \frac{20}{\hbar} \mathbf{p} \qquad (7.2)$$

After the measurement of x, the new state Ψ' (Equation 7.2) is completely different from the old state Ψ (Equation 7.1). The wave function has changed. In this extreme example, the particle state Ψ has changed from a single momentum eigenstate to a single position eigenstate.

To be sure that the maths is clear, here is another illustrative example. Consider an electron in a state Ψ that is a superposition of three position eigenstates. The electron is simultaneously at three different positions. Let's say for example that the wave function has amplitude A at $x = 1$, B at $x = 5$ and C at $x = 8$. The wave function expressed as a function of position is:

$$\Psi(x) \; = \; A\,\delta(x - 1) \; + \; B\,\delta(x - 5) \; + \; C\,\delta(x - 8) \qquad (7.3)$$

If we interpose something to constrain the electron's position (such as it hitting a screen), then the probability of the particle being at, say, $x = 5$ is:

$$Probability \; of \; electron \; interacting \; at \; x = 5 \qquad = \frac{|B|^2}{|A|^2 + |B|^2 + |C|^2} \qquad (7.4)$$

This is because the probability of the interaction occurring at $x = 5$ is related to the square of the amplitude at that point, and we must divide that by the total probability of the interaction occurring in the three positions because it must happen at one of them. After the interaction, the wave function has changed (I will call it Ψ') because now the electron is unambiguously at $x = 5$:

$$After \; electron \; interaction \; at \; x = 5 \qquad \Psi'(x) \; = \; \delta(x - 5) \qquad (7.5)$$

Had the interaction by chance occurred at a different position, such as $x = 1$, then Ψ' would be different reflecting that position. The superposition in terms of position has *collapsed* down to one specific value for *position*. However, this collapse changes the particle's superposition for values for *momentum*, so use the term *collapse of the wave function* with great care. Constraining the value of one particular variable will expand the superposition for another (note Box 7.3).

One reason this process causes such fascination is because it appears to be *irreversible* in time. Most of physics is built on processes that have a clear cause and effect. Typically this means that you can roll back the clock to figure out how things stood at some time in the past. For example, the maths of the orbit of the planets around the sun works forwards and backwards. You can figure out where they will be in a week's time *and* where they were a week ago. Measurement and interaction in quantum systems do not appear to allow this. You might know everything about Ψ' after the measurement (Equation 7.5), but that does not allow you to calculate the superposition Ψ of the electron before the measurement (Equation 7.3). This creates an asymmetry in time in quantum mechanics that has no obvious equivalent in general relativity and is one of the challenges for physicists in unifying the two theories.

Box 7.3   Zero, infinity and the uncertainty of eigenstates

Heisenberg's uncertainty principle states that $\Delta x \, \Delta \mathbf{p} \geq \frac{\hbar}{2}$. In a position eigenstate, $\Delta x = 0$ so you may be deceived into believing that $\Delta x \, \Delta \mathbf{p} = 0$ because anything multiplied by zero is zero. This is not correct because in a position eigenstate, momentum is *completely* undefined so $\Delta \mathbf{p} = \infty$ and infinity multiplied by zero is not always zero. An intuitive way to rationalise this is to think of zero as an infinitely small amount and infinity as an infinitely large amount – so multiply them together and they can produce a finite result.

7.3   When a Body Meets a Body …

One of the major achievements of quantum mechanics is to explain the structure of atoms. Electromagnetic attraction binds the negatively charged electrons to the positively charged protons in the atomic nucleus. We will discuss this in detail in a later chapter, but for now I want to highlight the difference in positional uncertainty of various particles and objects. The larger the mass of an object, the better defined its position (in a broad sense). Perhaps the easiest way to introduce this is with a simple theoretical question: can you collocate an electron and a proton, that is, can you constrain each one's position to the same limit?

The simple answer is *no* because the proton is about 2,000 times more massive than the electron. Let's compare the uncertainty relationship of an electron with that of a proton. The relationship $\Delta x\,\Delta \mathbf{p}_x \approx \hbar$ (as in Box 6.6), gives us the result that $\Delta x\,\Delta v_x \approx \frac{\hbar}{m}$ where v_x is velocity in the x direction. The different mass of the electron versus the proton means that $\Delta x\,\Delta v_x$ is about 2,000x higher for the electron:

Electron   *Proton*

$$\Delta x\,\Delta v_x \approx \frac{\hbar}{m_e} \qquad\qquad\qquad \Delta x\,\Delta v_x \approx \frac{\hbar}{m_p} \tag{7.6}$$

$$\approx \frac{10^{-34}}{10^{-32}} \approx 10^{-2} \qquad\qquad\qquad \approx \frac{10^{-34}}{2 \times 10^{-27}} \approx 5 \times 10^{-6} \tag{7.7}$$

With the usual apology to serious physicists, I am going to take a liberty and describe the value of the combined uncertainties $\Delta x\,\Delta v_x$ as a measure of the overall *fuzziness* of the particle position in the x dimension. How do I justify this? Δx is the uncertainty in position at a given moment. Δv_x is the uncertainty in how that position is changing over time ($v_x = \frac{dx}{dt}$). In layman's terms, Δx is the uncertainty of where the particle *is*, and Δv_x gives a measure of the uncertainty of where the particle *will be*. The greater the value of $\Delta x\,\Delta v_x$, the larger the overall uncertainty about the particle's present/future position. The same relationship holds for the other spatial dimensions ($\Delta y\,\Delta v_y$ and $\Delta z\,\Delta v_z$). The value of each is the same and depends only on the mass of the particle.

Let me stress that this is not a rigorous quantification. My only aim is to help you see intuitively that the smaller the mass of an object, the larger its positional fuzziness. The electron's lower mass means that its $\Delta x \Delta v_x$ is much higher than that of the proton, so using this as a measure (and my awful layman language) the electron's current/future positional definition is about 2,000 times fuzzier in the x direction, 2,000 times fuzzier in the y direction and 2,000 times fuzzier in the z direction. Overall that involves a volume about 10^{10} times larger than that of the proton. That is a whole lot fuzzier so, no, you cannot constrain an electron to the same current/future limits as you can a proton; the best you can do is constrain it to be nearby.

7.4   An Electron in a Box

It is time to take the training wheels off the bike and have some fun. Let's examine how the wave function of an electron changes when it is spatially constrained. This is the first step in explaining why atoms emit light of specific wavelengths and why atoms are stable. We can use an extremely primitive and simple model variously called the *electron-in-a-box* model or the *one-dimensional infinite potential well* to illustrate.

The model could not be simpler. We just decree that the electron is spatially constrained. As we are working in one spatial dimension, we will give the electron an allowable region in space of length L. The electron is allowed anywhere in this region, but is *never* allowed outside it. It is stuck in this one-dimensional box of length L.

Figure 7.1 illustrates the model. The electron can be anywhere in the box so the wave function can take various shapes and values – *but* – the value of the wave function must be zero outside the box because the electron cannot be there, *and* the wave function must be zero at either side of the box. If it were not zero at the sides there would have to be a jump up or down in value to zero at that point. Therefore, it would have two (or more) different values at the edge of the box. This is called a *discontinuity* and, if it happened, the derivative $\frac{\partial \Psi}{\partial x}$ would be infinite which means the electron would have infinite momentum and energy. Oops!

Therefore, the end points of the electron wave function at the edge of the box must be stationary zero-points, typically referred to as *nodes*. This type of wave is called a *standing wave* (see Box 7.4) . Consider the wave shown as $n = 1$ in Figure 7.1. Imagine it like a skipping rope held tight at each end with the centre vibrating up and down. It is a standing

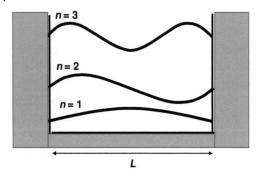

Figure 7.1 For a standing wave $\lambda = \frac{2L}{n}$ with n being an integer. All the waves have value $\Psi = 0$ at the edges. The waves are separated to reflect the differing energy level of each electron state. https://openstax.org/books/university-physics-volume-3/pages/7-4-the-quantumparticle-in-a-box#CNX_UPhysics_40_04_boxstates.

wave of wavelength $\lambda = 2L$ with two nodes (one at each end). Another possibility with some more vigorous waving of the skipping rope is shown as $n = 2$. You can see that the wavelength is L. In this case, it also has a stationary point in the middle so it has three nodes. Every step up in excitation adds another node. The pattern of stable wavelengths is $2L, L, \frac{2L}{3}, \frac{L}{2}$... more generally $\lambda = \frac{2L}{n}$ where n is any integer.

Box 7.4 Standing Waves: a mathematical refresher

For the electron constrained in the box to be stable, its spatial wave function must be a standing wave (see Figure 7.1) which fits in the box. A standing wave has fixed stationary nodes. Mathematically, it is constructed from two equivalent waves moving in opposite directions:

$$y_1 = A \sin(kx - \omega t) \quad \textit{moving to the right}$$

$$y_2 = A \sin(kx + \omega t) \quad \textit{moving to the left}$$

$$y = y_1 + y_2 = 2A \sin(kx) \cos(\omega t) \quad \textit{a standing wave} \tag{7.8}$$

Consider Equation 7.8^2. If $\sin(kx) = 0$ then $y = 0$ for all t. These points do not move. They are stationary zero-value nodes, unaffected by the $\cos(\omega t)$ term. The spatial shape of the wave function is created by the $\sin(kx)$ term. The $\cos(\omega t)$ term oscillates over time between $+1$ and -1 so the wave function oscillates from positive to negative values.

A standing wave is called a *time-independent* wave function. It oscillates over time, but its basic spatial shape does not change over time. If the x and t terms of the wave function can be separated as they are in Equation 7.8, then the equation is a standing wave and is spatially stable.

As only certain wavelengths are allowed, only certain values of momentum are allowed (refer back to Box 4.3 if needed). This, in turn, constrains the possible value of the kinetic energy of the electron as shown below (kinetic energy is labelled as $K.E$):

$$\mathbf{p} = \pm \frac{2\pi\hbar}{\lambda} \implies \mathbf{p} = \pm n \frac{\pi\hbar}{L} \tag{7.9}$$

$$K.E = \frac{\mathbf{p}^2}{2m} \implies K.E = n^2 \frac{\pi^2\hbar^2}{2mL^2} \implies K.E = n^2 C \ \textit{(where C is a constant)} \tag{7.10}$$

The energy levels have a pattern: $n^2 C$ where n is an integer and C is a constant that depends on the constraint L. When confined, the energy of the electron is *quantised*.

This model gives us a hint as to why atoms are stable. Classical physics (pre-quantum) cannot explain why electrons do not spiral down into the atomic nucleus. In the quantum world there is a ready explanation. The minimum kinetic energy of the constrained model electron in a box is $\frac{\pi^2\hbar^2}{2mL^2}$ (see Equation 7.10 with $n = 1$). Applying this logic more generally, if you very tightly constrain an electron, it must have a huge amount of energy.

It also gives us an underlying rationale for the distinct spectral emissions of atoms. The emissions of an electron must come from its switching between allowable energy levels. If the electron is constrained, these are quantised. Therefore we should not be surprised that the emissions of the constrained electrons in atoms have distinct spectral patterns.

One interesting feature of this model is that the standing wave function is a superposition of the constrained electron moving with momentum **p** to the left *and* to the right. The electron has momentum without physically moving! This sort of thing happens in quantum mechanics. How weird is that?

Let me end this discussion with the reminder that this is a *simple model*. In Chapter 15, we will discuss in more detail how the electron wave function is constrained in the atom by its attraction to the nucleus. The proper analysis must work in three spatial dimensions and account for the way potential energy changes with distance from the nucleus (the Coulomb potential).

7.5 Collapse of the Wave Function – a Twist

The physical act of measuring does not always have an impact on the wave function because the resulting state after measurements may leave the wave function in the same superposition and so unchanged. This is best explained with an example.

In the discussion on the Double-slit experiment (Section 4.4) we discussed the interference pattern that is created because the light (if we use light) is in a superposition of travelling through the two slits to the screen. I told you that the interference pattern disappears if you put a horizontal polarising filter over one slit and a vertical polarising filter over the other. This is shown as *A* in Figure 7.2. The different polarisation of the photons arriving at the screen could tell you which slit each photon had passed through. Therefore, in scenario *A* the superposition is destroyed. The light from the two photon streams creates two distinct spots on the screen.

Based on this result you understandably might conclude that the physical act of passing the photons through the polarisers destroys the superposition and collapses the wave function. However, this is not correct. For example, if both of the polarisers in *A* are horizontal, it is not possible to tell the photon streams apart, so the interference pattern of the Double-slit experiment does not disappear.

Scenario *B* on the right of Figure 7.2 shows this particularly clearly. After the top photon stream passes through the horizontal polariser, it is put through an angled polariser at 45 degrees. Half of the photon stream will pass through this angled filter and emerge polarised at 45 degrees. At the same time, the bottom photon stream passes through the vertical polariser and is then put through an identically angled polariser. This re-scrambles the polarisation of the two pathways before the photons hit the screen. It is no longer possible to tell which slit each photon has travelled through and the interference pattern returns.

In scenario *B* the photon streams pass through the same polarisers as scenario *A plus* the additional angled polarisers. Clearly it is not the physical process of measuring polarisation that is collapsing the superposition. The implication is clear. The change in the wave function does not depend just on physical measurement. It is important to evaluate any experiment *in its totality* to determine if it is possible to distinguish between different pathways. If you can distinguish between them, there is no superposition.

The next time you are having a soak in the bath tub, take a moment to think about this (and put on socks afterwards unlike Einstein in Box 7.5). It is fairly easy to understand *why* this happens. After all, if you can identify the particular path

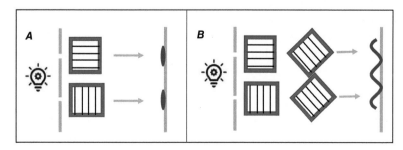

Figure 7.2 Schematic of Double-slit with polarisers.

of a photon, then it obviously cannot be in a superposition involving other possible pathways. However, it is a lot harder to understand *how* this happens.

Box 7.5   Einstein's socks

Einstein certainly was not a snappy dresser. He was well known for his dishevelled look. He refused to wear socks as he disliked how quickly they developed holes. Some historians have concluded that was because Einstein never cut his toenails!

7.6   Decoherence and the Measurement Problem

There are two connected but subtly different questions here. They concern *decoherence* and *the measurement problem* (often called the collapse of the wave function). Let me try to explain, this time with the example of an electron in the Double-slit experiment. Let's assume that the screen is phosphorescent and contains zinc sulphide (ZnS). The electron arrives with an independent wave function that is a broad spatial superposition. In hitting the screen, it is absorbed at a specific location by a ZnS molecule producing a flash of light:

1) *Decoherence*: What process leads to the wave function of the electron losing its independent nature and becoming entwined with the ZnS molecule and the screen?
2) *The measurement problem*: What process leads to the selection of one particular location on the screen for this interaction to occur?

Decoherence is an area of active research. Many measurements do not cause decoherence (spin measurement is an excellent example that we will discuss later). Avoiding decoherence is the key to constructing quantum computers so it is receiving a lot of attention. You will read all sorts of theories. For example, Roger Penrose has proposed that one source of decoherence may be the interaction of the wave function with gravity. The guy is a *genius*, but there are many differing views. Hopefully research will tell us more.

The measurement problem is another contentious subject. Many quantum physicists continue to accept what is called the *Copenhagen interpretation* proposed by Niels Bohr over a hundred years ago. This interpretation is best explained with an example. Consider again the Double-slit experiment. The electron can hit the screen in a number of places. The probability of each is determined by the square of the amplitude of the wave function. How does the electron transition from being in a number of possible places to just one? The response of the Copenhagen interpretation to the measurement problem is that this is a question you should not ask because, by definition, what is happening before the measurement is unmeasurable. This is sometimes described as *shut-up-and-calculate* or the *cop-out* interpretation. Personally, I find this a bit unfair. At least its advocates are open and honest about what they believe we can and cannot know.

There is also the *Bohmian interpretation* which is often called the *Pilot wave* theory. According to this, particles have a real and definite position prior to measurement. The wave function acts as a pilot wave guiding the movement of the particle. We can do no more than calculate the probability of a certain outcome because we do not know the exact position from which the particle started. This lack of exact information is often referred to as a *hidden variable*. The Bohmian interpretation met considerable resistance because it relies on the pilot wave being affected by all other particles, so it is *non-local*, which makes it inconsistent with special relativity. The current consensus (the Standard Model) assumes the opposite so the Bohmian interpretation faces an uphill struggle. Physicists get quite heated about this stuff. Reportedly, Pauli described it as *foolish simplicity... beyond all help*, and Oppenheimer disparaged it as *juvenile deviationism*, whatever that means.

Another almost infamous theory is Hugh Everett's *Many Worlds Interpretation* (see Box 7.6). Personally I find the logic very seductive, but it remains controversial. My wife, Rachel, gave me two brand new quantum mechanics books one Christmas with diametrically opposed views on the Many Worlds Interpretation so far be it from me to hazard too strong an opinion.

Box 7.6   Many Worlds Interpretation

Everett's idea is simple but mind-blowing. Consider the example of the electron hitting the screen and interacting with a molecule that leads to a flash of light. Imagine that the result is *not* a single flash of light at one location. Imagine instead that the result is a superposition of the interaction occurring at all locations (in line with the square of the amplitude of the electron wave function for each location). What would we observe? Everett argues persuasively that we would observe only one flash of light.

For simplicity let's model this assuming that the superposition is across only two locations so the flash will be at A (F_A) or B (F_B). I, Simon (S) observe the screen. The part of the superposition that is F_A creates a flash that hits my eye, interacts with my visual cortex and tells my brain that the flash is at A. At the same time the part of the superposition that is F_B tells my brain that the flash is at B. The overall superposition is $F_A S_A$ and $F_B S_B$. My *brain* is now part of the superposition! One version of me sees the flash at A and another version sees it at B.

The upside of this theory is that the measurement problem disappears because no particular value is favoured. All options are treated equivalently. The downside is that you have to get your head around a universe with an infinite and explosively increasing number of parallel scenarios, all coexisting in a massive superposition.

The short answer to the questions on decoherence and the measurement problem is that I don't know. Nobody knows for sure yet (though many think they do!). This is another one of those moments when if you know the answer you should ring Stockholm and collect your Nobel Prize.

However, it is far from obvious how you would test the difference between these various interpretations. They all make the same predictions with regards to everything you have read in this book so, it may not be easy to win that prize (check out Box 7.7 to see what Einstein did with his prize).

Box 7.7   Einstein's divorce

In 1903 Einstein married Mileva Maric, a female student who had been studying with him at Zurich Polytechnic. He was 24 years old at the time. As part of their divorce settlement in 1919, Einstein promised to hand over future money from the Nobel Prize that everyone expected him to win. He made good on this promise three years later. The prize was worth about US$30,000.

7.7   When a Body Leaves a Body – Entanglement at a Distance

In Section 7.3 we modelled what happens when particles come together, such as when an electron is constrained to be close to a proton. What about the opposite? What are the implications for objects or particles that split? You can build various scenarios such as blowing apart a macro object like a golf ball, the decay of a particle into two smaller particles or the emission of something like a photon from a particle. The result is that you start with one particle and end with two or more, and the resulting particles are *entangled*.

Consider a particle splitting into two resulting particles, A and B. We know that all particles are in some sort of superposition so let's say that particle A is in some sort of superposition of momentum states. As momentum has to be conserved, particle B must be in a correlated superposition of momentum states. The states of the two particles are entangled. They are not independent.

Now for the weird part. The particles fly apart. Far away, we measure the momentum of particle A (taking care that the particles do not interact with anything en route). The moment we measure the momentum of particle A, we know the momentum of particle B because of conservation of momentum. *But* if we had not measured the momentum of A, then both particles would remain in a superposition of coexisting momentum states. The measurement of the momentum of A means that we have a different wave function for B.

This *bugged* Einstein! In fact, he just could not accept it. He described it as *spooky action at a distance* yet it is where the combination of superposition and conservation laws clearly leads you. A lot of research has been done on this typically using entangled photons with correlated polarisation. Most famous is Alain Aspect's experiment (covered in more detail later in Section 17.4) which led to his winning the 2022 Nobel Prize. The result of it and subsequent work is that this spooky action at a distance is a reality. If you had trouble accepting that the particle state extends physically much further than the size of the particle itself, you now have to cope with the state including more than one particle. No longer can you think of particle A and particle B as separate entities. They are locked together as one even though they may physically be far apart.

What irked Einstein is that this action at a distance does not permit any delay. There is nothing in the maths we have done that allows it to be limited by the speed of light. Indeed, if it were, you could measure the momentum of particle A and then quickly measure the position of particle B in the time that the message is travelling to B about the measurement of A. Using conservation laws you could calculate the position *and* momentum of both particles breaking Heisenberg's uncertainty principle and driving a stake into the heart of all the maths of quantum mechanics!

It gets murkier. You may be tempted to think that measuring particle A *instantly* changes the wave function of B. But think back to relativity. For different observers, *instant* means different things as there is no absolute time. If one scientist measures the momentum of A and another measures that of B (far away) then their measurements will agree. But on one observer's clock the measurement of A might happen before that of B while on the clock of another observer (moving at a different speed) the measurement of B could happen before that of A. There is no *instant* change. We cannot choose a particular moment. There is no clear causal chain that measuring A affects B or vice versa. The link between the two ignores time. It appears somehow to tunnel back through it. The link just *is*. It is easy to understand why dear old Albert was so perplexed by this.

There is a saving grace for Einstein. Because quantum theory is probabilistic, the outcome of measurements is unpredictable. It turns out this means that you cannot communicate using entangled particles. The universal light speed limit still applies for messages and information.

7.8 Summary

At the start of this chapter we looked at how measurement can collapse the wave function, so that the particle has one value (or a more limited range of values) for a particular variable. Let me remind you to be cautious about the word *collapse*. Don't forget that decreasing the superposition for one variable increases the superposition for another. For example, if your measurement accurately defines the position of a particle, then the particle's momentum becomes less well defined. That is Heisenberg's uncertainty principle.

I defined as *quantum knowledge* the perfect information we might have of a quantum system that allows us to calculate the *probability* of each result. This is all we can know. The actual final result comes down to chance.

We compared the positional constraints of particles of different sizes. I used the combined uncertainty $\Delta x \Delta v_x$ and my layman language of fuzziness to try to illustrate how a smaller mass particle such as the electron is less well defined in terms of current/future position than a larger mass particle such as the proton. The two cannot be constrained in the same way (check out Box 7.8 for an amusing view of how this might affect a chicken).

Box 7.8   Just for laughs: Why did the chicken cross the road?
Charles Darwin: *For the species to survive, the chicken must reach the other side and expand its territory.*
Isaac Newton: *Chickens at rest stay at rest. Chickens in motion cross roads.*
Albert Einstein: *Whether the chicken crossed the road or the road moved beneath the chicken depends on your frame of reference.*
Werner Heisenberg: *We cannot be sure which side of the road the chicken is on, but it sure is moving fast.*
Simon Sherwood (your author): *Pass the gravy. This chicken is delicious.*

The electron-in-a-box scenario allowed us to model some of the implications if an electron is spatially constrained in the way it might be because of the electromagnetic attraction between electrons and protons in an atom. This simple model shows that constraining the electron quantises its energy levels. It gives us a flavour of why there is a distinct pattern to the emission spectra of atoms and why electrons do not spiral down into the atomic nucleus.

Section 7.6 introduced the measurement problem and decoherence, along with a brief introduction to the Copenhagen interpretation, Bohm's pilot wave theory and the seductively logical but mind-blowing idea of Everett's Many Worlds Interpretation (see Box 7.9 for background on Everett's tragic life).

The final twist came from considering what happens when particles split or emit, leading us to entanglement and what Einstein described as spooky action at a distance.

This chapter may have left you with rather more unanswered questions than you would like. That is the nature of quantum mechanics. It often seems easier to explain *why* things happen than *how* things happen.

In the following chapter we will summarise what we have covered so far and then start to ask the next big question: how does the wave function of a particle change over time and what can that tell us about the behaviour of particles? Read on!

Box 7.9   Hugh Everett III (1930–1982): a tale of scorn, ash and trash

Everett alighted on his *Many Worlds Interpretation* at the age of 24 after, in his words, *a slosh or two of sherry* with fellow students at Princeton. The theory is now widely respected, but was met with scorn when originally developed. Leon Rosenfeld, who was present at a key meeting with Niels Bohr on the subject, famously ridiculed Everett as *indescribably stupid... he could not understand the simplest things in quantum mechanics.*

For Everett, the meeting was *hell ... doomed from the beginning.* He was after all just a student, pushed by his advisor to work on the topic. In order to be published, Everett had to water down the theory. Exasperated, he dropped physics and took a government role assessing likely casualties in the event of a nuclear war. But in spite of all, Everett retained an unshakeable belief in his theory. Sadly, chain smoking, heavy drinking and obesity contributed to his early death from a heart attack aged 51, many years before his theory garnered any mainstream support. As a confirmed atheist, he instructed his ashes be thrown out with the trash.

Tragically, 14 years later, his daughter took her own life leaving similar instructions: *please sprinkle me in water... or the garbage, maybe that way I'll end up in the correct parallel universe with Daddy.* Everett believed in *quantum immortality...* that everyone's life continues in some version of the universe... I hope he is right and that he and his daughter are together.

Notes

1 A message for any reader who wants to point out that with perfect knowledge of the roll we might know more: *Stop being a pedant and get a life. You know what I am getting at!*

2 You can prove Equation 7.8 using $\sin(x \pm y) = \sin x \cos y \pm \cos x \sin y$.

Chapter 8

Module Summary and Schrödinger

CHAPTER MENU
8.1 Module Summary
8.2 Adding up the Implications...
8.3 The Path to Schrödinger's Equation
8.4 Module Memory Jogger

8.1 Module Summary

In the next chapter we need to step up the mathematics by introducing a complex form of the wave function. Don't panic! It will all be okay; but this is a good moment to review what you have learned. The following is a very brief summary of the book so far. Take a look through it to check the bits you understand (and the bits you don't). If you are confused or forgetful about a certain chapter or subject, take a quick look back as a refresher.

The first module in this book was *Special Relativity* which is built around two simple assumptions: that the speed of light is constant and the rules of physics are consistent for all observers. In Chapter 1, we looked at the implications of this for the fabric of spacetime. This gave us the invariant interval equation which compares the passage through time and space for an object when observed stationary versus when observed moving:

$$c^2 \, d\mathbf{T}^2 \; = \; c^2 \, dt^2 - dx^2 - dy^2 - dz^2 \qquad \textit{Invariant interval of Minkowski spacetime} \qquad (8.1)$$

Perhaps the most striking result is that time is no longer absolute. A moving clock ticks more slowly than a stationary clock by a factor of γ which is:

$$\gamma \; = \; \sqrt{\frac{1}{1 - \frac{v^2}{c^2}}} \qquad\qquad\qquad (8.2)$$

We analysed the various distortions to time and space that are essential for this spacetime (which is called Minkowski spacetime) to be mathematically consistent, specifically:

1) Time dilation so each second is longer and elapsed time is shorter: $\quad dt_{(moving)} \; = \; \gamma \, d\mathbf{T}_{(static)}$

2) Lengths appear shorter in the direction of motion: $\quad distance_{(moving)} \; = \; \dfrac{distance_{(static)}}{\gamma}$

3) Clocks appear out of sync as leading clocks lag by: $\quad \dfrac{distance_{(static)} \, v}{c^2}$

Chapter 2 looked at some of the paradoxes associated with special relativity. For example, if you compare your clock with that of a spaceship speeding by, you will see the spaceship's clock ticking more slowly. However, an observer on the spaceship would see your clock ticking more slowly than the spaceship's clock. This sort of paradox is resolved by considering all three distortions with the change in synchronisation of clocks playing a particularly important role, especially in the Twin Paradox.

Quantum Untangling: An Intuitive Approach to Quantum Mechanics from Einstein to Higgs, First Edition. Simon Sherwood.
© 2023 John Wiley & Sons Ltd. Published 2023 by John Wiley & Sons Ltd.

In Chapter 3 we switched our attention to energy and momentum by studying the rate of change of the invariant interval equation. We developed new relativistic definitions for momentum and energy. While the relativistic definition of momentum is almost identical to the classical version except at very high speeds, the relativistic definition of energy is radically different because it includes Einstein's famous term for rest mass energy, $E = mc^2$:

$$\mathbf{E} = mc^2\gamma \approx mc^2 + \frac{1}{2}mv^2 \qquad\qquad \mathbf{p} = mv\gamma \approx mv \qquad\qquad (8.3)$$

The analysis also revealed the relativistic relationship between energy, momentum and rest mass as shown below on the left in total and, on the right, separating the three spatial dimensions:

$$m^2c^4 = \mathbf{E}^2 - \mathbf{p}^2c^2 \qquad\qquad m^2c^4 = \mathbf{E}^2 - (\mathbf{p}_x^2 + \mathbf{p}_y^2 + \mathbf{p}_z^2)c^2 \qquad\qquad (8.4)$$

As a final flourish in the chapter we developed one further important relationship with the help of a Lorentz transformation. If different observers track the progress of an object moving through spacetime from a common point of origin ($t = x = y = z = 0$), then all will agree on the quantity ($\mathbf{p}x - \mathbf{E}t$). This relationship is shown below on the left in one spatial dimension and, on the right, using all three. The symbol for time τ, which I call particle time, is the time measured on the clock of an observer who is stationary relative to the particle. From the perspective of this observer, the particle has no momentum and its energy is simply its rest mass energy mc^2:

$$\mathbf{p}x - \mathbf{E}t = -mc^2\tau \qquad\qquad \mathbf{p}_x x + \mathbf{p}_y y + \mathbf{p}_z z - \mathbf{E}t = -mc^2\tau \qquad\qquad (8.5)$$

Box 8.1   Hendrik Lorentz (1853–1928)

Lorentz developed the mathematics for length contraction and time dilation although he thought of it more as a distortion than a fundamental reality. Einstein recognised the importance of his contribution, naming special relativity the Lorentz-Einstein theory. Lorentz won the Nobel Prize and was revered in the Netherlands. At the stroke of 12 on the day of his funeral, state telegraph and telephone services were suspended for three minutes in his honour.

We started this module on *Essential Quantum Mechanics* with a brief review in Chapter 4 of some problems that undermined classical physics including the stability of atoms (electrons don't spiral into the nucleus), atomic spectra (atoms emit radiation of specific wavelengths) and the ultraviolet catastrophe (black bodies emit unexpectedly small low amounts of short-wave radiation). The first step in resolving these issues was the acceptance, based on evidence such as the photoelectric effect, that the energy of light comes in quanta according to the formula $E = hf$ where h is Planck's constant and f is the frequency of the light wave.

De Broglie had the audacity to propose, correctly, that this applies to all particles. If $E = hf$ then, as $E = mc^2$ in the rest frame, the particle must have what he called some sinusoidal *internal periodic phenomenon* such as that shown on the left of Equation 8.6 (remember that $\hbar = \frac{h}{2\pi}$). Using the Lorentz transform in Equation 8.5, we can convert this to the coordinates for an observer who sees the particle moving (check out Box 8.1 on Lorentz). This is shown on the right side of Equation 8.6.

$$\phi = \sin\frac{1}{\hbar}(-mc^2\tau) \quad\Longrightarrow\quad \phi = \sin\frac{1}{\hbar}(\mathbf{p}x - \mathbf{E}t) \quad \textit{in moving frame} \qquad\qquad (8.6)$$

This is the free particle wave function. The convention is to show momentum as positive and energy as negative, but the reverse is legitimate as long as you are consistent. Also, the choice of a sine wave versus a cosine wave is arbitrary. The phase of the wave is unimportant because the physics is driven by the change of phase rather than the phase itself. Indeed, you will discover in later chapters that the phase of the wave function is undefined.

We discussed the Double-slit experiment that provides experimental evidence of the wave-like nature of particles. It also highlights that we must change our notion of what a particle is. In the Double-slit experiments the interference pattern results from the superposition of *two or more* distinct available pathways for each *single* particle. All available pathways affect the outcome. Schrödinger's mind experiment of a cat that is both dead and alive is an extreme example of how hard it is to match this up with the physical world we experience.

In Chapter 5, we took a closer look at the free particle wave function. We discussed the relationship of its partial derivatives with energy and momentum. This will prove very important in the Schrödinger equation. The derivatives are shown in Equation 8.7. I explained how this sort of relationship is driven by the time and space distortions of special relativity. Note that the notation $\phi_{\langle \frac{\pi}{2} \rangle}$ indicates that the phase of the derivatives has shifted by $\frac{\pi}{2}$. For example, if ϕ were a sine wave, then the derivatives would be cosine waves.

$$\frac{\partial \phi}{\partial t} = -\frac{\mathbf{E}}{\hbar}\,\phi_{\langle \frac{\pi}{2} \rangle} \qquad \frac{\partial \phi}{\partial x} = \frac{\mathbf{p}_x}{\hbar}\,\phi_{\langle \frac{\pi}{2} \rangle} \qquad \frac{\partial \phi}{\partial y} = \frac{\mathbf{p}_y}{\hbar}\,\phi_{\langle \frac{\pi}{2} \rangle} \qquad \frac{\partial \phi}{\partial z} = \frac{\mathbf{p}_z}{\hbar}\,\phi_{\langle \frac{\pi}{2} \rangle} \qquad (8.7)$$

The free particle wave function (see Equation 8.6) reflects an inherent uncertainty between the values of position and momentum. If the momentum of the particle has a precise value, then ϕ is a sinusoidal function in variable x with no peak so the value of x is undefined (and vice versa for $\mathbf{p}$ if the value of x is precisely known). This created a paradox. For an object to have a precise value for momentum, its position is undefined. It is in a superposition of being simultaneously at all possible positions.

Box 8.2 Heisenberg: The Catcher Was a Spy *(now a motion picture)*

During the Second World War, an American major league baseball player, Moe Berg, was sent to attend a lecture given by Heisenberg in Zurich. Berg's orders were to assassinate Heisenberg if he was helping the Germans develop an atomic bomb. Berg attended the lecture with a pistol, but did not act. He concluded that Germany's progress towards an atomic bomb was overstated.

We showed the paradox is solved provided we accept that particles are in a superposition of states rather than having a precise value of momentum and position. For a particle to have even loosely defined position, it must be in a superposition (which I label Ψ) of momentum eigenstates ($\phi_{\mathbf{p}}$). This results in a trade-off: the more precisely defined the momentum, the less accurately defined the position. This is Heisenberg's uncertainty principle (see Box 8.2 for an anecdote on his life). Following the rules of Fourier analysis, these superpositions allow virtually any overall shape for Ψ but there are some limits: Ψ must be reasonably well behaved in that it must be continuous (no breaks), single valued (not have two or more values for any single point); what I described tongue-in-cheek as *not too weird*.

We discussed the Born rule. The probability of an outcome depends on the value of $|\Psi|^2$ which is the square of the modulus of Ψ. No longer can we expect to predict the exact outcome of a measurement. In most cases, the best we can do is predict the probability of each outcome. Einstein hated this but it appears, in Einstein's words, that indeed *God plays dice*.

In Chapter 6 we took a detour into path integrals to see what is going on behind the scenes. Feynman thought about the Double-slit experiment. If there are two slits, then the result is the superposition of the two paths; if 20 slits, then it would be the superposition of 20 paths. So if there are an infinite number of slits (i.e. no barrier) it must be the superposition of every possible path to the screen. The mind-blowing implication is that the behaviour of a particle is the result of a superposition of every possible pathway, however crazy, however long.

We dipped lightly into the maths to show that the different phase of the wave function calculated along the pathways creates interference that results in the particle appearing to travel along the path of least (or more accurately, stationary) Action. We calculated this quantity and showed that, at anything but relativistic speeds, this means the path of the particle will be the one that minimises the difference between kinetic energy and potential energy. This quantity is called the Lagrangian and it, and the classical formula for Action (see Equation 8.8 noting that V is potential energy) have been known for hundreds of years.

$$\text{Classical Action:} \quad \int_0^t \left(\frac{1}{2}mv^2 - V \right) dt \qquad \text{Lagrangian:} \quad \frac{1}{2}mv^2 - V \qquad (8.8)$$

With the help of the Euler-Lagrange equation, this allowed us to tie things back to Newton's equations of motion and show why the mass m in mc^2 that is the rate of change of the wave function in the rest frame is the same m as in Newton's formula $F = ma$.

Chapter 7 provided an introduction to measurement and interaction. We looked at how measurement can collapse the wave function for a particular variable. However, decreasing the superposition for one variable (e.g. position) will increase it for another (e.g. momentum). Bear this in mind if you use the term *collapse*.

I used the quantity $\Delta x \, \Delta v_x$ to illustrate that smaller mass particles such as the electron cannot be spatially constrained in the same way as a larger mass particle such as the proton can. We then used the electron-in-a-box model to show how spatially constraining the electron might lead to distinct energy levels and a stable atom. This led into a discussion of decoherence, the measurement problem and the implication that measuring the property of one particle can affect the state of an entangled particle far away – the *spooky action at a distance* that Einstein so disliked.

8.2 Adding up the Implications...

We have covered a lot of ground. Anyone unversed in the world of complex numbers may shy away from the next chapters so I worry that I may lose some of you dear readers at this point. Fear not! We have already covered the most important points of quantum mechanics. Enough of the maths. Let's step back for a moment to consider the broader picture that has emerged.

The classical concept of a particle is wrong. Particles are wave-like. Waves are particle-like. And this relationship is defined by some strange *wave function*. What is this wave function? We know it has mathematical significance (I compared it in Section 5.8 with π). Does it physically exist? We don't know for sure, but we know that it defines the particle in the sense that it defines where it is, what it is, and how it behaves.

What does the form of the wave function tell us? It tells us that the universe is weirdly different from what we sense in our day-to-day lives. Particles do not have exact position and momentum. They exist in a mix of states with inbuilt uncertainty. I have called this *fuzzy* because there is no single precise value, but remember that everything is exactly and precisely defined. However, all we can know is the *probability* of the particle's being at a certain location or having a certain momentum.

How can we reconcile this uncertainty in position with the world we experience? The inbuilt position/velocity uncertainty of larger objects is minuscule compared to that of small particles. Our experience of these particles comes from their interactions with the macro world, that is, things that we can see. Interaction creates constraints, the wave function changes, and the particle's position is better defined. So to us, an object is an object, but behind the scenes, too small to observe, there is uncertainty.

And what about the Feynman path integral? Particles don't travel from one place to another along one route. Their behaviour is the outcome of combining all available options in a massive superposition as Feynman showed. However, the contribution of most of these pathways cancels out except for those close to the path of least Action, which is typically a straight line from *A* to *B*. This means that at the macro level we do not perceive the underlying quantum weirdness.

Sometimes I wonder at how far we have drifted away from our old comfortable world view. It is like Alice in Wonderland falling down the rabbit hole, and the hole goes on and on. There is a lot more to quantum mechanics, but this is as far as we can get without some more advanced mathematical tools. Before that, let me try to give you a taste of what is to come.

8.3 The Path to Schrödinger's Equation

What now? Where do we go from here? It is all very well knowing about particles and their wave functions, but how can we put that information to use? How can we use it to *predict* things? What is stable? What is not? We need to be able to calculate how wave functions *change over time and space*. That is the next challenge. For this, the most important tool is the Schrödinger equation.

Let me start with bad news for readers who struggle with maths. To use the Schrödinger equation we will need to introduce a complex formula for the free particle wave function – complex in the sense that it involves imaginary numbers and, in truth, the form is quite sophisticated. There is no way round this. However, before we get into this more advanced maths, I will try to give you a taste of what the Schrödinger equation is and why it makes sense, but it can only be a taste.

Our strategy is first to work with an energy-momentum eigenstate ϕ and, second, to extend to superpositions Ψ. The derivatives of ϕ are shown as Equations 8.9 and 8.10 (as derived in Section 5.1.2). These relationships are the consequence of the time and space distortions of special relativity (for a reminder, refer back to the rocket illustration in Figure 5.2).

For simplicity we will work with only one spatial dimension. Remember that the notation $\langle\frac{\pi}{2}\rangle$ means that the phase of the wave differs from that of ϕ by $\frac{\pi}{2}$ that is, 90 degrees (for a reminder, see Box 5.2). In order to calculate the second derivatives, the wave function is differentiated twice, which shifts the phase by π radians, resulting in the negative values shown. You easily can check on a calculator that $\sin(\theta \pm \pi) = -\sin\theta$ and that $\cos(\theta \pm \pi) = -\cos\theta$:

$$\frac{\partial\phi}{\partial t} = -\frac{\mathbf{E}}{\hbar}\phi_{\langle\frac{\pi}{2}\rangle} \qquad\qquad \frac{\partial^2\phi}{\partial t^2} = -\frac{\mathbf{E}^2}{\hbar^2}\phi \tag{8.9}$$

$$\frac{\partial\phi}{\partial x} = \frac{\mathbf{p}}{\hbar}\phi_{\langle\frac{\pi}{2}\rangle} \qquad\qquad \frac{\partial^2\phi}{\partial x^2} = -\frac{\mathbf{p}^2}{\hbar^2}\phi \tag{8.10}$$

8.3.1 The Klein-Gordon Equation

The first relationship we will construct is called the Klein-Gordon equation. We simply feed the second derivatives shown on the right of Equations 8.9 and 8.10 into the relativistic energy-momentum equation (as from Equation 8.4):

$$m^2c^4 = \mathbf{E}^2 - \mathbf{p}^2c^2 \tag{8.11}$$

$$\frac{m^2c^4}{\hbar^2}\phi = \frac{1}{\hbar^2}\mathbf{E}^2\phi - \frac{1}{\hbar^2}\mathbf{p}^2c^2\phi$$

$$= -\frac{\partial^2\phi}{\partial t^2} + c^2\frac{\partial^2\phi}{\partial x^2}$$

$$\implies \qquad \frac{\partial^2\phi}{\partial t^2} - c^2\frac{\partial^2\phi}{\partial x^2} + \frac{m^2c^4}{\hbar^2}\phi = 0 \quad \textit{Klein-Gordon} \tag{8.12}$$

Before we get too excited, we must check that this equation works for superpositions of multiple eigenstates. If not, it will be useless as it will work only for a particle with a theoretically perfect value for position or momentum. We can show it does indeed work for superpositions by considering a particle in a superposition of two eigenstates ($\Psi = \phi_a + \phi_b$).

As a reminder, differentiation can be done by parts and this works for both first *and* second derivatives as follows:

$$\Psi = \phi_a + \phi_b$$

$$\frac{\partial\Psi}{\partial t} = \frac{\partial(\phi_a + \phi_b)}{\partial t} = \frac{\partial\phi_a}{\partial t} + \frac{\partial\phi_b}{\partial t}$$

$$\frac{\partial^2\Psi}{\partial t^2} = \frac{\partial\left(\frac{\partial\phi_a}{\partial t} + \frac{\partial\phi_b}{\partial t}\right)}{\partial t} = \frac{\partial^2\phi_a}{\partial t^2} + \frac{\partial^2\phi_b}{\partial t^2} \tag{8.13}$$

Noting this, we add together the Klein-Gordon equations that we know work for eigenstates ϕ_a and ϕ_b (using Equation 8.12 for each). Remember that mass m is the same for both ϕ_a and ϕ_b as they are states of the same particle. We can show that the Klein-Gordon equation also works for all superpositions by adding together the results for the two separate eigenstates (Equation 8.14).

$$\frac{\partial^2 \phi_a}{\partial t^2} - c^2 \frac{\partial^2 \phi_a}{\partial x^2} + \frac{m^2 c^4}{\hbar^2} \phi_a = 0 \quad and \quad \frac{\partial^2 \phi_b}{\partial t^2} - c^2 \frac{\partial^2 \phi_b}{\partial x^2} + \frac{m^2 c^4}{\hbar^2} \phi_b = 0 \tag{8.14}$$

$$\frac{\partial^2 \phi_a}{\partial t^2} + \frac{\partial^2 \phi_b}{\partial t^2} - c^2 \left(\frac{\partial^2 \phi_a}{\partial x^2} + \frac{\partial^2 \phi_b}{\partial x^2} \right) + \frac{m^2 c^4}{\hbar^2} (\phi_a + \phi_b) = 0$$

$$\frac{\partial^2 (\phi_a + \phi_b)}{\partial t^2} - c^2 \frac{\partial^2 (\phi_a + \phi_b)}{\partial x^2} + \frac{m^2 c^4}{\hbar^2} (\phi_a + \phi_b) = 0$$

$$\implies \quad \frac{\partial^2 \Psi}{\partial t^2} - c^2 \frac{\partial^2 \Psi}{\partial x^2} + \frac{m^2 c^4}{\hbar^2} \Psi = 0 \qquad \textit{Klein-Gordon for } \Psi \tag{8.15}$$

The result is Equation 8.15, the Klein-Gordon equation for the superposition. The trick to ensure that any such equation works for superpositions is to be careful not to include any powers of a derivative such as squares (see Box 8.3).

Box 8.3   Superpositions abhor powers

If we represent $\mathbf{E}^2$ and $\mathbf{p}^2$ using $\frac{\partial^2 \Psi}{\partial t^2}$ and $\frac{\partial^2 \Psi}{\partial x^2}$, that works fine with superpositions because

$$if \quad \Psi = \phi_a + \phi_b \quad then \quad \frac{\partial^2 \Psi}{\partial x^2} = \frac{\partial^2 \phi_a}{\partial x^2} + \frac{\partial^2 \phi_b}{\partial x^2}$$

But this is *not* true if we try to represent $\mathbf{E}^2$ or $\mathbf{p}^2$ using $\left(\frac{\partial \Psi}{\partial t} \right)^2$ or $\left(\frac{\partial \Psi}{\partial x} \right)^2$ because

$$\left(\frac{\partial \Psi}{\partial x} \right)^2 = \left(\frac{\partial \phi_a}{\partial x} + \frac{\partial \phi_b}{\partial x} \right)^2 = \left(\frac{\partial \phi_a}{\partial x} \right)^2 + \left(\frac{\partial \phi_b}{\partial x} \right)^2 + 2 \frac{\partial \phi_a}{\partial x} \frac{\partial \phi_b}{\partial x} \neq \left(\frac{\partial \phi_a}{\partial x} \right)^2 + \left(\frac{\partial \phi_b}{\partial x} \right)^2$$

Check it out. Try to do the Klein-Gordon equation for a superposition using $\left(\frac{\partial \Psi}{\partial t} \right)^2$ and $\left(\frac{\partial \Psi}{\partial x} \right)^2$

Equations that work for superpositions must not contain powers because they must be linear.

If you are working in all three spatial dimensions, the Klein-Gordon equation is easily expanded to include derivatives for the y and z dimensions:

$$\frac{\partial^2 \Psi}{\partial t^2} - c^2 \left(\frac{\partial^2 \Psi}{\partial x^2} + \frac{\partial^2 \Psi}{\partial y^2} + \frac{\partial^2 \Psi}{\partial z^2} \right) + \frac{m^2 c^4}{\hbar^2} \Psi = 0 \qquad \textit{Klein-Gordon 3D} \tag{8.16}$$

The Klein-Gordon equation remains important in some area of physics (zero spin particles)[1], but there are problems. It is built from the energy-momentum relationship in Equation 8.11, which includes the term $\mathbf{E}^2$. The difficulty is that negative values of energy are equally valid solutions because $(\pm \mathbf{E})^2 = \mathbf{E}^2$. This created solutions with negative energy and probability that concerned the pioneers of quantum theory. Even more seriously, it does not give the right energy patterns for the electrons in an atom.

8.3.2   A Taste of Schrödinger's Equation

My objective is *not* to give a detailed derivation or proof of Schrödinger's equation. Its utility will become apparent in later chapters when we use it to analyse such things as the emission spectra of hydrogen. Rather, my aim is to explain the rationale behind his work and to show you why the approximations and assumptions built into his equation are intuitive and reasonable.

Schrödinger wanted a relationship based on $\mathbf{E}$ rather than $\mathbf{E}^2$ to avoid the problem of negative solutions, so he used the approximation that the relativistic energy of a free particle is equivalent to the particle's rest energy mc^2 plus its kinetic energy $\frac{p^2}{2m}$ as shown in Equation 8.17. Although not exact, it is an accurate approximation even at high speeds (see Table 3.1).

$$E \approx \frac{p^2}{2m} + mc^2 \qquad \textit{Approximate energy of a free particle} \tag{8.17}$$

Using Equation 8.17 we can multiply both sides by ϕ and then substitute for $E\phi$ and $p^2\phi$ using the derivatives of the wave function, in much the same way that we did for the Klein-Gordon equation (Equation 8.11). However, when substituting for $E\phi$, we must keep track of the $\frac{\pi}{2}$ phase shift that comes when you differentiate a sinusoidal function. Equation 8.18 adjusts the expression for the derivative so you can see how to substitute for $E\phi$. Don't forget that shifting the phase by π is the equivalent of multiplying a sinusoidal function by -1.

$$\frac{\partial \phi}{\partial t} = -\frac{\mathbf{E}}{\hbar}\phi_{\langle\frac{\pi}{2}\rangle} \qquad \Longrightarrow \qquad \frac{\partial \phi_{\langle\frac{\pi}{2}\rangle}}{\partial t} = -\frac{\mathbf{E}}{\hbar}\phi_{\langle\pi\rangle} = \frac{\mathbf{E}}{\hbar}\phi \tag{8.18}$$

We substitute for $p^2\phi$ using the second derivative $\frac{\partial^2 \phi}{\partial x^2}$ (see Equation 8.10). It is important that we use the second derivative because we want an equation that works for superpositions (see Box 8.3). These substitutions result in the equation shown as Equation 8.20 for a free particle.

$$E\phi = \frac{p^2}{2m}\phi + mc^2\phi \qquad \textit{accurate at non-relativistic speeds} \tag{8.19}$$

$$\hbar\frac{\partial \phi_{\langle\frac{\pi}{2}\rangle}}{\partial t} = -\frac{\hbar^2}{2m}\frac{\partial^2 \phi}{\partial x^2} + mc^2\phi \qquad \textit{this is for a free particle} \tag{8.20}$$

8.3.3 Incorporating Potential Energy

This is all well and good, but Schrödinger was not very interested in free particles. His aim was to understand the energy levels of electrons constrained by an electromagnetic field such as the electrons in atoms that are bound by their attraction to the positively charged nucleus. His proposal was simple and intuitive. He added to the energy balance in Equation 8.17 the *potential energy* that the electron has due to the presence of the electromagnetic field. Repeating the substitutions that we have just done results in Equation 8.21 using the total energy balance of an electron including its potential energy (labelled V).

$$E \approx \frac{p^2}{2m} + V + mc^2 \qquad \textit{approximate energy in an electromagnetic field}$$

$$\hbar\frac{\partial \phi_{\langle\frac{\pi}{2}\rangle}}{\partial t} = -\frac{\hbar^2}{2m}\frac{\partial^2 \phi}{\partial x^2} + (V + mc^2)\phi \tag{8.21}$$

We can simplify this expression. The potential energy V is a *variable* that may change depending on location and time. For instance, the potential energy of an electron in the atom varies with its distance from the nucleus. However, the rest mass energy mc^2 is a *constant*. We can ignore the mc^2 term providing we remember that any calculations will give the energy of the electron over and above its rest mass energy. In words, this is the classical energy balance that can be summarised as a particle's *classical energy* equals *kinetic energy* plus *potential energy*, and the result is:

$$E = \frac{p^2}{2m} + V \qquad \Longrightarrow \qquad \hbar\frac{\partial \phi_{\langle\frac{\pi}{2}\rangle}}{\partial t} = -\frac{\hbar^2}{2m}\frac{\partial^2 \phi}{\partial x^2} + V\phi \tag{8.22}$$

This is Schrödinger's equation, but there is one complicating glitch. The phase of the ϕ on the left side of the equation does not match that on the right. We have taken only the *first* derivative with respect to time shifting its phase by $\frac{\pi}{2}$. To handle this, we need to use a complex formula for the wave function that tracks the phase changes. We will introduce this in the next chapter. It allows us to account for the $\frac{\pi}{2}$ phase difference by multiplying with i (which is $\sqrt{-1}$). This gives the normal version of Schrödinger's equation that you will find in textbooks:

$$\phi_{\langle\frac{\pi}{2}\rangle} \textit{ becomes } i\phi \qquad \Longrightarrow \qquad i\hbar\frac{\partial \phi}{\partial t} = -\frac{\hbar^2}{2m}\frac{\partial^2 \phi}{\partial x^2} + V\phi \tag{8.23}$$

This equation has become the bedrock of quantum mechanics (summarised in Box 8.4). It is fairly simple to show[2] that it is true also for any linear superposition ($\Psi = A\phi_a + B\phi_b + C\phi_c...$), so it holds for all wave functions.

Box 8.4   Schrödinger's equation

$$i\hbar \frac{\partial \Psi}{\partial t} = -\frac{\hbar^2}{2m} \frac{\partial^2 \Psi}{\partial x^2} + V\Psi \qquad\qquad \textit{One spatial dimension}$$

$$i\hbar \frac{\partial \Psi}{\partial t} = -\frac{\hbar^2}{2m} \left(\frac{\partial^2 \Psi}{\partial x^2} + \frac{\partial^2 \Psi}{\partial y^2} + \frac{\partial^2 \Psi}{\partial z^2} \right) + V\Psi \qquad \textit{All spatial dimensions}$$

How do you use Schrödinger's equation? You insert the potential energy profile V that you want to investigate. For example, if you were investigating the hydrogen atom, the potential V would reflect the changing strength of the electromagnetic field at different distances from the proton. You also insert the mass m of the particle in question (an electron in this example). Then, you solve the equation to learn about the allowable energy levels and the spatial probability distribution of the electron. While this is easy in concept, the maths can be tricky.

If Schrödinger's equation is a baffling mystery of weird derivatives and complex numbers to you, try to remember its origin. All we have done is use the familiar classical definition of energy to tie together the derivatives of the wave function. Schrödinger proved the worth of his equation by calculating the energy levels of hydrogen and showing that his result is a good match to experimental data.[3] Throughout the next module, *Complex Quantum Mechanics*, we will use the Schrödinger equation. It is an accurate guide to the quantum behaviour of electrons.

I want to plant a minor warning in the back of your mind. The Schrödinger equation has flaws. First, it is built on the classical energy balance shown in Equation 8.17 which is an approximation (albeit an accurate one). Second, it equates the first derivative with respect to time with the second derivative with respect to space, which means it cannot fully conform to special relativity that treats time and space on a similar footing. Third, the electromagnetic potential energy is not quantised although we know the field consists of photons ($E = hf$). Later in the book, when we get to the module on *Relativistic Quantum Mechanics*, we will need more accurate tools such as the Dirac equation and quantum field theory. But for now, do not worry too much. The Schrödinger equation is accurate enough for most of the analysis you will ever need so sit back and enjoy. And if you are struggling with all of this, take comfort from Box 8.5.

Box 8.5   Are you struggling with these quantum concepts?

The aim of the book so far has been to introduce all the key concepts of quantum mechanics using as simple maths as possible. However, this does not make the concepts easy to grasp and accept. Even the founding experts in the field struggled.

Max Planck, who became a strong advocate, described introducing quanta into his equations as:
An act of despair.

Erwin Schrödinger reputedly quipped of quantum mechanics' probabilistic nature:
I don't like it and I am sorry I ever had anything to do with it.

Niels Bohr: *Those who are not shocked when they first come across quantum theory cannot possibly have understood it.*

Max von Laue, Nobel Laureate, reputedly reacted to de Broglie's idea of electron waves: *If that turns out to be true, I will quit physics.*

Albert Einstein: *Who would have thought around 1900 that in fifty years time we would know so much more and understand so much less?*

Richard Feynman: *I think I can safely say that nobody understands quantum mechanics.*

– so the good news is, if you are struggling, you are not alone!

78 | *8 Module Summary and Schrödinger*

8.4 Module Memory Jogger

Here is a list of some of the key topics we have covered so far. Hopefully, it can act as a memory-jogger if you look back at this book some day in the dim distant future (note that all the formulas include only one spatial dimension for simplicity). And for a few other resources, check out Box 8.6.:

- *A moving object's time changes (aging more slowly)*
- *A moving object's length contracts in the direction of motion*
- *A moving object's clocks fall out of sync (leading clocks lag)*
- *Relativistic energy:* $\mathbf{E} = mc^2\gamma \approx mc^2 + \frac{1}{2}mv^2$
- *Relativistic momentum:* $\mathbf{p} = mv\gamma \approx mv$
- *Mass-energy duality:* $m^2c^4 = \mathbf{E}^2 - \mathbf{p}^2c^2$ *so at rest* $\mathbf{E} = mc^2$
- *Lorentz transform shows all observers agree on:* $\mathbf{p}x - \mathbf{E}t$ *which at rest is* $-mc^2\tau$
- *More widely applying* $E = hf$ *leads to a sinusoidal wave function such as:* $\phi = \sin\frac{1}{\hbar}(\mathbf{p}x - \mathbf{E}t)$
- *Superpositions and Heisenberg's uncertainty principle:* $\Delta x\,\Delta\mathbf{p} \geq \frac{\hbar}{2}$
- *God plays dice – the Born rule that the probability of an outcome is proportional to* $|\Psi|^2$
- *Feynman path integral: everything that can happen does, kind of –*
- *Action and the Lagrangian linking the change in phase of the wave function with* $F = ma$
- *Electron-in-a-box illustrates why a constrained electron has quantised energy levels*
- *Decoherence and the measurement problem – an ongoing debate*
- *Entanglement leading to spooky action at a distance, which Einstein so disliked*
- *A brief introduction to Schrödinger's equation*

Box 8.6   Other Resources: Essential Quantum Mechanics

There are not many resources that give you an extensive description of quantum mechanics without the use of complex numbers. The classic book (and still one of the best) is:

In Search of Schrödinger's Cat by John Gribbin

For a deeper dive:

The Quantum Universe: Everything That Can Happen Does by Brian Cox and Jeff Forshaw plus the timeless masterpiece:

QED: the strange theory of light and matter by Richard Feynman

There are some very helpful and entertaining videos by 3Blue1Brown: *https://www.youtube.com/c/3blue1brown/featured* (as at December 2022)

I also recommend you look through the videos by PBS Spacetime: *https://www.youtube.com/c/pbsspacetime/videos* (as at December 2022)

Notes

1 In Chapters 24 and 25, we will use the Klein-Gordon solution in our discussion of the zero spin Higgs boson.
2 This is demonstrated later in Section 11.1.2.
3 This is covered in detail in Chapter 15.

Module III

Complex Quantum Mechanics

Chapter 9

Introducing Complex Numbers

CHAPTER MENU

9.1 Welcome to Complex Numbers
9.2 Representing the Wave Function with Complex Notation
9.3 Summary

9.1 Welcome to Complex Numbers

To dig deeper into quantum mechanics you are going to have to get used to working with the complex version of the particle wave function. Not only is it needed to simplify the maths, but without it, quantum mechanics is inconsistent with special relativity. If complex numbers are relatively alien to you, this may seem scary at first. I will do my utmost to take things slowly.

Box 9.1 Complex numbers: a brief refresher

Complex numbers are of the form $a + ib$ where $i = \sqrt{-1}$. a is called the *real* component and b the *imaginary* component. Addition and multiplication of complex numbers is straightforward using $i^2 = -1$:

$$(a + ib) + (c + id) = (a + c) + i(b + d)$$

$$(a + ib)(c + id) = ac - bd + i(ad + bc)$$

The *magnitude* of a complex number is calculated as:

$$|a + ib| = \sqrt{(a + ib)(a - ib)} = \sqrt{a^2 + b^2}$$

$(a - ib)$ is called the *complex conjugate* of $(a + ib)$ and can be written as $(a + ib)^*$

Some of you already will be familiar with the use of $e^{i\theta}$ to represent waves, but still I advise you to read this section unless you're completely at ease with the use of complex numbers in quantum mechanics. Again and again I hear people ask why i appears in Schrödinger's equation. Again and again I see answers that are confused.

We are going to limit our discussion at this stage specifically to the use of $e^{i\theta}$. This appears in every area of quantum mechanics and is essential for Schrödinger's equation. Towards the end of the book, complex numbers are used in a broader form to build the fields in quantum field theory, but let's leave that until later.

Why use complex numbers in the wave function? In broad terms the answer is quite simple. We use $e^{i\theta}$ and the maths that surrounds it as a *notational* aid to keep track of the change in phase of the particle state. It gives us flexibility to account for this extra variable (phase) in the equations. There is no magical mystery of the universe here although the maths fits so well that you do sometimes wonder.

Quantum Untangling: An Intuitive Approach to Quantum Mechanics from Einstein to Higgs, First Edition. Simon Sherwood.
© 2023 John Wiley & Sons Ltd. Published 2023 by John Wiley & Sons Ltd.

9.1.1 We Have a Problem

In this book I have generally used a sine wave to represent the wave function ϕ of a free (unconstrained) particle with momentum **p** and energy **E** in the form $\sin \frac{1}{\hbar}(\mathbf{p}x - \mathbf{E}t)$ but stressed repeatedly that we could equally have used a cosine wave. The difference between a sine wave and a cosine wave is simply where you put the point $x = 0$. The position of $x = 0$ relative to the wave is the *phase* of the wave. If $x = 0$ is at a peak, then it is a cosine wave. If $x = 0$ is at a zero point for the wave, then it is a positive or negative sine wave. A cosine wave is a sine wave shifted by one quarter of a wavelength which is $\frac{\pi}{2}$, so we can write $\cos(x) = \sin(x + \frac{\pi}{2})$. Shift another quarter of a wavelength and you have the negative of a sine wave, another quarter and you have a negative cosine wave and with another quarter you will be back to a sine wave as you will have shifted 2π in total which is a whole wavelength (if you need a refresher, refer back to Figure 5.1).

This creates a major headache. One observer's sine wave is another observer's cosine wave. The mathematical notation for the particle state ϕ is all over the place. One observer may see ϕ as $\cos \frac{1}{\hbar}(\mathbf{p}x - \mathbf{E}t)$ and another observer a little distance away sees it as $\sin \frac{1}{\hbar}(\mathbf{p}x - \mathbf{E}t)$ and what about an observer $\frac{\pi}{4}$ away who will see it as $\cos \frac{1}{\hbar}(\mathbf{p}x - \mathbf{E}t + \frac{\pi}{4})$ which can also be written approximately as $0.7(\sin \frac{1}{\hbar}(\mathbf{p}x - \mathbf{E}t) + (\cos \frac{1}{\hbar}(\mathbf{p}x - \mathbf{E}t))$? Is your head spinning yet? From relativity we know that all these viewpoints are equally valid. All the observers are seeing the *same* particle state. All of them are seeing the same ϕ.

It is impossibly confusing to have the notation for ϕ change so much. Earlier in the book I introduced notation such as $\phi_{(\frac{\pi}{2})}$ to represent its being out of phase with ϕ by $\frac{\pi}{2}$. This is helpful to show phase change, but does not help us manipulate or calculate. We need notation that clearly indicates that ϕ is ϕ whether it is a sine wave or a cosine wave. Whatever the phase, we want it to be obvious that we are dealing with the same particle state ϕ. But at the same time, we must be able to keep track of any phase changes as they are important if we get wave interference such as in the Double-slit experiment. Does this sound an impossible goal?

Step forward $e^{i\theta}$. Many of you will cringe at this. I mean *what*? But bear with me. It really is super-nifty.

$$e^{i\theta} = \cos\theta + i\sin\theta \tag{9.1}$$

Let's do a quick reality check on Equation 9.1. It may seem a bit bizarre, but the equation is true. You can search up a proof on the Internet if you feel so inclined. A little playing with numbers should get you comfortable with this expression. You can show with $\theta = 0$ that both the right side and left side equal 1 (remember that $e^0 = 1$). You can show both sides of the equation differentiate in the same way. The right side gives $-\sin\theta + i\cos\theta$ and the left side gives $ie^{i\theta}$, which is the same thing. For a bit of fun, set $\theta = \pi$ and you get Euler's beautiful identity $e^{i\pi} = -1$ (Box 9.4, later in this chapter, has some background on Euler if you are interested).

Enough playing. Why is $e^{i\theta}$ in Equation 9.1 useful for quantum mechanics? Let's start by looking at its ability to represent a wave of any phase.

9.1.2 Complex Notation for Phase

Our cunning plan is to take the wave part of the particle state that we might have written in the past as $\sin \frac{1}{\hbar}(\mathbf{p}x - \mathbf{E}t)$ or $\cos \frac{1}{\hbar}(\mathbf{p}x - \mathbf{E}t)$, and to express it, using the structure $e^{i\theta}$, as $e^{\frac{i}{\hbar}(\mathbf{p}x - \mathbf{E}t)}$ (this is actually a complex vector if you are interested in the technical details). Effectively we are using $e^{i\theta}$ by setting $\theta = \frac{1}{\hbar}(\mathbf{p}x - \mathbf{E}t)$ This will bring huge benefits. But first let's take a closer look at $e^{i\theta}$.

The structure of $e^{i\theta}$ is shown in Figure 9.1. The real component $\cos\theta$ is shown on the x axis and the imaginary component $i\sin\theta$ is shown on the y axis. Let's compare $e^{i\theta}$ with the plain vanilla sinusoid $\cos\theta$. If we change θ what happens? In the case of $\cos\theta$ that means a shift along the sinusoid, that is, a phase shift. In the case of $e^{i\theta}$, it is a rotation around the circle. First, consider a shift of 2π. This is a full wavelength shift for $\cos\theta$ so it is unchanged as $\cos\theta = \cos(\theta + 2\pi)$. Also a shift of 2π for $e^{i\theta}$ leaves it the same as it brings it full circle. Other shifts in θ have a similar effect on both functions. For example, a shift of $\frac{\pi}{2}$ is a quarter wavelength shift for $\cos\theta$ and a quarter of the way round the circle for $e^{i\theta}$, so the phase change reacts identically for both.

Now the clever bit. If we want to record a change in the phase of $\cos\theta$, we have no choice but to change θ or do something like switch from $\cos\theta$ to $\sin\theta$. In the case of $e^{i\theta}$, we have another, clearer way of altering the phase. We can multiply $e^{i\theta}$ by a complex number. We can track changes to the phase of $e^{i\theta}$ while leaving $e^{i\theta}$ intact. For example, the shift of $\frac{\pi}{2}$ that changes $\cos\theta$ to $-\sin\theta$ can be achieved for $e^{i\theta}$ by multiplying by i. A couple of examples are shown here:

Figure 9.1 The use of $e^{i\theta}$ allows us to preserve the form of the wave function while tracking changes to the phase.

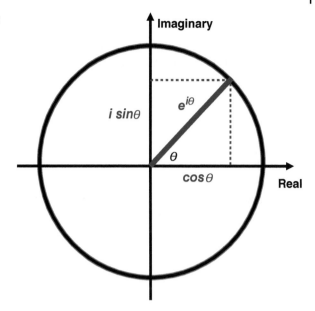

$$Add \; \frac{\pi}{2} \; to \; \theta \qquad e^{i(\theta+\frac{\pi}{2})} \; = \; i\,e^{i\theta} \qquad\qquad = \; -\sin\theta + i\,\cos\theta \qquad\qquad (9.2)$$

$$Add \; \frac{\pi}{4} \; to \; \theta \qquad e^{i(\theta+\frac{\pi}{4})} \; = \; \frac{(1+i)}{\sqrt{2}}\,e^{i\theta} \qquad\qquad\qquad (9.3)$$

The $\sqrt{2}$ that appears in Equation 9.3 *normalises* the expression so its magnitude remains 1.

9.1.3 Interference Calculations

So, we can use $e^{i\theta}$ to track phase changes. But how do we handle interference? Consider the Double-slit experiment (Section 4.4). Two different lengths of path mean different phases for ϕ and interference. Can the form $e^{i\theta}$ help us? Yes, it can! It could not be easier. Just add the amplitudes together. The real cosine components add or cancel, and the imaginary sine components add or cancel, and hey presto, you get the right answer.

Suppose we use our old version of ϕ in the form $\cos\theta$. If one cosine wave meets another cosine wave, negative or positive, they add up or cancel. Easy! But what if different amplitudes and phases of the wave meet? What if $a\cos\theta$ meets $b\sin\theta$ which is $\frac{\pi}{2}$ out of sync. The answer using the old notation is far from obvious:

$$a\cos\theta + b\sin\theta \; = \; ??? \qquad\qquad (9.4)$$

With the new notation it is easy. Let's label the first wave ϕ_1 and the second ϕ_2. We say $\phi_1 = ae^{i\theta}$. The change in phase is negative $\frac{\pi}{2}$ and so is represented in our new notation with multiplication by $-i$. So in our new notation the second wave is $\phi_2 = -ibe^{i\theta}$. Then you simply add the complex amplitudes together:

$$\phi_1 \; = \; ae^{i\theta} \; = \; a\cos\theta + ia\sin\theta$$

$$\phi_2 \; = \; -ibe^{i\theta} \; = \; b\sin\theta - ib\cos\theta$$

$$\phi_1 + \phi_2 \; = \; (a - ib)e^{i\theta} \qquad\qquad (9.5)$$

$$|\phi_1 + \phi_2| \; = \; \sqrt{(a-ib)(a+ib)} \; = \; \sqrt{a^2 + b^2} \qquad\qquad (9.6)$$

The new notation is easy to use and you can see at once that the result is the same $e^{i\theta}$ but with a different phase and with amplitude $\sqrt{a^2 + b^2}$.

9.1.4 A Friend with Benefits

There is one other important benefit of the complex notation. It sorts out a problem that we have when we use $\phi = \sin \frac{1}{\hbar}(p_0 x - E_0 t)$ for the state of an unconstrained particle with definite momentum p_0 and definite energy E_0. Let's look at ϕ at an instant in time and for convenience set $t = 0$. The result is $\phi = \sin \frac{p_0}{\hbar} x$ which is a sine wave over space stretching from $-\infty$ to $+\infty$. The problem is that this is not perfectly even over space. There are what we might call preferred locations. For example a sine wave is value zero at $x = 0$, π, 2π... but if the particle has no constraints, then this flies in the face of relativity. Without constraints there is no point of reference from which to define the point $x = 0$. I raised this problem back in Box 5.4.

One observer's sine wave is another observer's cosine wave if they stand a quarter of a wavelength apart. Who is right? Both. Special relativity tells us that all observers are equally right. The change in phase is important, but the starting phase of the wave function does not and cannot matter. Some sadistic physicist decided to give this sort of insensitivity to a variable the torturous name *gauge invariance*[1] and students have been scratching their heads ever since.

The problem is solved with the new complex vector notation $e^{i\theta}$ because it has a magnitude of 1 everywhere. Take a look back at Figure 9.1 and you can see that the amplitude of $e^{i\theta}$ is 1 for all θ. This is easy to show (remember that you must always multiply by the complex conjugate to calculate magnitude):

$$|e^{i\theta}|^2 = e^{i\theta}(e^{i\theta})^* = e^{i\theta}e^{-i\theta} = e^0 = 1 \tag{9.7}$$

For completeness this is also shown below with the expanded definition showing again that the amplitude of $e^{i\theta}$ is 1 for all θ.

$$|e^{i\theta}|^2 = e^{i\theta}(e^{i\theta})^* = (\cos\theta + i\sin\theta)(\cos\theta - i\sin\theta) = \cos^2\theta + \sin^2\theta = 1 \tag{9.8}$$

Consider Equation 9.9. I have taken the equation we developed for ϕ with a precise momentum p_0 at $t = 0$ and rewritten it in our new complex notation. The amplitude is the *same* for all x. It is always 1. There are no preferred spatial positions. Problem solved!

$$\phi = e^{\frac{i}{\hbar}p_0 x} \tag{9.9}$$

$$|\phi|^2 = e^{\frac{i}{\hbar}p_0 x} e^{-\frac{i}{\hbar}p_0 x} = e^0 = 1 \tag{9.10}$$

The switch to complex vector notation gives clearer equations (where ϕ stays ϕ), makes calculating interference simpler *and* smooths the amplitude of ϕ so that its magnitude is the same whatever the phase θ, and so makes it consistent with the requirements of special relativity. This complex notation truly is a friend with benefits (check out Box 9.2 for a bit of complex number history).

Box 9.2   The complex life of Gerolamo Cardano (1501–1576)

Back in the 1500s, Cardano used complex numbers to help solve cubic equations. In a colourful life he earned much of his income through gambling, helped by his treatises on probability and, perhaps more usefully, by his study on how best to cheat. His lifestyle was not a stable family environment. Of his three children, one was beheaded for murdering his wife and another disinherited for constantly stealing to cover gambling debts. Cardano spent several months in jail at the hands of the Inquisition having been declared *vehemently suspect of heresy*. To gain his freedom, he had to promise never to teach or publish again before his death.

9.1.5 Not a Free Lunch

As you can see, the benefits of using complex vector notation for phase are enormous. In fact, it is unrealistic to try to tackle a serious problem without it. But as the Americans say, there ain't no such thing as a free lunch, or as the English say, you have to pay the piper. There are some mathematical consequences. You must always multiply a complex number by its complex conjugate to calculate its magnitude (refer back to the refresher in Box 9.1 if you are unclear on this).

For example, if you need to work out the probability of a particle being at location x_0 for a wave function Ψ, then as part of the calculation you will need to calculate the square of the magnitude of Ψ at x_0. This *must* be done using $\Psi^*\Psi$ for the point $x = 0$ where Ψ^* is the complex conjugate of Ψ. Again and again you will see this structure appearing in the maths.

We will introduce later some notations to help keep track of what is the wave function and what is the complex conjugate, such as the *bra* and *ket* system invented by Dirac. The *ket* is shown as $|\Psi\rangle$ and represents the wave function Ψ (or, on occasions, some function such as integration of Ψ). The *bra* is shown as $\langle\Psi|$ and represents the equivalent but using the complex conjugate Ψ^*. The *bra* and *ket* generally have to be multiplied together to determine the magnitude. Hence Dirac's rather charming names for them that give *bra-ket* for the combination $\langle\Psi|\Psi\rangle$.

9.2 Representing the Wave Function with Complex Notation

Shown below on the right side of Equation 9.11 is the free particle wave function expressed in complex form, using one spatial dimension for simplicity. Its derivatives are also shown. For reference, Box 9.3 shows how to calculate the derivatives of this sort of exponential function:

$$\phi = \sin\frac{1}{\hbar}(\mathbf{p}x - \mathbf{E}t) \implies e^{\frac{i}{\hbar}(\mathbf{p}x - \mathbf{E}t)} \tag{9.11}$$

$$\frac{\partial\phi}{\partial t} = -\frac{i}{\hbar}\mathbf{E}\phi \qquad \frac{\partial^2\phi}{\partial t^2} = -\frac{1}{\hbar^2}\mathbf{E}^2\phi \qquad \frac{\partial\phi}{\partial x} = \frac{i}{\hbar}\mathbf{p}\phi \qquad \frac{\partial^2\phi}{\partial x^2} = -\frac{1}{\hbar^2}\mathbf{p}^2\phi \tag{9.12}$$

Box 9.3 Derivatives of exponential functions: a refresher

This type of exponential function differentiates as follows (where a and b are constants):

$$\frac{\partial}{\partial x}e^{(ax+by)} = a\,e^{(ax+by)} \qquad \frac{\partial^2}{\partial x^2}e^{(ax+by)} = a^2\,e^{(ax+by)}$$

$$\frac{\partial}{\partial y}e^{(ax+by)} = b\,e^{(ax+by)} \qquad \frac{\partial^2}{\partial y^2}e^{(ax+by)} = b^2\,e^{(ax+by)}$$

9.3 Summary

I hope this chapter has convinced you of the benefits of using the complex vector notation for the wave function. There is no mysticism in using i. It simply is a notational tool. The use of the structure $e^{i\theta}$ tracks the change of phase of the wave function in a beautifully clear way. We can multiply it by complex numbers to represent any phase or change of phase without distorting the underlying wave function itself. It makes interference calculations simpler and, as an (essential) added bonus, it generalises the wave function of a free particle to avoid peaks and troughs that would contradict the principle of relativity.

Box 9.4 Leonhard Euler (1707–1783)

Euler (pronounced *oiler* not *yuler*) achieved his Masters at only 17 years old. He was a prolific writer, producing over 500 books during his lifetime, many during the last 17 years of his life when he was blind (plus another 400 books published posthumously). It is a rather sad sign of the times that of his 13 children only 5 survived into adulthood. Euler himself lived to the ripe old age of 76, working until the day he died. Reportedly, he was discussing the orbit of Uranus with a colleague when he collapsed of a brain hemorrhage. In the words of Pierre-Simon Laplace: *Read Euler, read Euler, he is the master of us all.*

Euler's identity: $\mathbf{e^{i\pi} = -1}$

Note

1 Gauge invariance is an important feature of quantum physics and is covered in detail in Chapter 21.

Chapter 10

Superpositions and Fourier Transforms

CHAPTER MENU

10.1 The Maths of Fourier Transforms
10.2 Heisenberg's Uncertainty Principle and the Gaussian Distribution
10.3 The Quantum Footprint
10.4 Time and Energy
10.5 Summary

How does the complex notation work for superpositions? Consider a particle state that we might have expressed as a superposition of momentum eigenstates as shown in Equation 10.1. This is just as we did earlier with Equation 5.18 and again I have chosen a moment in time at $t = 0$ to simplify the expression (see Box 10.1 for details on notation).

$$\Psi = A_1 \phi_{p_1} + A_2 \phi_{p_2} + A_3 \phi_{p_3}.... \tag{10.1}$$

The momentum eigenstates generally will have phase differences between them so where ϕ_{p_1} is at a minimum, ϕ_{p_2} might be at a maximum. With complex notation, the weighting of each eigenstate component in the superposition is a *complex number* to account for this. In Equation 10.2, the weightings are expanded so you can see them expressed as complex numbers such as $(a + ib)$ where a and b are real constants.

$$\Psi = A_1 e^{\frac{i}{\hbar} p_1 x} + A_2 e^{\frac{i}{\hbar} p_2 x}... = (a+ib) e^{\frac{i}{\hbar} p_1 x} + (c+id) e^{\frac{i}{\hbar} p_2 x}... \tag{10.2}$$

The phase of each component is set by the value of the complex number. For example, if $a = 1$ and $b = 0$, then the real part is a cosine wave and the imaginary part is a sine wave. If we set $a = 0$ and $b = 1$, the real part is a negative sine wave and the imaginary part is a cosine wave as shown below. So, A_1 and A_2 record the phase of each component:

$$\text{if } a = 1 \text{ and } b = 0 \quad A_1 e^{\frac{i}{\hbar} p_1 x} = e^{\frac{i}{\hbar} p_1 x} = \cos \frac{1}{\hbar} p_1 x + i \sin \frac{1}{\hbar} p_1 x \tag{10.3}$$

$$\text{if } a = 0 \text{ and } b = 1 \quad A_1 e^{\frac{i}{\hbar} p_1 x} = i e^{\frac{i}{\hbar} p_1 x} = -\sin \frac{1}{\hbar} p_1 x + i \cos \frac{1}{\hbar} p_1 x \tag{10.4}$$

Using complex coefficients is very efficient because one complex coefficient such as A_1 above codes for both the phase and magnitude of the relevant eigenstate. By the way, don't forget that if you need to calculate the magnitude of the component, you must multiply A_1 by A_1^* (the complex conjugate of A_1) and take the square root. In the example above, the magnitude of the ϕ_{p_1} momentum eigenstate therefore is the square root of $(a + ib)(a - ib)$ which is $\sqrt{a^2 + b^2}$.

Box 10.1   Notation detail

From now on, the state of the particle will be expressed as the complex wave function. Ψ will be used generally for these wave functions, including for superpositions, and ϕ reserved for specific individual eigenstates.

Quantum Untangling: An Intuitive Approach to Quantum Mechanics from Einstein to Higgs, First Edition. Simon Sherwood.
© 2023 John Wiley & Sons Ltd. Published 2023 by John Wiley & Sons Ltd.

Any particle with even broadly defined position must be in a superposition of momentum eigenstates. You can describe the particle as being in such-and-such a superposition of momentum eigenstates *or* as being in such-and-such a superposition of position eigenstates. They are two ways of describing the same thing. We say that both momentum eigenstates and position eigenstates form a *basis* for describing Ψ.

If you know a composite particle state in terms of position x, that is, you know $\Psi(x)$, how do you calculate its composition in terms of momentum eigenstates, i.e. $\Psi(p)$? The answer is that you need to use the Fourier transform of $\Psi(x)$. A Fourier transform decomposes any waveform or function into the sinusoid components that make it up. You may have heard of its use in the processing of audio signals when it allows one to isolate and then remove certain frequencies of noise that may be distorting the signal. We need the same process because, in order to calculate $\Psi(p)$, we need to know the momentum eigenstates that add up to make Ψ. Each of these momentum eigenstates is a sinusoid when expressed in terms of x. The mathematical process is the same as is needed when analysing the sinusoidal components in an audio signal: Fourier analysis.

10.1 The Maths of Fourier Transforms

Box 10.2 Mathematics warning!

The mathematics in Fourier transforms is challenging. There are only a few key transforms that you might need and most of these can be found in tables. Do not panic if you struggle with this section. Just try to follow the gist.

Equation 10.5 shows the mathematical formula for calculating a Fourier transform. If you are like me, your initial reaction will be a visceral feeling of horror (see Box 10.2). If you want to get a feel of what is happening in the Fourier transform equation and process, I strongly recommend you to take a look at the video by the group 3Blue1Brown[1]. It is on YouTube and does a much better job of explaining than I could ever manage.

$$\Psi(k) = \int_{-\infty}^{+\infty} \Psi(x)\, e^{ikx} dx \tag{10.5}$$

The first thing to notice in Equation 10.5 is that I have shown the Fourier transform to $\Psi(k)$ rather than $\Psi(p)$. The relationship between momentum **p** and the wave number k is $\mathbf{p} = \hbar k$ (see the wave refresher in Box 4.2 if you need a reminder of how the relationship between **p** and k works). The advantage of using k is that there is complete symmetry between x and k in the free particle wave function, which keeps things simple[2]:

$$\phi = e^{\frac{i}{\hbar}\mathbf{p}x} = e^{ikx} \tag{10.6}$$

Note also that most formulas for Fourier transform include a multiplying factor of $\frac{1}{\sqrt{2\pi}}$ at the front. This is because converting from $\Psi(x)$ to $\Psi(k)$ and then back again to $\Psi(x)$ increases the value of the resulting $\Psi(x)$ by 2π. On this occasion, we don't need to keep track of this because we are interested in the distribution (or shape) of Ψ. Multiplying Ψ by a constant does not change the relative probability of the particle being in a certain position or having a certain momentum.

There is good news and bad news. The bad news is that (in my opinion) Fourier transforms are tricky to calculate from scratch. Personally, I am no fan! The good news is that you really don't need much more than a passing familiarity with what is happening. I will limit myself to two examples: one trivial, and one slightly harder but illuminating. We will also use the Fourier transform of the Gaussian distribution to illustrate Heisenberg's uncertainty principle.

Do *not* worry if you cannot cope with these. The important thing is to understand the concept. There are tables of common Fourier transforms and even on-line Fourier transform programmes to help you, should you need them (but be sure to use the right transform, see Box 10.3 for a warning). And I doubt that you will need them unless some sadistic teacher decides to put you through your paces.

Box 10.3   Choose the right Fourier transform formula

If you look in Wikipedia (and many textbooks) a mysterious 2π appears in the exponential factor such as:

$$\Psi(f) = \int_{-\infty}^{+\infty} \Psi(t)\,e^{i2\pi ft}\,dt \qquad \textit{versus} \qquad \Psi(k) = \int_{-\infty}^{+\infty} \Psi(x)\,e^{ikx}\,dx$$

This is because Fourier transforms are often used for switching between time and frequency, which is not the same as $x \Longleftrightarrow k$ or $t \Longleftrightarrow \omega$ because: $f = 2\pi\omega$. Warning: I once wasted a long time working with the wrong Fourier formula so I now always transform $x \Longleftrightarrow k$ and $t \Longleftrightarrow \omega$ using the formula from Equation 10.5 to avoid confusion.

10.1.1   Example 1: Fourier Transform of a Position Eigenstate

We will start with the trivial example (it is trivial because we already know the answer). Let's calculate the Fourier transform of a free particle in a position eigenstate. The mathematical representation for a particle precisely at $x = x_0$ is the Dirac delta function $\delta(x - x_0)$, which by definition is zero everywhere except at $x = x_0$ so the wave function has amplitude only at that position. We take this and apply the formula from Equation 10.5.

$$\Psi(x) = \delta(x - x_0) \tag{10.7}$$

$$\Psi(k) = \int_{-\infty}^{+\infty} \delta(x - x_0)\,e^{ikx}\,dx = e^{ikx_0} = e^{\frac{i}{\hbar}x_0\mathbf{p}} \tag{10.8}$$

$$\Psi(x) = \delta(x - x_0) \qquad \Longleftrightarrow \qquad \Psi(p) = e^{\frac{i}{\hbar}x_0\mathbf{p}} \tag{10.9}$$

The result in Equation 10.9 should not come as a surprise. For a position eigenstate at $x = x_0$ we simply replace x in the free particle wave function by the value x_0. Let me step through the workings carefully. The integral in Equation 10.8 looks nasty, but it is actually very straightforward. The integral is adding all the values of e^{ikx} for all values of x from $-\infty$ to $+\infty$. But the delta function means that the only piece of the summation that has value is the e^{ikx} when $x = x_0$, which gives e^{ikx_0}. All the other parts of the integral summation are zero. Simple! The last part of Equation 10.8 switches from k to $\mathbf{p}$ using $\mathbf{p} = \hbar k$.

10.1.2   Example 2: Fourier Transform of $\frac{\partial \Psi}{\partial x}$

The second example of a Fourier transform is more illuminating. Before you read the solution you may want to stop and think, because you already know the answer. What is the Fourier transform of the differential of Ψ with respect to x, that is, what is the Fourier transform of $\frac{\partial \Psi}{\partial x}$?

The easiest way to calculate this is to switch x and k in the Fourier transform formula (x and k are symmetrical) and take the derivative:

$$\Psi(x) = \int_{-\infty}^{+\infty} \Psi(k)\,e^{ikx}\,dk \tag{10.10}$$

$$\frac{\partial \Psi(x)}{\partial x} = \frac{\partial}{\partial x} \int_{-\infty}^{+\infty} \Psi(k)\,e^{ikx}\,dk$$

$$= \int_{-\infty}^{+\infty} ik\,\Psi(k)\,e^{ikx}\,dk \tag{10.11}$$

$$\left[\frac{\partial \Psi(x)}{\partial x}\right] = \int_{-\infty}^{+\infty} [ik\,\Psi(k)]\;e^{ikx}\,dk \tag{10.12}$$

$$\frac{\partial \Psi(x)}{\partial x} \qquad \Longleftrightarrow \qquad ik\,\Psi(k) = \frac{i}{\hbar}\,\mathbf{p}\,\Psi(p) \tag{10.13}$$

Table 10.1 Switching between space and momentum coordinates.

Description	$\Psi(x)$	$\Psi(p)$
Position eigenstate	$\delta(x - x_0)$	$e^{\frac{i}{\hbar} x_0 \mathbf{p}}$
Momentum eigenstate	$e^{\frac{i}{\hbar} p_0 x}$	$\delta(\mathbf{p} - p_0)$
First derivative (*by x*)	$\frac{\partial \Psi}{\partial x}$	$\frac{i}{\hbar} \mathbf{p} \Psi$
Second derivative (*by x*)	$\frac{\partial^2 \Psi}{\partial x^2}$	$-\frac{1}{\hbar^2} \mathbf{p}^2 \Psi$

The trick is to recognise that Equation 10.11 has exactly the same structure as the Fourier transform formula in Equation 10.10. I have highlighted this using square brackets in Equation 10.12. The square bracket part on the right must be the Fourier transform of that on the left because that is the very definition of a Fourier transform. And what pops out? It is the result we derived for the derivative of the free particle wave function (you can compare with the derivatives in Equation 9.12).

Let me stress again that you should not be disheartened if this maths is a bit much for you. I am showing you this only to demonstrate that switching between the x and p coordinate systems involves a Fourier transform (with an adjustment for the difference between k and p). A few examples are summarised in Table 10.1.

10.2 Heisenberg's Uncertainty Principle and the Gaussian Distribution

Heisenberg's uncertainty principle states that the uncertainty (standard deviation) in the position of a particle multiplied by the uncertainty (standard deviation) in the value of momentum of the particle must always equal or exceed the minimum value $\frac{\hbar}{2}$. To illustrate how this is baked into the Fourier transform process, I want to introduce the Gaussian distribution (see Box 10.4 and Box 10.5 for background on Carl Gauss and his correspondence with the equally amazing Sophie Germain). This distribution is frequently used as a model in quantum mechanics so it is worth taking a moment to get to grips with it:

$$\text{Gaussian distribution:} \quad e^{-\frac{x^2}{2\sigma^2}} \qquad \text{where } \sigma \text{ is its standard deviation} \tag{10.14}$$

The Gaussian distribution is perhaps better known as the *normal distribution* – yes that bell-shaped curve beloved of statisticians. It is of interest to us because its Fourier transform is another Gaussian (normal) distribution. In terms of quantum mechanics, this means if $\Psi(x)$ is a Gaussian, then $\Psi(k)$ is also a Gaussian. The Gaussian in Equation 10.15 is particularly special as its Fourier transform is the same Gaussian. It transforms to exactly the same function so the shapes and standard deviations of $\Psi(x)$ and $\Psi(k)$ are identical. The standard deviation of both is 1, as you can check by setting $\sigma = 1$ in Equation 10.14. If you feel a strange urge to see the details of this Fourier transform, you can find it in Box 10.6. The degree of difficulty in that calculation may give you an inkling as to why I dislike them!

$$\Psi(x) = e^{-\frac{x^2}{2}} \quad \Longleftrightarrow \quad \Psi(k) = e^{-\frac{k^2}{2}} \qquad \text{Fourier transform} \tag{10.15}$$

$$\sigma_x = 1 \qquad\qquad\qquad \sigma_k = 1 \tag{10.16}$$

What does this mean for the relationship between the uncertainty in position and wave number? We need to be careful with notation so as not to confuse two different things. We have labelled the standard deviation of $\Psi(x)$ as σ_x. This is different from the standard deviation in the position of the particle that we show as Δx. The probability of the particle being in a certain position depends on $|\Psi|^2$ (if you need a refresher, refer back to Section 5.6) so we need the expressions for $\Psi^2(x)$ and $\Psi^2(k)$ shown in Equation 10.18.

$$\Psi(x) = e^{-\frac{x^2}{2}} \quad \Longleftrightarrow \quad \Psi(k) = e^{-\frac{k^2}{2}} \qquad \text{Fourier transform} \tag{10.17}$$

$$\Psi^2(x) = e^{-x^2} \qquad\qquad \Psi^2(k) = e^{-k^2} \tag{10.18}$$

Δx is the standard deviation of $\Psi^2(x)$ while Δk is the standard deviation of $\Psi^2(k)$. The standard deviation of each is $\frac{1}{\sqrt{2}}$. You can check this by inserting that value as σ to the Gaussian equation (Equation 10.14) to see it matches Equation 10.18. The final step is to convert from k to $\mathbf{p}$ using $\mathbf{p} = \hbar k$. The result for $\Delta x \, \Delta \mathbf{p}$ is Heisenberg's minimum value of $\frac{\hbar}{2}$ as shown below.

$$\Delta x = \frac{1}{\sqrt{2}} \qquad\qquad \Delta k = \frac{1}{\sqrt{2}} \qquad \textit{then use } \mathbf{p} = \hbar k \qquad\qquad (10.19)$$

$$\Delta x = \frac{1}{\sqrt{2}} \qquad\qquad \Delta \mathbf{p} = \frac{\hbar}{\sqrt{2}} \qquad \implies \qquad \Delta x \, \Delta \mathbf{p} = \frac{\hbar}{2} \qquad\qquad (10.20)$$

Interestingly, all particle states that are Gaussians are at the Heisenberg minimum of uncertainty, and any particle state that is at the Heisenberg minimum must be a Gaussian distribution. As I noted earlier, you will come across Gaussians fairly frequently. The ground state of the Quantum Harmonic Oscillator is an example we will look at later.

Box 10.4   Carl Friedrich Gauss (1777–1855), child prodigy

Gauss was born to a poor family. His mother did not even record his date of birth. One story, perhaps just legend, is that at 7 years old, he added the integers 1 to 100 in seconds by noticing the first and last repeatedly add to 101 so the answer is $50 \times 101 = 5050$.

His first mathematical discoveries, while still a teenager, were noticed by the Duke of Brunswick who paid for him to have a university education. Gauss made major contributions to algebra, number theory and, of course, the famous distribution that is named after him.

He was yet another genius who suffered from depression, brought on initially by the death of his first wife and a child within five years of marriage. A perfectionist, he banned his children from studying maths, fearing they could not compare with him and would stain the family name.

10.3   The Quantum Footprint

I find the Gaussian distribution a useful model in thinking about the quantum world because the impact of measurement is so clear. If we start with a particle state that is Gaussian and has Δx uncertainty in position and $\Delta \mathbf{p}$ uncertainty in its momentum, what happens when we measure position more accurately? The measurement constrains the position of the particle. It reduces Δx. Suppose we halve Δx. For this to happen, the uncertainty in momentum $\Delta \mathbf{p}$ must double. Why? Because we know that for all Gaussians $\Delta x \Delta \mathbf{p} = \frac{\hbar}{2}$, so if Δx falls, $\Delta \mathbf{p}$ must correspondingly increase.[3]

I like to think of each particle state as having a *quantum footprint* of uncertainty. For a Gaussian, this is an irreducible size. If we say this quantum footprint is width Δx and height $\Delta \mathbf{p}$, then its area is always $\frac{\hbar}{2}$. The top of Figure 10.1 shows a starting particle state. If you measure position accurately, you reduce Δx and this changes the particle state to that shown at the bottom of the figure. As you squeeze the width Δx of this irreducible footprint, then its height $\Delta \mathbf{p}$ must expand. You may prefer to imagine it as a lump of soft clay. If you force the clay through a small hole (reducing Δx) it will be squeezed and therefore lengthen on the other measure (increasing $\Delta \mathbf{p}$) and vice versa. The quantum footprint is a helpful way to visualise Heisenberg's uncertainty principle.

The vast majority of particle states are not Gaussians so generally we cannot expect that $\Delta x \Delta \mathbf{p}$ is exactly $\frac{\hbar}{2}$. However, we can apply the broader approximation: $\Delta x \, \Delta \mathbf{p} \approx \hbar$ or h (see Box 6.6 if you need a reminder). This tells us that, more generally, if a measurement better defines the particle's position, it typically will increase the uncertainty in the particle's momentum, and vice versa.

We will stick with the analogy of a quantum footprint for a little longer. Let's suppose that we measure the particle's position and then we measure its momentum. Compare this with measuring the particle's momentum first and then its position. Is the resulting particle state the same in the two cases? No! Forgive me if I give a detailed example. This is a very important point.

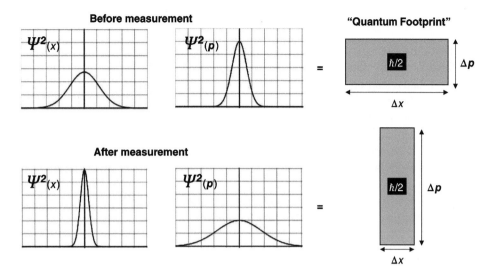

Figure 10.1 Measuring the position of a Gaussian particle state accurately will reduce the uncertainty in position Δx but must correspondingly increase the uncertainty in momentum Δ**p**.

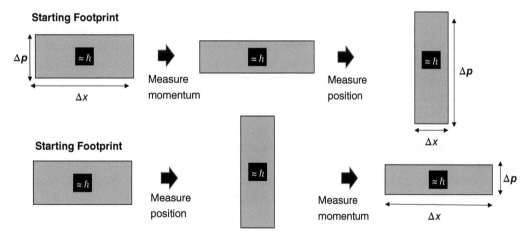

Figure 10.2 Measuring momentum then position is not the same as measuring position then momentum. The footprints of the resulting particle states are very different.

Consider the scenario shown in Figure 10.2. The starting quantum footprint is shown on the left. I have applied the broadly applicable relationship using $\Delta x \, \Delta \mathbf{p} \approx \hbar$. The top line of the figure shows the impact of first measuring momentum (reducing Δ**p** and increasing Δx) and then measuring position (reducing Δx and increasing Δ**p**). The bottom shows the same measurements in reverse order. The resulting particle state is very different depending on the order of measurement. This is true even if the results of the measurements happen to be the same.

In mathematics this behaviour is described as *non-commutative*. If we have two operations such as a and b, then applying them in a different order makes a difference: $ab \neq ba$. In quantum mechanics we have to take care to account for the non-commutative relationship of some operations.

With all this discussion of non-commutative measurements, I need to offer a reminder. The non-commutative nature between position x and momentum $\mathbf{p}_x$ is also true between y and $\mathbf{p}_y$, and between z and $\mathbf{p}_z$. However no such relationship applies to other mixes of these variables such as x and $\mathbf{p}_y$ or indeed x and y or say $\mathbf{p}_y$ and $\mathbf{p}_z$. If you measure position in the z direction then measure momentum in the x direction ($\mathbf{p}_x$) the order of measurement does not matter because the impact your measurement has on $\Psi(z)$ does not affect $\Psi(\mathbf{p}_x)$. Please refer back to Section 5.4 if needed.

Box 10.5   Monsieur LeBlanc – a tale of prejudice

Gauss corresponded with the mathematician Monsieur LeBlanc. During the Napoleonic Wars, the French occupied the town where Gauss lived. A military detachment arrived at his home to check on his welfare, sent by a certain Sophie Germain. To his surprise, Gauss learned that his long time correspondent, Monsieur LeBlanc, was actually a woman.

Battling huge prejudice, Sophie had studied maths secretly by candlelight in her bedroom while her disapproving parents slept. In 1816, she became the first woman to win a Paris Academy Science prize. As a woman she was unable to attend her own award ceremony!

Gauss praised her: *When a woman, because of her sex, our customs and prejudices, encounters infinitely more obstacles than men..., yet overcomes these fetters..., she doubtless has the most noble courage, extraordinary talent, and superior genius.*

Having been denied a university education, Gauss recommended she be awarded an honorary degree. Sadly, Sophie died of breast cancer before she could receive it. She was 55 years old.

10.4   Time and Energy

I suspect it will not have escaped your attention that the free particle wave function also suggests that some sort of non-commutative relationship and/or uncertainty principle should apply between t and $\mathbf{E}$ as these variables sit side by side in the free particle wave function in the same way as x and $\mathbf{p}$ (as you can see, for example, in Equation 9.11).

Um, er, well, um, cripes. Here, we run into some problems. So far, I have tried as best I can to respect the interplay of time and space in special relativity. For example, I have used where possible the relativistic definitions of momentum and energy (written as bold $\mathbf{p}$ and $\mathbf{E}$). However, in this module we will be using Schrödinger's equation which is an accurate practical tool, but is *not* relativistic (see discussion at the end of Section 8.3.3).

A kind teacher would tell you not to worry about this too much. That is probably right because it is a thorny issue which is likely to confuse rather than help you, but I would like you to harbour a small nagging discomfort about this at the back of your mind. We will soon be discussing momentum operators ($\hat{p}$), energy operators ($\hat{E}$) and position operators ($\hat{x}$), but quantum mechanics does not allow for a corresponding time operator. We will address this later in the book in the section on *Relativistic Quantum Mechanics*.

With this caveat, let's look at the relationship between time and energy. We can use the same analogy of a quantum footprint. Consider two different scenarios. In the first $\Delta\mathbf{E}$ is small and Δt is large. In the second it is the reverse (see Figure 10.3).

How do we interpret the first scenario? Energy is well defined so $\Delta\mathbf{E}$ is small and Δt is large. We can take this to the extreme by considering a particle state with a precise value of energy, an energy eigenstate. In this case Δt is infinite and the quantum footprint of the particle state is spread equally across all time. What does this mean? It means that the particle state remains the same at all points in time. It means that a single energy eigenstate does *not change* over time (unless you mess with it). The longer Δt, the more stable the particle state. The shorter Δt, the faster the particle state changes.

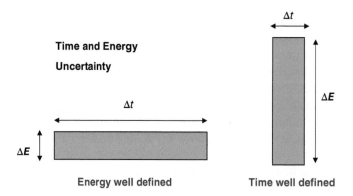

Figure 10.3   The uncertainty relationship between time and energy.

Let's turn to the second scenario where Δt is small and $\Delta \mathbf{E}$ is large. This tells us that there is great uncertainty in the state's energy or, as I prefer to say, the particle state is a superposition of a lot of different energy eigenstates. The particle state has a very short Δt. What does this mean? It means that the state changes very rapidly.

Looking at the same thing the other way round, transitory states can exist with short Δt and very uncertain energy. These states can include high energy options in the superposition. The more transitory the state, the higher the uncertainty in energy level so the higher that energy might possibly be. This leads to all sorts of quantum weirdness: particles that can tunnel through energy barriers because for a short instant they have higher energy than the classical world allows, and virtual particles that pop into and out of existence momentarily. We will discuss this in depth in coming chapters.

10.5 Summary

I hope that you can forgive me for this rather dry chapter. Fourier analysis is not the most mood-brightening topic. It plays an important role in quantum mechanics, but don't worry if you cannot follow the maths. The key thing is to understand the concept. Position and momentum are dependent variables. $\Psi(x)$ is fully defined by $\Psi(p_x)$ and vice versa. You can toggle between the two with Fourier transforms.

We discussed the Gaussian distribution. Its Fourier transform is another Gaussian and the combined uncertainty of the two is the Heisenberg minimum at $\Delta x \Delta \mathbf{p}_x = \frac{\hbar}{2}$. We used what I called the quantum footprint to illustrate why the order of measurements is important – quantum mechanics entails some non-commutative mathematics.

We touched on the uncertainty relationship between time and energy. If the particle has precise energy, it is in an energy eigenstate so $\Delta \mathbf{E} = 0$ and $\Delta t = \infty$. This means that it spreads over all time. It is stable. Where does this take us? It signals the importance of *energy eigenstates*. For example, if we can find the energy eigenstates of electrons in an atom, then we will have found the atom's stable configurations. Much of the analysis in the following chapters will be a hunt for these energy eigenstates.

The other thing it highlights is that any particle state that is not a single energy eigenstate must and will change over time. It is important that we understand how it changes. This brings us back to Schrödinger's equation with Box 10.7 added just for a laugh.

Box 10.6 Detailed calculation of the Fourier transform for $e^{-\frac{x^2}{2}}$ **Optional!**

If you really want to torture yourself, if you are fascinated by Fourier transforms, or you are just plain weird, here it is in all its glory:

$$f(x) = e^{-\frac{x^2}{2}}$$

$$f(k) = \frac{1}{\sqrt{2\pi}} \int_{-\infty}^{\infty} f(x)e^{ikx}dx$$

$$= \frac{1}{\sqrt{2\pi}} \int_{-\infty}^{\infty} e^{-\frac{x^2}{2}} e^{ikx}dx \qquad = \frac{1}{\sqrt{2\pi}} \int_{-\infty}^{\infty} e^{-\frac{x^2}{2}+ikx}dx$$

$$= \frac{1}{\sqrt{2\pi}} \int_{-\infty}^{\infty} e^{-\frac{1}{2}(x-ik)^2 - \frac{k^2}{2}}dx \qquad = \frac{1}{\sqrt{2\pi}} e^{-\frac{k^2}{2}} \int_{-\infty}^{\infty} e^{-\frac{1}{2}(x-ik)^2}dx$$

Substitute with $z = \frac{1}{\sqrt{2}}(x - ik)$ so that $z^2 = \frac{1}{2}(x-ik)^2$ noting $dz = \frac{dx}{\sqrt{2}}$

$$f(k) = \frac{1}{\sqrt{2\pi}} e^{-\frac{k^2}{2}} \int_{-\infty}^{\infty} e^{-z^2}\sqrt{2}\,dz$$

$$= \frac{1}{\sqrt{\pi}} e^{-\frac{k^2}{2}} \int_{-\infty}^{\infty} e^{-z^2}dz \qquad = \frac{1}{\sqrt{\pi}} e^{-\frac{k^2}{2}}\sqrt{\pi}$$

$$f(k) = e^{-\frac{k^2}{2}}$$

Box 10.7   Just for laughs:
What do you call a scientist who steals energy?   *A joule thief.*

Notes

1 See *But what is the Fourier transform?* at https://www.youtube.com/watch?v=spUNpyF58BY (as at December 2022).
2 Technical point: if you are working with all three spatial dimensions, remember that both **p** and k are *vector* quantities.
3 Technical point: this assumes for simplicity that the resulting particle state is also a Gaussian.

Chapter 11

Schrödinger's Equation

CHAPTER MENU
11.1 Understanding Schrödinger's Equation
11.2 Operators, Eigenstates and Eigenvalues
11.3 Commutation Relations
11.4 Expectation Values and Dirac Notation
11.5 Energy Eigenstates are Stationary
11.6 Time-independent Schrödinger Equation

11.1 Understanding Schrödinger's Equation

I gave you a taste of this equation in Subsection 8.3.2. You may remember the challenge that surfaced because of the phase difference between the derivatives used in the Schrödinger equation. Using complex numbers we can address this more formally. Schrödinger's aim was to find an equation that describes how particle states *change over time*, that is, an equation that describes $\frac{\partial \Psi}{\partial t}$. We are going to need the first derivative with respect to time and the second derivative with respect to space of the complex wave function of an eigenstate with precise value for energy and momentum, shown below as Equation 11.1 using one spatial dimension. Equation 11.2 shows these expressions manipulated to make them easier to substitute later.

$$\phi = e^{\frac{i}{\hbar}(\mathbf{p}x - \mathbf{E}t)} \quad \Longrightarrow \quad \frac{\partial \phi}{\partial t} = -\frac{i}{\hbar} \mathbf{E} \phi \qquad \frac{\partial^2 \phi}{\partial x^2} = -\frac{1}{\hbar^2} \mathbf{p}^2 \phi \tag{11.1}$$

$$\mathbf{E} \phi = i\hbar \frac{\partial \phi}{\partial t} \qquad \mathbf{p}^2 \phi = -\hbar^2 \frac{\partial^2 \phi}{\partial x^2} \tag{11.2}$$

The Schrödinger equation is so important that I will run through exactly the same logic as covered in Subsection 8.3.2. The relativistic energy for a free particle is approximately equivalent to its rest energy mc^2 plus its kinetic energy $\frac{p^2}{2m}$, as shown in Equation 11.3. This is extremely accurate at non-relativistic speeds. We multiply both sides of the equation by ϕ and substitute E with $\frac{\partial \phi}{\partial t}$. We are careful to substitute p^2 with $\frac{\partial^2 \phi}{\partial x^2}$ so it works for superpositions (see Box 8.3). This gives Equation 11.4 for a free particle which has precise values of momentum and energy.

$$E \approx \frac{p^2}{2m} + mc^2 \qquad \textit{accurate at non-relativistic speeds} \tag{11.3}$$

$$E \phi = \frac{p^2}{2m} \phi + mc^2 \phi$$

$$i\hbar \frac{\partial \phi}{\partial t} = -\frac{\hbar^2}{2m} \frac{\partial^2 \phi}{\partial x^2} + mc^2 \phi \qquad \textit{this is for a free particle} \tag{11.4}$$

Quantum Untangling: An Intuitive Approach to Quantum Mechanics from Einstein to Higgs, First Edition. Simon Sherwood.
© 2023 John Wiley & Sons Ltd. Published 2023 by John Wiley & Sons Ltd.

11.1.1 Incorporating Potential Energy

Schrödinger was not very interested in free particles. He wanted to understand the energy levels of electrons constrained by an electromagnetic field such as the electrons in atoms. To account for this, he proposed that the *potential energy* of the electron from the electromagnetic field, labelled V, be added to the energy balance. Repeating the substitutions that we have just done results in Equation 11.5.

$$E \approx \frac{p^2}{2m} + V + mc^2 \quad \textit{approximate energy in an electromagnetic field}$$

$$i\hbar \frac{\partial \phi}{\partial t} = -\frac{\hbar^2}{2m} \frac{\partial^2 \phi}{\partial x^2} + (V + mc^2)\phi \tag{11.5}$$

The potential energy V is a *variable* that may change depending on location and time. For instance, the potential energy of an electron in the atom varies with its distance from the nucleus. However, the rest mass energy mc^2 is a *constant*. We can ignore the mc^2 term providing we remember that any calculations will give the energy of the electron over and above its rest mass energy. In words, this gives the classical energy balance that can be summarised as a particle's *classical energy* equals *kinetic energy* plus *potential energy*. This gives 11.6 which is Schrödinger's equation for a particle in eigenstate ϕ which has precise values of energy and momentum.

$$E = \frac{p^2}{2m} + V \quad \Longrightarrow \quad i\hbar \frac{\partial \phi}{\partial t} = -\frac{\hbar^2}{2m} \frac{\partial^2 \phi}{\partial x^2} + V\phi \tag{11.6}$$

11.1.2 Superpositions

For much that follows, it is crucial that Schrödinger's equation holds for all particle states. Therefore we need to show that the equation holds for superpositions. As a simple example, consider a superposition with two elements that I label $\Psi = A\phi_a + B\phi_b$ where A and B are complex numbers. The first step is to calculate the derivatives of Ψ as shown in Equation 11.7.

$$\Psi = A\phi_a + B\phi_b \quad \textit{where A and B are complex numbers}$$

$$\frac{\partial \Psi}{\partial t} = A\frac{\partial \phi_a}{\partial t} + B\frac{\partial \phi_b}{\partial t} \qquad \frac{\partial^2 \Psi}{\partial x^2} = A\frac{\partial^2 \phi_a}{\partial x^2} + B\frac{\partial^2 \phi_b}{\partial x^2} \tag{11.7}$$

We multiply the left of Equation 11.7 by $i\hbar$ to get 11.8. Equation 11.9 substitutes using Schrödinger's equation which we know holds individually for both ϕ_a and ϕ_b. With a bit of reorganising, this results in Schrödinger's equation for Ψ. Any allowable particle state can be expressed as a superposition of eigenstates. Therefore, Schrödinger's equation holds for *all* particle states.

$$i\hbar \frac{\partial \Psi}{\partial t} = A\,i\hbar \frac{\partial \phi_a}{\partial t} + B\,i\hbar \frac{\partial \phi_b}{\partial t} \tag{11.8}$$

$$= A\left(-\frac{\hbar^2}{2m} \frac{\partial^2 \phi_a}{\partial x^2} + V\phi_a\right) + B\left(-\frac{\hbar^2}{2m} \frac{\partial^2 \phi_b}{\partial x^2} + V\phi_b\right) \tag{11.9}$$

$$= -\frac{\hbar^2}{2m}\left(A\frac{\partial^2 \phi_a}{\partial x^2} + B\frac{\partial^2 \phi_b}{\partial x^2}\right) + V\left(A\Psi_a + B\phi_b\right)$$

$$i\hbar \frac{\partial \Psi}{\partial t} = -\frac{\hbar^2}{2m} \frac{\partial^2 \Psi}{\partial x^2} + V\Psi \quad \textit{Schrödinger's equation for } \Psi \tag{11.10}$$

11.1.3 Schrödinger's Equation in Words

This is called the *time-dependent* Schrödinger equation (Box 11.1 plus Box 11.3 if you want the intriguing story behind the equation's discovery). It is so fundamental to our studies that I want to pause and step through it again in words rather than numbers. After all, the equation is not the sort of thing you run across in schoolroom maths. Equation 11.11 shows the equation. What has Schrödinger done? The equation is an energy balance. The left side relies on the relationship between total energy (classically defined) and the first derivative of Ψ with respect to time. The right side is kinetic energy plus potential energy. The kinetic energy term relies on the relationship between momentum squared and the second derivative of Ψ with respect to space.

Consider a free particle with definite energy and momentum, so it is a simple sinusoidal wave function. Each differentiation shifts it in phase by $\frac{\pi}{2}$ (see Subsection 5.1.1). The term on the left side is a *first* derivative, so Ψ is multiplied by i to account for the $\frac{\pi}{2}$ phase change. The first term on the right side is a *second* derivative, so it is multiplied by -1 to account for the π phase change (see Section 8.3 if you cannot remember why).

Box 11.1   Schrödinger equation   **Time-dependent version**

In one spatial dimension for $\Psi(x,t)$:

$$i\hbar\frac{\partial\Psi}{\partial t} = -\frac{\hbar^2}{2m}\frac{\partial^2\Psi}{\partial x^2} + V\Psi \tag{11.11}$$

In all three spatial dimensions for $\Psi(x,y,z,t)$:

$$i\hbar\frac{\partial\Psi}{\partial t} = -\frac{\hbar^2}{2m}\left(\frac{\partial^2\Psi}{\partial x^2} + \frac{\partial^2\Psi}{\partial y^2} + \frac{\partial^2\Psi}{\partial z^2}\right) + V\Psi \tag{11.12}$$

The potential energy term V on the far right of the equation is very important because it sets the scenario that is under study. For example, if $V = 0$, then the scenario is an unconstrained particle and the solution is the free particle wave function.

We will look at a number of scenarios with different patterns to V. Most fundamental to Schrödinger's search was the hunt for the energy levels of the hydrogen atom. In this case the potential energy varies with $\frac{1}{r}$ where r is the distance between the electron and the proton. It will take some time to build up to this as it involves taking into account angular momentum and using spherical coordinates. Before that we will look at the Quantum Harmonic Oscillator that is a scenario similar to a spring or pendulum. This model will become very familiar to anyone who takes an advanced course in quantum mechanics as it is at the core of quantum field theory.

In the examples above, potential energy is set to vary only over space $V(x)$, but it can also be varied over time $V(x,t)$. Changing the pattern of potential energy V allows us to build scenarios to model how particles behave under different constraints. This is how we use the Schrödinger equation to explore the quantum world.

11.2   Operators, Eigenstates and Eigenvalues

Operators play an important role in the maths of quantum mechanics. An *operator* is anything that *acts on* the wave function. Among other benefits, this allows us to shorten the notation in equations because we can specify a sequence of actions with a single label. This is best explained with some examples. By the way, I will no longer label energy **E** and momentum **p** in bold because in building the Schrödinger equation we have switched to the classical definitions rather than the relativistic ones. We again will need some of the important derivative relationships, so they are shown below (in one spatial dimension).

$$E\phi = i\hbar\frac{\partial\phi}{\partial t} \qquad E^2\phi = -\hbar^2\frac{\partial^2\phi}{\partial t^2} \qquad p\phi = -i\hbar\frac{\partial\phi}{\partial x} \qquad p^2\phi = -\hbar^2\frac{\partial^2\phi}{\partial x^2} \tag{11.13}$$

Consider a free particle that has a specific value of momentum p_0 and energy E_0. The free particle wave function for this is shown in Equation 11.14. If we want to know the outcome of measuring the momentum of this particle, we can calculate it using the derivative relationship for momentum. We can act on ϕ in the way shown in Equation 11.15 to generate the result $p_0\phi$.

$$\phi(x,t) \ = \ e^{\frac{i}{\hbar}(p_0 x - E_0 t)} \tag{11.14}$$

$$\hat{p}\,\phi \ = \ -i\hbar\,\frac{\partial\phi}{\partial x} \ = \ -i\hbar\,\frac{i}{\hbar}\,p_0\,e^{\frac{i}{\hbar}(p_0 x - E_0 t)} \ = \ p_0\,\phi \tag{11.15}$$

$$\hat{p}\,\phi \ = \ p_0\,\phi \qquad\qquad \textit{Defining... } \ \hat{p} \ = \ -i\hbar\,\frac{\partial}{\partial x} \tag{11.16}$$

We define this action as the *momentum operator* and it is usually labelled $\hat{p}$. I will endeavour always to put a little hat over the symbol of any operator. It is much simpler to write $\hat{p}$ than the longer version, but don't forget that the operator $\hat{p}$ is just shorthand notation for the action $-i\hbar\,\frac{\partial}{\partial x}$.

The example shown in Equations 11.14 to 11.16 is a special case because particle state ϕ has a single precise value of momentum. This is why we call ϕ an *eigenstate* of the momentum operator. The value for momentum p_0 is called the *eigenvalue*. The nomenclature comes from the world of matrices and vectors. If you are unfamiliar with it, Box 11.2 on eigenvectors may help.

What happens if we apply the momentum operator to a particle state that is *not* a momentum eigenstate? The calculation below shows its action on a particle state $\Psi(x,t)$ that is a superposition of two component eigenstates:

$$\Psi(x,t) \ = \ b_1\,\phi_1 \ + \ b_2\,\phi_2 \qquad \textit{where } b_1 \textit{ and } b_2 \textit{ are complex numbers} \tag{11.17}$$

$$\Psi \ = \ b_1\,e^{\frac{i}{\hbar}(p_1 x - E_1 t)} \ + \ b_2\,e^{\frac{i}{\hbar}(p_2 x - E_2 t)}$$

$$\hat{p}\,\Psi(x,t) \ = \ -i\hbar\,\frac{\partial\Psi}{\partial x} \ = \ p_1\,b_1\,\phi_1 \ + \ p_2\,b_2\,\phi_2 \ \neq \ \textit{multiple of } \Psi \tag{11.18}$$

Again, the important point is that it draws out the momentum of each component of the superposition. But note that in this context momentum p is a *variable*, not a single value. The result will not be a simple multiple of Ψ because there are two possible values of momentum ($p_1 \neq p_2$).

We are particularly interested in operators that give us information about the likely result of measurements. Another important example is the energy operator $\hat{E}$ that we can create from the derivative relationships just as we did for the momentum operator. It draws out energy from each energy eigenstate in exactly the same way as the momentum operator $\hat{p}$ draws out momentum:

$$\phi(x,t) \ = \ e^{\frac{i}{\hbar}(p_0 x - E_0 t)}$$

$$\hat{E}\,\phi \ = \ i\hbar\,\frac{\partial\phi}{\partial t} \ = \ i\hbar\left(-\frac{i}{\hbar}\right)E_0\,e^{\frac{i}{\hbar}(p_0 x - E_0 t)} \ = \ E_0\,\phi \tag{11.19}$$

The action of $\hat{E}$ will draw out a single value E for the energy of the state if and *only* if the particle state Ψ is an energy eigenstate.

A list of commonly used operators is shown below in Table 11.1. Note that the operator $\hat{p}^2$ that draws out p^2 is based on the *second* derivative. This is because it acts to get the first derivative and then acts again to get the second. Note also that the last two operators give the same result. This is because the Schrödinger equation tells us that $\hat{E} = \hat{H}$. Compare the actions of $\hat{E}$ and $\hat{H}$ with the Schrödinger equation in Equation 11.11 if you want to check.

Table 11.1 Widely used operators (in one spatial dimension).

Operator	Symbol	Action on $\Psi(x,t)$	Action on Eigenstate
Momentum operator	$\hat{p}$	$-i\hbar\,\dfrac{\partial\Psi}{\partial x}$	$\hat{p}\,\phi_p \ = \ p\,\phi_p$
Momentum squared	$\hat{p}^2$	$-\hbar^2\,\dfrac{\partial^2\Psi}{\partial x^2}$	$\hat{p}^2\,\phi_p \ = \ p^2\,\phi_p$
Position operator	$\hat{x}$	$x\Psi$	$\hat{x}\,\phi_x \ = \ x\,\phi_x$
Energy operator	$\hat{E}$	$i\hbar\,\dfrac{\partial\Psi}{\partial t}$	$\hat{E}\,\phi_E \ = \ E\,\phi_E$
Hamiltonian operator	$\hat{H}$	$-\dfrac{\hbar^2}{2m}\,\dfrac{\partial^2\Psi}{\partial x^2} + V\Psi$	$\hat{H}\,\phi_E \ = \ E\,\phi_E$

The table shows the operators in one spatial dimension. If you are working in all three spatial dimensions, there are $\hat{p}$ and $\hat{p}^2$ operators for each spatial dimension:

$$\hat{p}_x \Psi = -i\hbar \frac{\partial \Psi}{\partial x} \qquad \hat{p}_y \Psi = -i\hbar \frac{\partial \Psi}{\partial y} \qquad \hat{p}_z \Psi = -i\hbar \frac{\partial \Psi}{\partial z} \qquad (11.20)$$

$$\hat{p}_x^2 \Psi = -\hbar^2 \frac{\partial^2 \Psi}{\partial x^2} \qquad \hat{p}_y^2 \Psi = -\hbar^2 \frac{\partial^2 \Psi}{\partial y^2} \qquad \hat{p}_z^2 \Psi = -\hbar^2 \frac{\partial^2 \Psi}{\partial z^2} \qquad (11.21)$$

The operator for position $\hat{x}$ works the same way as the other operators, drawing out the x value. As $\Psi(x, t)$ is already expressed as a function of x, you need only multiply $\Psi(x, t)$ by x to generate the desired result $x\,\Psi$. For work in three spatial dimensions there are identical $\hat{y}$ and $\hat{z}$ operators.

As shown in Table 11.1, the Hamiltonian operator $\hat{H}$ is an abbreviated form of the actions on the right side of the Schrödinger equation. As the result is the same, you will find $\hat{H}$ and $\hat{E}$ used interchangeably on many occasions. In three dimensions, $\hat{H}$ needs to take into account the square of total momentum and becomes:

$$\hat{H}\Psi = -\frac{\hbar^2}{2m}\left(\frac{\partial^2 \Psi}{\partial x^2} + \frac{\partial^2 \Psi}{\partial y^2} + \frac{\partial^2 \Psi}{\partial z^2}\right) + V\,\Psi \qquad (11.22)$$

All of the operators in Table 11.1 are associated with measurable values. This means all of their eigenvalues must be *real* (not complex). After all, the value of say momentum cannot be imaginary! Operators that have only real eigenvalues are called *Hermitian*, another of those confusing terms that you may stumble upon in textbooks.

Box 11.2   Eigenvectors of matrices and their eigenvalues

The terminology comes from the world of matrices. The *eigenvectors* of a matrix are those vectors that, when multiplied by the matrix, are unchanged except in magnitude.

For the randomly selected 2x2 matrix below $\begin{bmatrix} 2 \\ 1 \end{bmatrix}$ is an eigenvector but $\begin{bmatrix} 1 \\ 2 \end{bmatrix}$ is not:

Eigenvector: $\qquad \begin{bmatrix} 4 & 0 \\ 1 & 2 \end{bmatrix} \begin{bmatrix} 2 \\ 1 \end{bmatrix} = \begin{bmatrix} 8 \\ 4 \end{bmatrix} = 4 \begin{bmatrix} 2 \\ 1 \end{bmatrix} \qquad$ Eigenvalue $= 4$

Not an eigenvector: $\qquad \begin{bmatrix} 4 & 0 \\ 1 & 2 \end{bmatrix} \begin{bmatrix} 1 \\ 2 \end{bmatrix} = \begin{bmatrix} 4 \\ 5 \end{bmatrix} \neq C \begin{bmatrix} 1 \\ 2 \end{bmatrix}$

The same terminology is widely used in quantum mechanics. If the effect of an operator on a particle state is to change only its magnitude for example, $\hat{p}\,\Psi = p_0\,\Psi$ where p_0 is a number, then we say that Ψ is an *eigenstate* of the operator $\hat{p}$, and the number p_0 is the *eigenvalue*.

We can recast the Schrödinger equation using operator symbols. It helps clarify how and why the equation works. All we do is to substitute in the operator expressions from Table 11.1. It is also convention to describe potential energy V as an operator because in the term $V\Psi$, V is acting on the wave function. Of course, the form of V varies with the scenario under study.

$$i\hbar \frac{\partial \Psi}{\partial t} = -\frac{\hbar^2}{2m}\frac{\delta^2 \Psi}{\delta x^2} + V\,\Psi$$

$$\hat{E}\,\Psi = \frac{\hat{p}^2}{2m}\,\Psi + \hat{V}\,\Psi \qquad \textit{or even} \quad \hat{E}\,\Psi = \hat{H}\,\Psi \qquad (11.23)$$

You may feel that we have not achieved much with this new notation. However, when things get more complicated I assure you that you will appreciate it!

11.3 Commutation Relations

Operators give us a way to check where and how Heisenberg's uncertainty principle applies. Measuring position affects the uncertainty in momentum and vice versa, so the order of measurement matters (this is illustrated in Figure 10.2). Where there is an uncertainty relationship, the operators show the same sensitivity. They do *not commute* in the sense that $\hat{x}\hat{p}_x\Psi \neq \hat{p}_x\hat{x}\Psi$.

We can show this effect by calculating the *commutator* of the two operators: $[\hat{x}, \hat{p}_x]\Psi = \hat{x}\hat{p}_x\Psi - \hat{p}_x\hat{x}\Psi$. If it is *not* zero, then there is an uncertainty relationship between the measures. The calculation of $[\hat{x}, \hat{p}_x]\Psi$ is shown here and the impact is $i\hbar\Psi$:

$$\hat{x}\hat{p}_x\Psi = \hat{x}\left(-i\hbar\frac{\partial\Psi}{\partial x}\right) = -i\hbar x\frac{\partial\Psi}{\partial x}$$

$$\hat{p}_x\hat{x}\Psi = \hat{p}_x(x\Psi) = -i\hbar\frac{\partial(x\Psi)}{\partial x} = -i\hbar x\frac{\partial\Psi}{\partial x} - i\hbar\Psi$$

$$\hat{x}\hat{p}_x\Psi - \hat{p}_x\hat{x}\Psi = i\hbar\Psi \qquad \Longrightarrow \qquad [\hat{x}, \hat{p}_x]\Psi = i\hbar\Psi \qquad (11.24)$$

The same is true of course for $[\hat{y}, \hat{p}_y]\Psi$ and $[\hat{z}, \hat{p}_z]\Psi$. In contrast, position and momentum in different directions do not sit side by side in the free particle wave function, so commutators such as $[\hat{x}, \hat{p}_y]\Psi$ are zero and there is no uncertainty relationship:

$$\hat{x}\hat{p}_y\Psi = \hat{x}\left(-i\hbar\frac{\partial\Psi}{\partial y}\right) = -i\hbar x\frac{\partial\Psi}{\partial y}$$

$$\hat{p}_y\hat{x}\Psi = \hat{p}_y(x\Psi) = -i\hbar\frac{\partial(x\Psi)}{\partial y} = -i\hbar x\frac{\partial\Psi}{\partial y}$$

$$\hat{x}\hat{p}_y\Psi - \hat{p}_y\hat{x}\Psi = 0 \qquad \Longrightarrow \qquad [\hat{x}, \hat{p}_y]\Psi = 0 \qquad (11.25)$$

Turning back to the case in which the commutator does have a value such as $[\hat{x}, \hat{p}_x]\Psi = i\hbar\Psi$, what does the i mean? It means that if you compare the action of $\hat{p}_x$ on Ψ followed by the action of $\hat{x}$ with the effect of applying the two operators in reverse order, the difference is Ψ multiplied by $\hbar$ with the *phase* of Ψ shifted by $\frac{\pi}{2}$. The phase shift is shown by i. All the i is telling us is that the sinusoid components in the result are out of phase with the original.

Box 11.3 Schrödinger's inspiration

Although Erwin Schrödinger's (1887–1961) name is most popularly associated with his cat paradox (see Section 4.5), his most famous scientific discovery is Schrödinger's equation. Schrödinger had what you might describe as a very active love life. He struggled to settle in at several universities because his required living arrangements included both a wife and a mistress. Schrödinger revealed his equation in early 1926 after a trip to Arosa in Switzerland. He had a history of tuberculosis and Arosa boasted a famous sanatorium so he left his wife to spend a few weeks recuperating. But it appears he took an old girlfriend from Vienna along for the trip.

So what was the inspiration behind his equation? Sources differ. Was it bed rest and relaxation in a sanatorium? Or was it the attentions of his old flame from Vienna? You decide. Or perhaps, like his famous cat, Schrödinger was in a superposition of sickness and ardour so we will never know the answer because it is now too late to open the box and peep!

One word of warning. Do not confuse the actions of these operators with actually making measurements. That is a very different thing. If you act on Ψ with say $\hat{p}_x$, it draws out the possible values of p_x and their relative probabilities. Actual measurement would completely change Ψ to one particular momentum eigenstate. The action of the operator tells us about the possible outcomes. It is a mathematical manipulation, like stretching or rotating a vector, which can tell us something about the wave function, whereas a measurement is a physical process.

The operators and commutators shown above are fairly straightforward, but on lots of occasions the situation is less clear. For example, when we get to angular momentum you will see that it is a composite measure that involves position and momentum so things get more complicated. Uncertainty relationships abound in quantum mechanics: between position and momentum, between different directions of angular momentum, between spins – and the commutator relationships are surprisingly important and helpful.

One use is that you can derive the Heisenberg uncertainty relationship of two measures easily because the minimum is half the absolute value of the commutator of the two operators. In Equation 11.26, this calculation is shown for $\Delta x \Delta p_x$. Don't forget the the magnitude of i is 1. This is because $|i| = \sqrt{i(-i)} = 1$.

$$\Delta A \Delta B \geq \frac{1}{2} |[\hat{A}, \hat{B}]| \qquad \textit{for example} \quad \Delta x \Delta p_x \geq \frac{1}{2} |[\hat{x}, \hat{p}_x]| = \frac{1}{2} |i\hbar| = \frac{\hbar}{2} \tag{11.26}$$

11.4 Expectation Values and Dirac Notation

If a particle is in a superposition of position eigenstates, the exact result of a measurement of position is down to chance, but we can calculate the average result. This is called the *expectation value* of position and is labelled $\langle\Psi|\hat{x}|\Psi\rangle$ for reasons that will soon become clear. To illustrate, consider a particle that is in a superposition of three different locations: a, b or c. To calculate the expectation value (average) we must multiply the value of x at each position by the probability of the particle being found there and then sum across the three. The probability depends on the square of the absolute value of Ψ at each position, so the calculation is:

$$\textit{Expectation value} = \frac{a|\Psi(a)|^2 + b|\Psi(b)|^2 + c|\Psi(c)|^2}{|\Psi(a)|^2 + |\Psi(b)|^2 + |\Psi(c)|^2} = \frac{\sum x|\Psi(x)|^2}{\sum |\Psi(x)|^2} \tag{11.27}$$

Frequently we will be working with continuous variables such as the position of a particle. We handle all possibilities by switching from summation to integration:

$$\textit{Expectation value of } x = \frac{\int_{-\infty}^{+\infty} x|\Psi(x)|^2 \, dx}{\int_{-\infty}^{+\infty} |\Psi(x)|^2 \, dx} \tag{11.28}$$

Typically we don't have to bother with the denominator in Equation 11.28 because the coefficients of wave functions are usually normalised (i.e. scaled) so that the integral of $|\Psi(x)|^2$ equals 1 but it is best to use the full form just in case. Don't forget always to multiply Ψ with its complex conjugate Ψ^* when calculating the value of $|\Psi(x)|^2$. This is important because changing the phase of the wave function should not change the magnitude of anything measurable (see Box 11.4).

Also, don't forget that for calculations such as $\langle\Psi|\hat{x}|\Psi\rangle$ in 11.28, you must use the amplitude of the wave function using the *same* variable, in this case $\Psi(x)$. To calculate the expectation value for momentum, you would need to use $\Psi(p)$, not $\Psi(x)$, so that you combine each value of p with the amplitude of the wave function for that particular level of momentum p.

That bright spark Dirac came up with some nifty notation to keep things clear: bras and kets. The *ket* $|\Psi\rangle$ indicates that you must work with Ψ while the *bra* $\langle\Psi|$ indicates the need for its complex conjugate Ψ^*. In Dirac's notation, the top of Equation 11.28 becomes $\langle\Psi|\hat{x}|\Psi\rangle$ and the bottom is $\langle\Psi|\Psi\rangle$. The integration is just assumed rather than explicitly shown. The whole of Equation 11.28 simply becomes:

$$\textit{Expectation value of } x = \langle\Psi|\hat{x}|\Psi\rangle \quad \textit{or if } \Psi \textit{ is not normalised} = \frac{\langle\Psi|\hat{x}|\Psi\rangle}{\langle\Psi|\Psi\rangle} \tag{11.29}$$

This may seem a bit complicated at first, but you will soon get used to it. It applies for all operators so for momentum it is $\langle\Psi|\hat{p}|\Psi\rangle$ and for energy $\langle\Psi|\hat{E}|\Psi\rangle$.

Box 11.4 Changing the phase of the wave function does not affect measurables

Multiplying a wave function by $e^{i\theta}$ is the equivalent of shifting its phase by angle θ because $e^{i\theta}e^{ix} = e^{i(\theta+x)}$. Adding θ to the phase does *not* change anything measurable. All measurables depend on the value of $\Psi^*\Psi$ and the θ phase change cancels out:

$$\textit{if } \Psi \rightarrow e^{i\theta}\Psi \quad \textit{then} \quad \Psi^*\Psi \rightarrow e^{-i\theta}\Psi^* e^{i\theta}\Psi = \Psi^*\Psi$$

11.5   Energy Eigenstates are Stationary

Energy eigenstates are stationary. This is *big* news! Typically we use the Schrödinger equation to hunt for energy eigenstates. Why? Because they are stable. I introduced the stability of energy eigenstates in Section 10.4 when we discussed the quantum footprint of time and energy. This will prove very important so let's look at why and how energy eigenstates are stable.

Consider a particle state Ψ that is a superposition of three different eigenstates as shown below, each with a different energy. How will this wave function Ψ change over space and time?

$$\Psi = A\, e^{\frac{i}{\hbar}(p_a x - E_a t)} + B\, e^{\frac{i}{\hbar}(p_b x - E_b t)} + C\, e^{\frac{i}{\hbar}(p_c x - E_c t)} \tag{11.30}$$

There is no obvious pattern. The value of the first component in the superposition is oscillating over time at a frequency determined by E_a, that of the second is oscillating at a different frequency determined by E_b, and that of the third is oscillating at yet another frequency. They are not in sync so the relative values of the components will fluctuate over time.

Now let's assume instead that the particle state Ψ is an energy eigenstate. I will label this ϕ_E to reflect that the particle is in a state with a precise value of energy that I will label E_0. Any measurement of energy will always give E_0, so the components in the superposition all must have energy E_0. As you can see in Equation 11.31, in this case the energy-time term can be factored out because it is the same for all parts of the superposition. You can separate ϕ_E into two independent parts, one varying only with time and the other varying only with space as shown in Equation 11.32. This is a standing wave (see earlier Box 7.4 if you want an overview of standing waves).

$$\phi_E = A\, e^{\frac{i}{\hbar}(p_a x - E_0 t)} + B\, e^{\frac{i}{\hbar}(p_b x - E_0 t)} + C\, e^{\frac{i}{\hbar}(p_c x - E_0 t)}$$

$$= \left(A\, e^{\frac{i}{\hbar} p_a x} + B\, e^{\frac{i}{\hbar} p_b x} + C\, e^{\frac{i}{\hbar} p_c x} \right) e^{-\frac{i}{\hbar} E_0 t} \tag{11.31}$$

$$\phi_E(x, t) = \phi_E(x)\, \phi_E(t) = \phi_E(x)\, e^{-\frac{i}{\hbar} Et} \qquad \textit{if energy E is a single value} \tag{11.32}$$

What does this mean? The value of the three components still oscillate, but they do so at the same frequency. In layman's terms, they are in sync. Their values relative to each other do *not* change over time. Energy eigenstates are spatially stable standing waves and so are called *stationary states*. To be clear, the probability of finding a particle in a location is driven by $|\phi_E|^2$ and, in the case of an energy eigenstate, this does *not* change over time:

$$|\phi_E|^2 = \phi_E(x)\, e^{-\frac{i}{\hbar} Et}\, \phi_E^*(x)\, e^{+\frac{i}{\hbar} Et} = |\phi_E(x)|^2 \qquad \textit{no change over time} \tag{11.33}$$

11.6   Time-independent Schrödinger Equation

There is a wonderful shortcut that takes advantage of the spatial stability of energy eigenstates. It is a simpler version of the Schrödinger equation. Aarrgghh! I hear you cry. Another one! But this is powerful. It works *only* for energy eigenstates and is called the *Time-independent* Schrödinger equation.

Let's take the expression for an energy eigenstate as shown in Equation 11.32 and feed it into the Schrödinger equation:

$$i\hbar \frac{\partial \Psi}{\partial t} = -\frac{\hbar^2}{2m} \frac{\delta^2 \Psi}{\delta x^2} + V\Psi \qquad \textit{Schrödinger equation}$$

$$i\hbar \frac{\partial}{\partial t} \left(\phi_E(x)\, e^{-\frac{i}{\hbar} Et} \right) = -\frac{\hbar^2}{2m} \frac{\delta^2}{\delta x^2} \left(\phi_E(x)\, e^{-\frac{i}{\hbar} Et} \right) + V\, \phi_E(x)\, e^{-\frac{i}{\hbar} Et} \tag{11.34}$$

The left side of 11.34 can be solved by simple differentiation. Remember that $\phi_E(x)$ is *not* time-dependent:

$$i\hbar \frac{\partial}{\partial t} \left(\phi_E(x)\, e^{-\frac{i}{\hbar} Et} \right) = E\, \phi_E(x)\, e^{-\frac{i}{\hbar} Et} \tag{11.35}$$

Turning to the right side of Equation 11.34, the time component is unaffected when differentiated with respect to space and so can be separated out. Combining this with the result we have just derived for the left side gives:

$$E\,\phi_E(x)\ e^{-\frac{i}{\hbar}Et} = \left(-\frac{\hbar^2}{2m}\frac{\delta^2\phi_E(x)}{\delta x^2} + V\,\phi_E(x)\right) e^{-\frac{i}{\hbar}Et}$$

$$\implies \quad E\,\phi_E(x) = -\frac{\hbar^2}{2m}\frac{\delta^2\phi_E(x)}{\delta x^2} + V\,\phi_E(x) \quad \textit{Time-independent version} \tag{11.36}$$

The result is the Time-independent version of the Schrödinger equation (Box 11.5 plus the summary in Box 11.6). There are a couple of important things to note. The first is that E is a *number*. It is a *single value*. It is the energy of the particle *over* its rest mass energy (as explained in Subsection 11.1.1). Having one single value for E is what makes it an energy eigenstate. The second is that this works only if the potential V does *not* change over time.

I should finish with a health warning. The maths I have used is sloppy (that is the word used by a well-informed friend). My aim is to help you understand intuitively *why* the time-independent version works for energy eigenstates (I promise it does). If you need a full proof including boundary conditions, you must look elsewhere.

Box 11.5   Time-independent Schrödinger equation   **Only for energy eigenstates**

Energy eigenstates are stationary states. E, a *number*, is the energy of the eigenstate.

$$E\,\phi_E(x) = -\frac{\hbar^2}{2m}\frac{\partial^2\phi_E(x)}{\partial x^2} + V\,\phi_E(x)$$

In all three spatial dimensions the following equation applies for $\phi_E(x,y,z)$:

$$E\,\phi_E(x,y,z) = -\frac{\hbar^2}{2m}\left(\frac{\partial^2\phi_E}{\partial x^2} + \frac{\partial^2\phi_E}{\partial y^2} + \frac{\partial^2\phi_E}{\partial z^2}\right) + V\,\phi_E$$

Using operator notation this can be written as shown below. Remember that E is a number:

$$E\,\phi_E = \frac{\hat{p}^2}{2m}\phi_E + \hat{V}\,\phi_E \qquad \textit{or sometimes abbreviated as} \qquad E\,\phi_E = \hat{H}\,\phi_E$$

Box 11.6   Schrödinger's equations summarised

$$\hat{E}\,\Psi \equiv i\hbar\frac{\partial\Psi}{\partial t} \qquad \hat{H}\,\Psi \equiv -\frac{\hbar^2}{2m}\frac{\partial^2\Psi}{\partial x^2} + V\,\Psi$$

$$\hat{E}\,\Psi(x,t) = \hat{H}\,\Psi(x,t) \qquad \textit{Schrödinger equation}$$

$$E\,\phi_E(x) = \hat{H}\,\phi_E(x) \qquad \textit{Time-independent Schrödinger equation}$$

Chapter 12

Schrödinger's Equation in Action

CHAPTER MENU

12.1 Free Particle Wave Function (E > V)
12.2 Creeping into Forbidden Places (E < V)
12.3 The Finite Potential Well
12.4 Quantum Tunnelling and the Sun
12.5 Dodging Potential Obstacles (E > V)
12.6 Quantum Biology
12.7 Wave Packets: A Model for Localised Particles
12.8 Summary

In the last chapter we discussed the time-dependent and time-independent versions of Schrödinger's equation and we looked at operators, eigenstates, eigenvalues, commutation relations, expectation values, Dirac notation and energy eigenstates. This is all important material, but I can imagine readers screaming for something that these equations tell us, so let's get more practical right now. For example, hidden in these equations is the secret of quantum tunnelling that is fundamental to what powers the sun. It involves a bit more maths to show how it works but worth a look don't you think?

12.1 Free Particle Wave Function (E > V)

Let's warm up with a simple question. What wave function will a particle have in a region where its energy E is *more* than the potential energy V (i.e. $E > V$)? Let's assume that the particle energy E and the potential V are each a single constant value (no variation across space or time). E is defined as the energy above rest mass energy so the particle has kinetic energy of $E - V$. In classical physics the particle has enough energy to be in this region. What does the Schrödinger equation say about it?

As we are dealing with an energy eigenstate we can use the time-independent Schrödinger equation, which can be rearranged as in Equation 12.2 because E and V are simple real numbers. The result is that the second derivative is a *negative constant* as shown in Equation 12.3.

$$E \phi_E(x) = -\frac{\hbar^2}{2m} \frac{\partial^2 \phi_E(x)}{\partial x^2} + V \phi_E(x) \tag{12.1}$$

$$\frac{\partial^2 \phi_E(x)}{\partial x^2} = \frac{2m}{\hbar^2} (V - E) \phi_E \tag{12.2}$$

$$E > V \quad \frac{\partial^2 \phi_E(x)}{\partial x^2} = -C \phi_E \qquad -C, \text{ a negative constant} \tag{12.3}$$

The solution is, of course, a sinusoid. It is our old friend the free particle wave function for a particle with momentum p such that its kinetic energy $\frac{p^2}{2m}$ is $(E - V)$. Take the free particle wave function and separate out the component that is

Quantum Untangling: An Intuitive Approach to Quantum Mechanics from Einstein to Higgs, First Edition. Simon Sherwood.
© 2023 John Wiley & Sons Ltd. Published 2023 by John Wiley & Sons Ltd.

x dependent as shown in Equation 12.4. The second derivative is shown on the right of Equation 12.5. The final step is to equate this with the expression for the second derivative in Equation 12.2 and show the relationship between $(E-V)$ and p.

$$\phi_E = e^{\frac{i}{\hbar}(px-Et)} = e^{\frac{i}{\hbar}px}\, e^{-\frac{i}{\hbar}Et} \implies \phi_E(x) = e^{\frac{i}{\hbar}px} \tag{12.4}$$

$$\frac{\partial \phi_E(x)}{\partial x} = \frac{i}{\hbar} p\, \phi_E \qquad\qquad \frac{\partial^2 \phi_E(x)}{\partial x^2} = -\frac{p^2}{\hbar^2}\, \phi_E \tag{12.5}$$

$$\frac{2m}{\hbar^2}(V - E) = -\frac{p^2}{\hbar^2} \qquad \implies \qquad E - V = \frac{p^2}{2m} \tag{12.6}$$

This should come as no surprise. We have discussed in detail the sinusoidal nature of the wave function of a free particle. We have discussed superpositions. And we used the definition of kinetic energy to explain the structure of the Schrödinger equation.

– but one thing we have never discussed, and a classical physicist would not even consider worth discussing, is what happens if a particle is in a region where its energy (above rest mass energy) is *less* than the required potential energy $(E < V)$; prepare yourself for a shock!

12.2 Creeping into Forbidden Places (E < V)

What does the Schrödinger equation tell us about the wave function when E is *less* than V? Let me give a classical example of what this means. I will use gravity because it is so familiar. Imagine that you throw a ball up in the air. It has kinetic energy. At it goes up, its kinetic energy drops and it gains potential energy. At the highest point it has no kinetic energy left. It can go no higher. At this point, all of its energy (over rest mass energy) is potential energy. Any higher would mean $E < V$ so the ball would have to have *negative* kinetic energy.

Any region where $E < V$ is a classically disallowed region. Negative kinetic energy sounds impossible, but what does the Schrödinger equation tell us? In the Schrödinger equation, kinetic energy is expressed as a second derivative. If $(E < V)$ then the second derivative must be *positive* as shown in Equation 12.9, and that is not impossible – it is very possible.

$$E\,\phi_E(x) = -\frac{\hbar^2}{2m}\frac{\partial^2 \phi_E(x)}{\partial x^2} + V\,\phi_E(x) \tag{12.7}$$

$$\frac{\partial^2 \phi_E(x)}{\partial x^2} = \frac{2m}{\hbar^2}(V - E)\,\phi_E \tag{12.8}$$

$$E < V \qquad \frac{\partial^2 \phi_E(x)}{\partial x^2} = +C\,\phi_E \qquad + C,\ a\ positive\ constant \tag{12.9}$$

If $E < V$, the second derivative must be a *positive* multiple of ϕ. This means the wave function is not sinusoidal. It is exponential. Equation 12.10 shows how such an exponential wave function results in the required positive second derivative. There are two solutions: a positive (growing) exponential and a negative (decaying) exponential.

$$E < V \qquad \phi = e^{\pm\sqrt{C}\,x} \implies \frac{\partial \phi}{\partial x} = \pm\sqrt{C}\,\phi \implies \frac{\partial^2 \phi}{\partial x^2} = +C\,\phi \tag{12.10}$$

Let me quickly dismiss the positive exponential $e^{+\sqrt{C}x}$ which is not a legitimate solution . The wave function explodes to infinity as x increases. Why is this a problem? The probability of finding the particle must sum to 100% (the particle must be *somewhere*). As the probability is proportional to $|\Psi|^2$, the integral $\int |\Psi|^2 dx$ must be finite. In the case of $e^{+\sqrt{C}x}$ it is infinite. We describe it as not *normalisable* so $e^{+\sqrt{C}x}$ is not a valid wave function.

However, the negative exponential $e^{-\sqrt{C}x}$ falls to zero as x increases, so it is normalisable. It is the only valid solution. With emphasis, let me repeat what the Schrödinger equation tells us. In this scenario, the second derivative *must* be a *positive* multiple of ϕ so the wave function in this classically disallowed region cannot be zero. It must be the decaying exponential $e^{-\sqrt{C}x}$.

This means that the wave function does not disappear when it encounters a region which, in classical terms, the particle does not have enough energy to reach. The wave function penetrates the region and fades away exponentially. If you measure

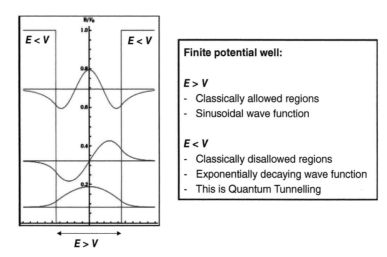

Figure 12.1 The finite potential well: the wave function reaches into areas where $E < V$ so there is a chance of finding the particle in classically disallowed regions.

the particle's position, there is a small but real probability of finding it in the classically disallowed region. The value of constant C that determines the rate of exponential decay of the wave function depends on the height of the potential barrier $(V - E)$ and the mass of the particle, as shown in Equation 12.11:

$$E < V \ exponential \ decay: \quad \phi = e^{-\sqrt{C}x} \quad where \ C = \frac{2m}{\hbar^2}(V - E) \tag{12.11}$$

12.3 The Finite Potential Well

The *finite potential well* is a neat model that illustrates this. In the one-dimensional model, a particle is placed in a potential that is zero in a central region and a constant potential V outside the region as shown in Figure 12.1. The diagram shows the three lowest energy eigenstates for the particle. In classical terms the particle should not be able to leave the well because it does not have the energy to surmount the potential energy barrier V on either side. But you can see from the figure that the quantum wave functions extend outside the well, decaying away exponentially. Remember that the wave functions are standing waves and are all vibrating up and down. It does not matter if the wave function is positive or negative because it is the square of the magnitude of the wave function at any point that drives the probability of finding the particle there.

We looked at the *infinite* potential well earlier in the book (see Section 7.4). In that case we simply forbade the wave function from existing outside the well. In terms of the Schrödinger equation, the exponential decay is instant for the infinite potential well $(V - E = \infty)$. It is worth comparing Figure 12.1 with the infinite well in Figure 7.1. You can see that the patterns are very similar except that in the finite potential well, the wave functions creep into classically forbidden areas. The lower the energy gap $(V - E)$, the further the creep. This is *quantum tunnelling* (for a detailed explanation of the underlying mechanism, I beg your patience until Section 19.4 on quantum field theory).

12.4 Quantum Tunnelling and the Sun

Quantum tunnelling allows particles to penetrate short distances through seemingly impassable barriers. The process is illustrated in Figure 12.2. A particle with energy E approaches from the left a potential barrier of energy V, which it classically cannot pass $(E < V)$. The incident wave function decays exponentially, but there is some residual wave function at the other side of the barrier. This means that there is a reduced, but real, probability that the particle will pass the barrier. If it does then it will continue on its way with energy E because the reduced amplitude of the wave function affects the probability of the particle's passing, but does not change its energy.

Figure 12.2 Quantum tunnelling: even if the barrier potential V is higher than the particle's kinetic energy E, there is a chance of it tunnelling through.

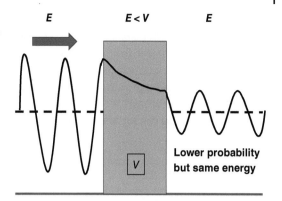

Lower probability but same energy

What is the probability that the particle will pass through the barrier? The probability of finding a particle somewhere is proportional to the *square* of the wave function, so the relationship between the transmission probability T and the length of the barrier L is:

$$\phi = e^{-\sqrt{C}L} \implies T \approx e^{-2\sqrt{C}L} \quad where \quad C = \frac{2m}{\hbar^2}(V - E) \tag{12.12}$$

Why don't we see quantum tunnelling by larger objects? The maths gives us the answer. The mass of the particle has an exponential impact on the decay rate of the wave function and therefore on the probability of transmission. Consider a barrier that an electron has a 50% chance of passing. If you confront a ZnS molecule with the same barrier, the extra mass of the molecule (180,000 times that of the electron) reduces the transmission probability of quantum tunnelling from 50/50 to about one in 10^{130}. Objects that we can see are much larger (even a grain of sand is about 10^{24} times the mass of an electron) so the probability of transmission from quantum tunnelling is such a vanishingly small number that your calculator would not begin to be able to handle it.

Quantum tunnelling plays a major role in the sun. The basic fusion reaction (hydrogen to helium) requires two protons to approach close enough for the attraction of the strong force to overcome their electrical repulsion. The range of the strong force is very short, about 4 femtometres (a femtometre is 10^{-15}m). Fusion in the sun occurs at about 10^7 Kelvin, but it should take a temperature of over 10^9 Kelvin to get the protons close enough. The protons have only one hundredth of the energy they need so how does fusion happen? The answer is quantum tunnelling. We can illustrate this in a fun way (well, I think it is fun) with a *very* crude model. The model is of a proton that runs into a simple rectangular potential energy barrier exactly like that shown in Figure 12.2. How far can a proton tunnel in?

Only about one in 10^{28} approaches by a proton results in a reaction, but the sun is so massive that this is enough to sustain stable fusion. The crude model suggests the proton can tunnel in and fuse at this rate from about 10^{-12} metres. So the protons tunnel in from over 100 times the range of the strong force. Hence the required energy is 100 times less and the sun's fusion can occur at about 10^7 Kelvin instead of 10^9 Kelvin. The model calculation is shown in Box 12.1.

Box 12.1   Quantum tunnelling: a model of fusion in the sun

The first step is to estimate the energy V of the barrier, that is, the energy a proton needs to get within 4×10^{-15}m of another, so the strong force can kick in. This is called the Coulomb potential and is about 6×10^{-14} joules. In the calculation, q is the particle charge and r the distance apart:

$$Coulomb\ potential \approx (9 \times 10^9)\frac{q_1 q_2}{r} \approx (9 \times 10^9)\frac{(1.6 \times 10^{-19})^2}{4 \times 10^{-15}} \approx 6 \times 10^{-14}$$

In the sun, about one in 10^{28} proton interactions results in fusion. What length L of barrier can a proton tunnel through and have this transmission probability? We can use Equation 12.12 and solve for L:

$$C = \frac{2m_p}{\hbar^2}(V - E) = \frac{2(2 \times 10^{-27})}{(10^{-34})^2}(6 \times 10^{-14}) \approx 2 \times 10^{26} \implies \sqrt{C} \approx 10^{13}$$

$$T\ probability: T = 10^{-28} \approx e^{-64} \qquad T\ formula: T \approx e^{-2\sqrt{C}L} \approx e^{-2 \times 10^{13}L}$$

$$\implies (2 \times 10^{13})L = 64 \qquad \implies L \approx 10^{-12}\ metres$$

I love using this sort of crude model. Of course it is inaccurate (for example, the Coulomb potential that resists the protons decreases rapidly with distance *r* so it is not rectangular), but the model gives a feel of what is happening. Back here on earth, controlled nuclear fusion could give mankind a limitless source of renewable energy. Current efforts such as the huge superconducting magnets called *tokamaks* have to operate at temperatures over 10 times those of the sun in order to achieve fusion, at huge detriment to their energy balance. Might there be a way to tune quantum tunnelling in order to reduce these temperatures? We don't know, but perhaps nothing is more important, and the hunt is on (see Box 12.2 for a corny joke). Now do you see why quantum mechanics is worth studying?

Box 12.2 Just for laughs: A physicist's diet

Which take-away food gives you the most energy? *Fission chips.* But why do physicists yearn for mixed cuisine? *They would prefer fusion.*

12.5 Dodging Potential Obstacles (E > V)

Quantum tunnelling is not always good news. The backbone of computer memory is Direct Random Access Memory (DRAM) integrated circuit chips. Over the years, miniaturisation has increased the speed of memory access and lowered its cost. I worked once on a project to speed up the development cycle of the vapour-deposition and etching machines essential in integrated circuit production. It truly is amazing technology. But quantum tunnelling poses a threat because at some stage of miniaturisation electrons (and therefore current) will start to leak across junctions. There is much active research in this area and there is hope that we might even be able to turn the quantum tunnelling problem to our advantage!

In *classical* physics, you either have enough kinetic energy to pass a potential barrier (in which case you do pass it) or you do not have enough kinetic energy (in which case you do not). We have seen that *quantum* physics is not so simple and that the fuzzy wavelike nature of quantum particles allows them sometimes to find a way through barriers that classically would stop them. The corollary should not surprise you. This same fuzzy wavelike nature means that on occasion, quantum particles bounce back off potential barriers that, classically, they should be able to pass.

Let's remind ourselves what classical physics says should happen. Figure 12.3 shows the scenario of a ball approaching a hill from the left. If it has enough kinetic energy it will reach the plateau on top. It will roll along the plateau with a reduced velocity as kinetic energy is no longer *E* but is now *E − V* because of the work done in climbing the hill. It will roll on and on along the plateau finally descending down the other side with kinetic energy *E* restored. What determines whether or

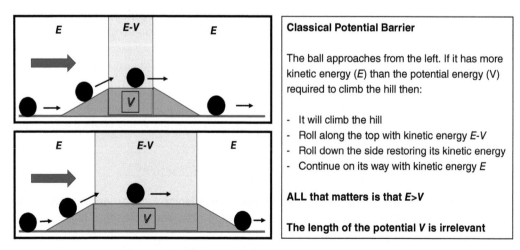

Classical Potential Barrier

The ball approaches from the left. If it has more kinetic energy (*E*) than the potential energy (V) required to climb the hill then:

- It will climb the hill
- Roll along the top with kinetic energy *E-V*
- Roll down the side restoring its kinetic energy
- Continue on its way with kinetic energy *E*

ALL that matters is that *E>V*

The length of the potential *V* is irrelevant

Figure 12.3   The classical view: the ball must have sufficient kinetic energy to climb the barrier.

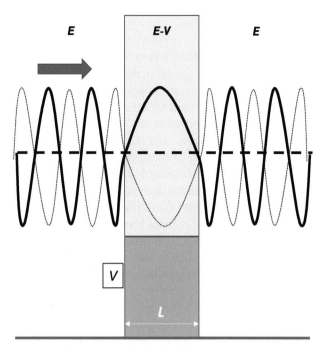

Figure 12.4 Quantum transmission is affected by the length of the barrier.

not the ball passes the potential? The *only* thing that matters is that $E > V$. As shown in the lower part of the figure, the length of the top of the potential (the plateau) is irrelevant.

The same scenario is very different for a quantum mechanical particle. The quantum particle may not pass the potential barrier even if $E > V$. What is more, the *length* of the barrier (the plateau) is a major factor in determining whether or not the particle will pass, especially if the particle's kinetic energy E is only slightly higher than the potential barrier's energy V.

Take a look at the model of a simple one-dimensional rectangular potential barrier in Figure 12.4. A particle travels in from the left with energy E and encounters the barrier of potential V. As $E > V$, the particle classically should slow down in the region of the potential, but always make it through and continue on its way. Why is this not true for a quantum particle and why is the length of the barrier important?

In Figure 12.4 the particle approaches from the left with kinetic energy E. While it is in the region of potential V, the wavelength of the particle state must increase to reflect its lower kinetic energy $(E - V)$ and hence lower momentum. The length of barrier that gives perfect transmission is when there can be a stable resonating standing wave in the region of potential. This is the situation shown in the figure. It is an easy exercise to calculate the length of barrier that leads to maximum (perfect) transmission:

$$L = \frac{n}{2}\lambda \qquad \text{where n is an integer so there is a standing wave} \tag{12.13}$$

$$L = \frac{n}{2}\frac{2\pi}{k} = \frac{n\pi\hbar}{p} = \frac{n\pi\hbar}{\sqrt{2m(E-V)}} \tag{12.14}$$

If E is only very slightly higher than V, the length of the barrier can have a significant impact on transmission, that is, there is a significant chance that the particle will reflect back off the potential. And you may actually improve transmission by *lengthening* the barrier. This rectangular potential barrier model is fairly crude (I am not expecting you to find many rectangular barriers in nature) but it, and the quantum tunnelling model, give insight into how quantum particles behave around potential barriers. This raises a big question:

Can we facilitate chemical reactions by tuning the shape and length of potential barriers?

12.6 Quantum Biology

Some scientists believe that evolution has already harnessed this tactic and that quantum mechanics plays a fundamental role in how biological systems work. This is the focus of Quantum Biology.

Enzymes are obvious candidates for research. These protein molecules facilitate the biochemical reactions that are essential for life. The classical explanation is that an enzyme binds with the reactant molecules and forms a transition state that lowers the reaction *activation energy* (the reaction's equivalent of V in Figure 12.4). But quantum biologists argue that enzyme activity goes much further than this and can also alter the *length* and *shape* of the reaction's potential barrier, effectively *quantum tuning* the potential barrier to maximise transmission and quantum tunnelling.

It is difficult to find experiments to separate the various mechanisms that an enzyme might employ: lowering V; shaping the potential barrier and tuning L for maximum transmission; shortening L for quantum tunnelling – how can you tell them apart? One trick is to replace some of the reactant hydrogen atoms (each has one proton) with deuterium (one proton, one neutron) and tritium (one proton, two neutrons). These isotopes have the same chemical characteristics as hydrogen but with two-fold and three-fold higher mass. In theory this should make little difference to the classical chemical pathways, but will dramatically reduce any quantum tunnelling because the extra mass has an exponential impact on tunnelling transmission (see the formula in Equation 12.12). There is growing evidence from these isotope experiments that quantum tunnelling does indeed play an important role in enzyme activity.

The action of some enzymes increases reaction rates by factors of up to a billion or even a trillion fold. Enzymes are beautifully tuned for efficiency. To me, it is only logical that evolution has latched on to quantum mechanisms during its random but steady search for improvement over the last few billion years.

Quantum Biology is a relatively new field, but if you are interested, there are a few books on the subject such as Al-Khalili and McFadden's *Life on the Edge*. In it they discuss evidence for much more sophisticated quantum tuning. In photosynthesis there are several routes to a successful reaction. They argue that the potential barriers of these different reaction pathways are quantum tuned to be in sync. The amplitudes of the transmission wave functions from each and every possible pathway add up constructively. If true, the probability of a successful outcome is then the *square* of this sum, vastly increasing the overall efficiency of the reaction. What a beautiful, amazing and exciting thought that is!

Box 12.3   Schrödinger versus Heisenberg: a quantum bust-up

Schrödinger's quantum theory, based on wave mechanics, was preceded a year earlier by that of Heisenberg, which was built on matrix mechanics. After much contention and deliberation, the two approaches were finally proved to be mathematically equivalent, but in the earlier years, how do you think Schrödinger and Heisenberg reacted to each other's theories? A spirit of mutual endeavour and respect? Not quite!

Schrödinger on Heisenberg (1926): *I know of [Heisenberg's] theory, of course, but I feel intimidated, not to say **repelled**, by what seem to me the very difficult methods of matrix mechanics, and by the lack of clarity.*

Heisenberg on Schrödinger (1926): *The more I think about the physical portion of Schrödinger's theory, the more **repulsive** I find it... What Schrödinger writes about the visualisability of his theory... I consider **crap**.*

How did the Nobel committee sensitively handle the competing achievements of these two sparring partners? They awarded Heisenberg the 1932 Nobel Prize and Schrödinger the 1933 Nobel Prize on the same day, 10 December 1933.

12.7   Wave Packets: A Model for Localised Particles

We will now switch topic to discuss a particle that is spatially constrained so it has a well-defined position (Δx is small). We can use the work of Schrödinger and Heisenberg (see Box 12.3 for how the two got on with each other) to figure out what will happen.

Let's start with a different question: *How do localised particle states form?* Based on the Schrödinger equation, the answer is that they don't! Or, at least, it is extraordinarily unlikely that a wave function will progress from an undefined spatial state to a well-defined spatial state (it is possible in the same way that all the particles of gas around you theoretically could rush into the same corner of your room). So, how do localised particle states form? It is the result of *particle interactions*. Consider an electron being emitted from an atom. Because of the interaction with the atom, it starts with a well-defined location (within about 10^{-10}m) so it leaves the atom as a free particle starting from a constrained localised position. The aim of this section is to understand how it evolves over time.

Localised particle states are spread over a limited spatial area. They are described as *wave packets*. The model typically used for a localised particle is a Gaussian (the normal distribution). The classic bell-shaped curve has a clear peak, is symmetrical and its standard deviation is easy to calculate. Also, very importantly, we know its Fourier transform (another Gaussian). As you will see, this makes the maths much easier.

Let's set the starting time at $t = 0$ with $\Psi(x)$ as a Gaussian distribution that we will centre on $x = 0$ for simplicity. What does this look like in terms of momentum? The Fourier transform of a Gaussian is another Gaussian (see Section 10.2 if you need a reminder) so $\Psi(p)$ is a Gaussian that we also will centre on $p = 0$ for simplicity. This is illustrated in the top half of Figure 12.5. What happens? The particle state spreads as we will show.

Before we do the maths, let me share with you an intuitive explanation. We discussed in depth in Section 5.3 and again in Section 10.3 that for a free particle to be constrained in a small volume of space (so Δx is small) it must be in a superposition of momentum eigenstates. Some of these will be momentum $p < 0$ (particle states moving to the left) and others with $p > 0$ (particle states moving to the right). Over time this variation in momentum spreads out the spatial distribution of Ψ as shown in the bottom half of Figure 12.5. This sounds sensible and is not completely without merit, but please beware of this seductive sort of argument. You must be very cautious. Classical analogues do not always apply. In this case the particle state does indeed spread, but you will come across many particle states with uncertainty in momentum that do not. For example, in the infinite well model the particle state is stationary while being in a superposition of having momentum to the right and left. Proceed with care!

Let's get into the mathematics. We will work with the variables x and k because they are symmetrical in the free particle wave function e^{ikx} (remember that $p = \hbar k$). At $t = 0$, both $\Psi(x)$ and its Fourier transform $\Psi(k)$ are Gaussians. How can we calculate the time evolution of Ψ? The trick is to recognise that, for a free particle there is no potential ($V = 0$), so the time evolution can be expressed in terms of k as follows:

$$e^{-\frac{i}{\hbar}Et} = e^{-i\hbar\frac{k^2}{2m}t} \qquad \text{With no potential: } E = \frac{p^2}{2m} = \frac{\hbar^2 k^2}{2m} \qquad (12.15)$$

Starting point:

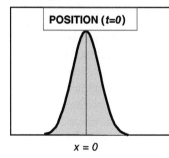

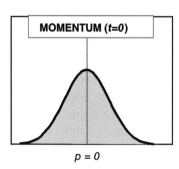

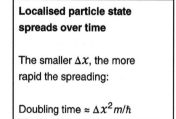

Change in spatial distribution over time:

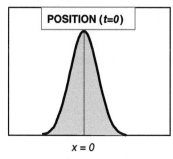

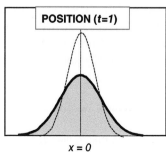

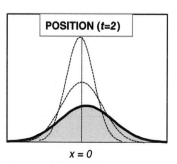

Figure 12.5 Localised particle states spread over time.

Therefore, we can apply the time evolution to $\Psi(k)$ expressing E in terms of k. This tells us how $\Psi(k)$ changes over time. It turns out that it remains a Gaussian, but its standard deviation changes over time. Because $\Psi(k, t)$ is still a Gaussian, we can use the Fourier transform relationship again to get us back to $\Psi(x, t)$. The details of this calculation are shown in Box 12.4. The result is that the standard deviation of $\Psi(x)$ grows over time and the uncertainty in position Δx grows.

Box 12.4   Calculation of spread of Gaussian wave packet (with standard deviation $\sigma = a$)

$$\Psi(x, 0) = e^{-\frac{x^2}{2a^2}} \implies \Psi(k, 0) = e^{-\frac{a^2 k^2}{2}} \qquad \textit{Fourier transform at time = 0}$$

$$\Psi(k, t) = e^{-\frac{a^2 k^2}{2}} e^{-\frac{i}{\hbar} Et} = e^{-\frac{a^2 k^2}{2}} e^{-\frac{i}{\hbar} \frac{\hbar^2 k^2}{2m} t} = e^{-\frac{k^2}{2}\left(a^2 + \frac{i\hbar}{2m} t\right)}$$

$$\Psi(k, t) = e^{-\frac{k^2}{2}\left(a^2 + \frac{i\hbar}{2m} t\right)} \implies \Psi(x, t) = e^{-\frac{x^2}{2} \frac{1}{a^2 + \frac{i\hbar}{2m} t}} \qquad \textit{Fourier transform at time = t}$$

$$\sigma^2(x, 0) = a^2 \qquad \sigma^2(x, t) = \left| a^2 + \frac{i\hbar}{2m} t \right|$$

The variance σ^2 of the spatial Gaussian increases over time. The standard deviation σ and thus the uncertainty in position Δx double approximately every $\frac{(\Delta x)^2 m}{\hbar}$ seconds.

We can quantify this. The smaller the uncertainty in position, Δx, the faster the spread. This means that a quantum particle with a well defined location changes very quickly. How quickly? If an electron is emitted from an atom then its size (standard deviation of spread) is about 10^{-10} metres and this will double in only 10^{-16} seconds. That is very fast indeed:

$$t_{double} \approx \frac{(\Delta x)^2 m}{\hbar} \approx \frac{(10^{-10})^2 \, 10^{-30}}{10^{-34}} \approx 10^{-16} \ seconds \tag{12.16}$$

This seems to create a paradox (Aarrgghh - another one!). Why doesn't everything dissipate into a sludgy soup of nothing? Why do we have this impression that electrons, atoms and molecules have any sort of position or are *real* in any sense of the word?

The first and most important part of the answer is that most particles are not free. They are constantly interacting with each other. Imagine an electron state Ψ dissipating as shown in the bottom of Figure 12.5. What happens as the electron interacts with other particles? Every interaction is in effect a measurement that resets the spatial distribution of Ψ. The interactions stop Ψ spreading. Is there a way to better describe this? Perhaps this is the domain of a philosopher, but I like to imagine particle states born of interactions. They are localised so perhaps more particle than wave. These states spread steadily into looser wave forms until they are jerked backed to a more localised particle form by the next interaction. The spatial uncertainty of the state is like an accordion tightening with each interaction, and spreading in between. But perhaps I am falling into the very trap that I warned of, trying to use familiar terms and words for superpositions that are quantum and defy classical description (for an example of this weirdness check out the Zeno effect in Box 12.5).

The second part of the answer is that the dissipation speed for macro objects would be very very slow even if they started tightly spatially constrained and were not interacting. What if you managed completely to constrain and isolate a grain of sand with a mass of about 10^{-6} kg? How long would it take for there to be a noticeable spatial spread in its wave function? For example how long would it take in theory for the positional uncertainty to double in size from say 0.1mm to 0.2mm? You would have to be extremely patient. The answer is 10^{13} years so you would need somehow to isolate the grain of sand for a thousand times the age of the universe.

$$t_{double} \approx \frac{(\Delta x)^2 m}{\hbar} \approx \frac{(10^{-4})^2 \, 10^{-6}}{10^{-34}} \approx 10^{20} \ seconds \approx 10^{13} \ years \tag{12.17}$$

While we don't need to worry about this effect in the macro world, it does illustrate how fast things can change at the level of quantum particles. This in turn highlights the importance of energy eigenstates. Their spatial stability helps us to develop a picture of the quantum world.

Box 12.5   The quantum Zeno effect

The time evolution of particles is slowed by frequent measurement: the quantum Zeno effect. One of the first to propose the Zeno effect was Alan Turing (1912–1954) who rose to fame leading the UK team that cracked the Enigma code and shortened the Second World War. The UK rewarded his efforts with chemical castration while persecuting him as a homosexual.

As an example of the Zeno effect, consider a particle that is tightly spatially constrained. Normally, over time, its wave function would spread. However, frequent measurement collapses the wave function again and again so the particle remains spatially constrained. This is often described in layman's terms as *the-watched-pot-never-boils* effect of quantum mechanics.

12.8   Summary

In this chapter we have discussed some of the implications that flow from the Schrödinger equation. We used the time-independent version to look at the behaviour of quantum particles when confronted with a potential. In a classically forbidden zone where the particle energy is lower than the potential ($E < V$), the wave lingers exponentially creating the possibility of quantum tunnelling that is so important in powering the sun (check out Box 12.6 about Sakharov and the tokamak, one of the most promising approaches to generating fusion power on earth).

In regions where the particle energy is higher than the barrier ($E > V$), there is still a chance that the particle will not pass. If E is only slightly higher than V then much depends on the length and shape of the potential barrier. Does nature exploit this? Do enzymes function in part by quantum tuning the potential barriers of the reactions they promote? Quantum biologists believe so.

We looked at the time evolution of a spatially constrained particle using a Gaussian model. The spatial spread of a constrained small particle increases rapidly, but this is kept in check by continuous interaction with other particles. The rate at which the spatial spread increases is inversely proportional to mass so this spread is insignificant for larger objects (even if it were possible to isolate them).

The calculations in this chapter show that the quantum world can change extremely quickly. We will spend the next few chapters studying energy eigenstates, a beacon of stability in our mad frantic quantum universe.

Box 12.6   The tokamak, and a spokesperson for the conscience of mankind

The tokamak is, at time of writing, the most promising prototype fusion reactor. It uses a powerful magnetic field to confine a hydrogen plasma, under extreme heat and pressure, in the form of a torus (the shape of a doughnut, if you prefer a tastier description). The idea of a tokamak was first proposed in the 1950s by Andrei Sakharov (1921–1989) in the USSR. Sakharov played a pivotal role in the development of the Soviet atomic bomb and was nicknamed the father of the Soviet hydrogen bomb.

In spite of this, in the 1960s Sakharov became a vocal advocate against nuclear proliferation: *Something new and terrible had entered our lives, a product of the greatest of the sciences, of the discipline I revered.* He grew disillusioned with the USSR and published essays calling for democratisation and better human rights. In 1975 Sakharov was awarded the Nobel Peace Prize, being described as *a spokesman for the conscience of mankind*. The USSR authorities described him as *Domestic Enemy Number One* and refused him permission to travel to Oslo.

In 1980 the American Humanist Society named him *Humanist of the Year*. At the same time, the USSR confined him for six years under police surveillance. A few years later, his wife was imprisoned and he undertook several hunger strikes to expedite her release. Sakharov was finally freed in 1986 as part of Mikhail Gorbachev's policy of *perestroika* and *glasnost*. He continued as an activist for the remaining three years of his life.

If mankind is to get away from the brink, it must overcome its divisions. Andrei Sakharov

Chapter 13

Quantum Harmonic Oscillator

CHAPTER MENU
13.1 Introduction 13.2 Penetration Model for the QHO 13.3 Schrödinger's Equation for the QHO 13.4 The QHO in Three Dimensions 13.5 Formal Definition of the Creation and Annihilation Operators 13.6 The Path to Quantum Field Theory (QFT)

13.1 Introduction

The Quantum Harmonic Oscillator is the quantum version of the Simple Harmonic Oscillator. It describes an object caught in a potential that is similar in effect to that of a spring or pendulum. If it strays from the point of equilibrium ($x = 0$) there is a restoring force directly proportional to the displacement x. This force accelerates back towards the equilibrium point. If you displace the object and then release it, it oscillates around the equilibrium point. This is *simple harmonic motion*. An example is the motion of a pendulum that oscillates to and fro.

13.1.1 The Simple Harmonic Oscillator

Let's briefly review the classical Simple Harmonic Oscillator (SHO). Springs or pendulums are typical examples such as shown in Figure 13.1. A well-known feature of simple harmonic motion is that, whatever the initial displacement from the equilibrium point, the frequency of oscillation is the same. This is why pendulums are used in clocks. Wherever you release the pendulum from, each oscillation takes the same time so the clock always ticks at the same rate.

If you have studied physics you will be familiar with the maths of the SHO, but for those who are not, the system is simple (hence the name). When it moves out of equilibrium, the force is $F = -Kx$ where K is a constant from Hooke's law of elasticity based originally on a spring (see Box 13.1 for a warning on variables and Box 13.2 if you want to know more about Hooke). This leads to a sinusoidal motion of angular frequency ω that characterises the system as shown in Equation 13.1. We can show that the acceleration of the object/pendulum at any point is $-\omega^2 x$ (towards the central point $x = 0$) by using the wave function for angular frequency ω and taking its second derivative with respect to time.

$$x(t) = A \sin \omega t \quad \implies \quad Acceleration: \ \frac{d^2 x}{dt^2} = -\omega^2 A \sin \omega t = -\omega^2 x \tag{13.1}$$

In the SHO, the force is $F = -Kx$ and we know from Newton that $F = ma$ (where a is the acceleration). Using the expression for acceleration from Equation 13.1 gives $F = -m\omega^2 x$, showing that the constant K in the force is $m\omega^2$:

$$F = -m\omega^2 x \quad \implies \quad K = m\omega^2 \quad as \ F = -Kx \ is \ the \ restoring \ force \tag{13.2}$$

Quantum Untangling: An Intuitive Approach to Quantum Mechanics from Einstein to Higgs, First Edition. Simon Sherwood.
© 2023 John Wiley & Sons Ltd. Published 2023 by John Wiley & Sons Ltd.

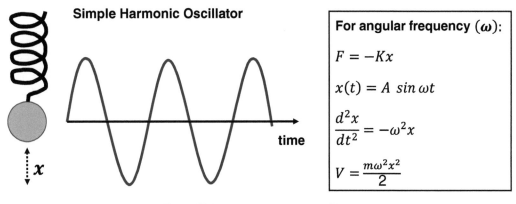

Figure 13.1 Simple Harmonic Oscillator.

To develop the Quantum Harmonic Oscillator we need the SHO term for potential energy (V) so we can plug it into the Schrödinger equation. This is derived by integrating the restoring force over x and gives the potential $V = \frac{m\omega^2 x^2}{2}$. These relationships are summarised in Figure 13.1.

Box 13.1   A warning on variables: K and ω appear in analysis of the QHO

K is often called the spring constant and generally refers to the constant of proportionality between F and displacement x. Do *not* confuse it with the wave number k in the free particle wave function. ω generally refers to the angular frequency of a comparative SHO system (pendulum or such). Do *not* confuse it with the angular frequency ω of particle wave functions.

13.1.2   The SHO and QHO: Why Do We Care?

The SHO in classical physics and the QHO in quantum physics are useful in modelling the behaviour of almost any system that oscillates slightly away from an equilibrium. For the QHO this includes atoms in lattices, vibrating molecules, indeed any trapped particle.

This is because the restoring force of virtually any equilibrium system can be approximated with the SHO/QHO restoring force of Kx (with K a constant) provided that the displacement x is small. To explain I need to introduce the Maclaurin series. This is named after Colin Maclaurin although it is a variant of the well known Taylor series that is named after Brook Taylor who didn't invent that either; such are the mysteries of nomenclature. Anyway, the point of the Maclaurin series is that any function $f(x)$ can be written as a series in the way shown below, with A_0, A_1, A_2... all being constants:

$$f(x) = A_0 + A_1 x + A_2 x^2 + A_3 x^3 + A_4 x^4 + A_5 x^5... \tag{13.3}$$

The Maclaurin series tells you how to calculate A_0, A_1, A_2... but we do not need to worry about that. The important thing is that it can be done, so any relationship between the force on an object and its distance from equilibrium, however complicated, can be written in this form. Now the trick. We consider a very small displacement Δx. This means that the higher power components (Δx^2 and higher) will be insignificant and can be ignored (Equation 13.4). Furthermore, at equilibrium both x and F are zero, so therefore $A_0 = 0$. This leaves us with $K\Delta x$ as an approximation of the restoring force for a small displacement for *any* equilibrium system.

$$\begin{aligned} Restoring\,force &= A_0 + A_1\,\Delta x + A_2\,\Delta x^2 + A_3\,\Delta x^3 + A_4\,\Delta x^4 + A_5\,\Delta x^5... \\ &\approx A_0 + A_1\,\Delta x \tag{13.4} \\ &\approx A_1\,\Delta x \implies restoring\,force \approx K\,\Delta x \; for\, small\, \Delta x \tag{13.5} \end{aligned}$$

If you understand the dynamics of the QHO, you understand the dynamics of small oscillations in *all* quantum systems that are in equilibrium. When Schrödinger developed his equation, the first thing he did was to model the hydrogen atom. The next was to use his equation for the QHO. It is that fundamental.

Box 13.2   Hooke's Law of Elasticity: ceiiinossssttuv

The life of Robert Hooke (1635–1703) was filled with arguments. He was described variously as irritable, suspicious, despicable, cantankerous, mistrustful and vengeful. His biggest dispute was with Newton. Hooke maintained that he shared with him the idea of an inverse square law for gravity, a claim that Newton vehemently denied. To protect his discovery of Hooke's law ($F = Kx$), he published it first as *ceiiinossssttuv* revealing it two years later to be an anagram of *ut tensio sic vis*, the Latin for *as the extension, so the force*.

These were turbulent times. Hooke lived through the execution of Charles I, through England's time as a republic under Cromwell and the subsequent restoration of Charles II to the throne. He died a wealthy man having helped Christopher Wren rebuild London after the Great Fire of 1666. Indeed, Hooke devised the method used to construct the dome of St Paul's cathedral.

13.2   Penetration Model for the QHO

I need you to cast your minds back to the infinite potential well (electron-in-a-box) model in Section 7.4 which showed that a constrained particle can only have specific amounts of energy. In the model, we constrained the particle in a one-dimensional space of length L. For the wave function to fit as a standing wave, its wavelength must be $\lambda = \frac{2L}{n}$ where n can be any positive integer. The wavelength tells us the momentum of the particle which, in turn, tells us its energy as shown again in Equation 13.6. Using more precise terminology, this tells us the possible *energy eigenstates* of the constrained particle.

$$p = \pm\frac{2\pi\hbar}{\lambda} \implies E = \frac{p^2}{2m} = \frac{n^2\pi^2\hbar^2}{2mL^2} \quad \textit{Particle constrained in length L} \tag{13.6}$$

What if we constrain a particle in a QHO potential? Our challenge is to find its energy eigenstates. The correct approach is to apply the Schrödinger equation, but before we do that, let me give you a flavour of what is happening by using another simple model. The easiest way to visualise a particle in a QHO potential is to imagine it on the end of a spring as in Figure 13.1.

How tightly is the particle constrained? We can make an *approximation* using the maths of the classical SHO. As the particle moves out from the centre of the potential, the spring pulls it back. The distance x that it can drift from the centre depends on its energy E. At the limit, all its energy is potential energy so the maximum possible value of x is $E = m\omega^2 x^2$ (see equation of SHO potential energy in Figure 13.1). The particle can be a maximum distance x either side of the centre of the potential so the constraining length is $L = 2x$. We can use this to give an approximation for L in terms of E as shown in Equation 13.7.

$$E = \frac{m\omega^2 x^2}{2} = \frac{m\omega^2 L^2}{8} \implies L^2 = \frac{8E}{m\omega^2} \tag{13.7}$$

We can feed this expression for L^2 into Equation 13.6 and solve for E:

$$E = \frac{n^2\pi^2\hbar^2}{2m}\frac{m\omega^2}{8E}$$

$$E^2 = \frac{n^2\pi^2\hbar^2}{2m}\frac{m\omega^2}{8} = \frac{n^2\pi^2\hbar^2\omega^2}{16} \implies E = n\frac{\pi}{4}\hbar\omega \tag{13.8}$$

This gives the possible values of particle energy E in the simple model. Remember that n can be any positive integer. The possible values come in regular steps separated by $\frac{\pi}{4}\hbar\omega$ which is about $0.8\hbar\omega$. In the actual QHO each step is $\hbar\omega$, so the pattern is very similar. Figure 13.2 illustrates what is happening in the model.

Let's compare with the original infinite potential well (electron-in-a-box) model. In that model, the constraining length L is *fixed*. Each value of n is an energy eigenstate (an energy level that the particle can have), but the energy of the particle does not change L. In more realistic scenarios, the amount that the particle is constrained depends on its energy. This is the case with the QHO. As the energy of the particle increases, L increases.

Consider the change from energy level $n = 1$ to $n = 2$ as illustrated in Figure 13.2. If we give the electron an energy kick up to $n = 2$, it has more energy and can drift further from the centre, penetrating deeper into the QHO potential. If L had not changed, we would have expected the electron energy to quadruple in the jump from $n = 1$ up to $n = 2$, but the increase in L partially offsets this. The net result is that the energy eigenstates of the QHO are equally spaced.

Figure 13.2 Simplified model: particle constrained when the potential exceeds its kinetic energy E.

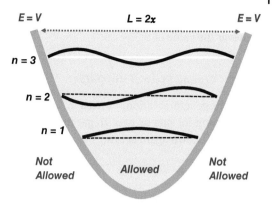

Please remember that this is just an *approximate model*. It ignores all sorts of things such as quantum tunnelling and the effect on the wave function of the change in value of the potential inside the QHO. It is a simple tool to help you understand why the energy eigenstates of the QHO are *equally* spaced, which is an important feature of the QHO because it means that the same is true for all quantum fields near equilibrium. Each step up or down in excitation absorbs or releases the same amount of energy. This will prove very important when we come to discuss quantum field theory later in the book. For some of you that may be as much as you want to know about the QHO and you can relax and skim the rest of this chapter. But the QHO is important, so any physicists need to read on.

13.3 Schrödinger's Equation for the QHO

Now we have set the stage with the simple model, let's calculate the QHO energy eigenstates correctly. The time-independent Schrödinger equation works for energy eigenstates (see Box 11.5) so we can use that, plugging in the definition of potential energy V from the Simple Harmonic Oscillator: $V = \frac{m\omega^2 x^2}{2}$. This gives the result in Box 13.3.

Box 13.3   Quantum Harmonic Oscillator using time-independent Schrödinger equation

$$E\,\phi_E \;=\; -\frac{\hbar^2}{2m}\frac{\partial^2 \phi_E(x)}{\partial x^2} + V\,\phi_E(x) \;=\; -\frac{\hbar^2}{2m}\frac{\partial^2 \phi_E}{\partial x^2} + \frac{m\omega^2 x^2}{2}\,\phi_E \qquad \textit{for energy eigenstates}$$

This is the first time we have used Schrödinger's equation, so let me pause and step back through the logic. Energy eigenstates are spatially stable so we can ignore change over time and focus only on $\phi(x)$. Energy E is a precise value and is the total of kinetic energy plus potential energy. The term for potential energy depends on the scenario under study and in this case is $\frac{m\omega^2 x^2}{2}$, the potential energy of the QHO.

We can rearrange the QHO equation to get the second derivative on the left side and simplify it by introducing two constants which I will call A^2 and B. I have used A^2 to avoid squares in the expression for A in Equation 13.10:

$$E\phi \;=\; -\frac{\hbar^2}{2m}\frac{\partial^2 \phi}{\partial x^2} + \frac{m\omega^2 x^2}{2}\phi \tag{13.9}$$

$$\frac{\partial^2 \phi}{\partial x^2} \;=\; \left(\frac{m\omega^2 x^2}{2}\phi - E\phi\right)\frac{2m}{\hbar^2} \;=\; \frac{m^2\omega^2}{\hbar^2}x^2\phi - \frac{2m}{\hbar^2}E\phi$$

$$\frac{\partial^2 \phi}{\partial x^2} \;=\; A^2 x^2\phi - B\phi \qquad where: \quad A = \frac{m\omega}{\hbar} \qquad B = \frac{2m}{\hbar^2}E \tag{13.10}$$

The Schrödinger equation for the energy eigenstates of QHO simplifies to Equation 13.10 where A and B are constants. Any particle state ϕ that complies with this equation is an energy eigenstate of the QHO. Why? Because the kinetic energy and potential energy add to exactly the same total E everywhere in the particle state, so there is only one precise value of E whatever the value of x.

13.3.1 Ground State of the QHO

To find the general solution, we need to find every ϕ that solves this equation. Tricky, very tricky; so we will content ourselves with hunting for one or two. To get the first takes a spirit of trial and error, plus a little common sense. The simplified equation is shown again below:

$$\frac{\partial^2 \phi}{\partial x^2} = A^2 x^2 \phi - B\phi \qquad \text{\textit{Simplified Schrödinger equation for the QHO}} \qquad (13.11)$$

What type of ϕ is likely to solve this equation? Consider the first term $A^2 x^2 \phi$. The second derivative of ϕ is throwing off the same ϕ multiplied by $A^2 x^2$. What is the simplest function that does this? To mathematicians amongst you it will be obvious. To others, perhaps less so, but all will soon be clear. It is:

$$\phi = e^{-A\frac{x^2}{2}} \qquad \Longrightarrow \qquad \frac{\partial \phi}{\partial x} = -A x e^{-A\frac{x^2}{2}} \qquad (13.12)$$

$$\frac{\partial^2 \phi}{\partial x^2} = A^2 x^2 e^{-A\frac{x^2}{2}} - A e^{-A\frac{x^2}{2}} = A^2 x^2 \phi - A\phi \qquad (13.13)$$

You can see how the exponential generates the $A^2 x^2 \phi$ term in the second derivative. And hey presto, we have a solution! This ϕ fits the equation so it is indeed an energy eigenstate of the QHO. And comparing Equation 13.11 with Equation 13.13, you can see that $B = A$ so using the definitions of A and B from Equation 13.10 we get:

$$B = A \qquad \textit{where} \qquad A = \frac{m\omega}{\hbar} \qquad B = \frac{2m}{\hbar^2} E \qquad (13.14)$$

$$\frac{2m}{\hbar^2} E = \frac{m\omega}{\hbar} \qquad \Longrightarrow \qquad E = \frac{\hbar\omega}{2} \qquad (13.15)$$

The solution that we have derived turns out to be the lowest possible energy that the QHO can have (we will prove this later). It is called the ground state and is typically labelled as ϕ_0:

$$\phi_0 = e^{-A\frac{x^2}{2}} = e^{-\frac{m\omega}{2\hbar} x^2} \qquad E_0 = \frac{\hbar\omega}{2} \qquad \text{\textit{Ground state of QHO}} \qquad (13.16)$$

This is a Gaussian (normal distribution) so it is a bell-shaped curve centred on the equilibrium point $x = 0$. It is a standing wave flicking up and down at a rate determined by its energy. For the full equation including time dependency, we can add the sinusoidal term that depends on energy (refer back to Equation 11.32 if you find this confusing). I am ignoring what is called the normalisation factor (see Box 13.4):

$$\phi_0(x,t) = e^{-\frac{m\omega}{2\hbar} x^2} e^{-i\frac{Et}{\hbar}} = e^{-\frac{m\omega}{2\hbar} x^2} e^{-i\frac{\omega}{2} t} \qquad (13.17)$$

Box 13.4   Normalisation and the QHO ground state

The full version of the QHO ground state that you will see in textbooks is:

$$\phi_0 = \left(\frac{m\omega}{\pi\hbar}\right)^{\frac{1}{4}} e^{-\frac{m\omega}{2\hbar} x^2}$$

The front term is a normalisation factor that is set to ensure that the total probability of finding the particle adds up to exactly 1 that is,

$$\int_{-\infty}^{+\infty} |\phi_0(x)|^2 \, dx = 1 \qquad \textit{or in Dirac notation} \quad \langle\phi_0|\phi_0\rangle = 1$$

In this book, I will tend not to include these normalisation factors unless needed.

The big message from the QHO ground state is that, in sharp contrast to the classical Simple Harmonic Oscillator, there is no zero energy solution (see Box 13.5 for another weak joke). A classical pendulum can sit stationary at the equilibrium point but a quantum particle cannot. This is because the uncertainty in both position and momentum of the quantum particle (Δx and Δp) would be zero which is not allowed by Heisenberg's uncertainty principle. Incidentally, as the QHO ground state is a real Gaussian, it is at a Heisenberg minimum so the uncertainty relationship is: $\Delta x \Delta p = \frac{\hbar}{2}$.

Box 13.5 Just for laughs:

Why does a burger have less energy than a steak?   *Because it is in its ground state.*

13.3.2 A Trick to Find the Other Energy Eigenstates of the QHO

$$\frac{\partial^2 \phi}{\partial x^2} = A^2 x^2 \phi - B\phi \qquad\qquad where: \quad A = \frac{m\omega}{\hbar} \quad B = \frac{2m}{\hbar^2} E \qquad\qquad (13.18)$$

You *might* be able to guess the next energy eigenstate by trial and error, but the maths gets very complicated. Fortunately there is a cunning trick. You take a known energy eigenstate such as the ground state ϕ_0, multiply it by Ax and subtract its derivative. And abracadabra, this always gives you another energy eigenstate. Let me demonstrate by doing this to the ground state solution that we have found:

$$\phi_0 = e^{-A\frac{x^2}{2}} \qquad\Longrightarrow\qquad \phi_1 = Ax\,\phi_0 - \frac{\partial \phi_0}{\partial x} \qquad\qquad (13.19)$$

$$\phi_1 = Ax\,e^{-A\frac{x^2}{2}} + Ax\,e^{-A\frac{x^2}{2}} = 2Ax\,e^{-A\frac{x^2}{2}} \qquad\Longrightarrow\qquad \phi_1 = x\,e^{-A\frac{x^2}{2}} \qquad\qquad (13.20)$$

We can ignore the constants at the front as they have no effect on the differentials and we are not worrying about normalisation. You can see from Equation 13.22 that the second derivative of ϕ_1 is indeed of the form $A^2 x^2 \phi - B\phi$. This is the first energy level above the ground state. Its value for B is $3A$ meaning that its energy is $\frac{3\hbar\omega}{2}$ which is $\hbar\omega$ higher than that of the ground state.

$$\phi_1 = x\,e^{-A\frac{x^2}{2}} \qquad\Longrightarrow\qquad \frac{\partial \phi_1}{\partial x} = -A x^2 e^{-A\frac{x^2}{2}} + e^{-A\frac{x^2}{2}} \qquad\qquad (13.21)$$

$$\frac{\partial^2 \phi_1}{\partial x^2} = A^2 x^3 e^{-A\frac{x^2}{2}} - 2Ax\,e^{-A\frac{x^2}{2}} - A\,e^{-Ax\frac{x^2}{2}} = A^2 x^2 \phi_1 - 3A\,\phi_1 \qquad\qquad (13.22)$$

The trick in Equation 13.19 always works. The super-smart Paul Dirac discovered it (you will hear much more about him later). We now have a fairly simple *operation* that we can perform on one QHO energy eigenstate to *create* another. For obvious reasons, this in its full form (which includes normalisation) is called the *creation operator* and is labelled $\hat{a}^+$.

Each time that you perform the $\hat{a}^+$ operation, you get a new energy eigenstate with energy $\hbar\omega$ higher than the previous one. The eigenstates are like rungs in an *energy ladder*, each separated in energy by $\hbar\omega$. The creation operation takes you up one rung.

$$E = \left(n + \frac{1}{2}\right)\hbar\omega \qquad where\ n\ is\ an\ integer \qquad\qquad (13.23)$$

To go down a rung there is a similar trick except that you add, rather than subtract, the derivative. In its full form, with normalisation built in, it is amusingly called the *annihilation operator* ($\hat{a}^-$). The best way to show the annihilation operation is to use it on the QHO first energy level ϕ_1 to generate the ground state ϕ_0:

$$\phi_1 = x\,e^{-A\frac{x^2}{2}} \qquad\Longrightarrow\qquad \phi_0 = Ax\,\phi_1 + \frac{\partial \phi_1}{\partial x} \qquad annihilation\ operation \qquad\qquad (13.24)$$

$$\phi_0 = Ax^2 e^{-A\frac{x^2}{2}} - Ax^2 e^{-A\frac{x^2}{2}} + e^{-A\frac{x^2}{2}} \qquad\Longrightarrow\qquad \phi_0 = e^{-A\frac{x^2}{2}} \qquad\qquad (13.25)$$

The creation operation $\hat{a}^+$ always generates the next energy level up, $\hbar\omega$ higher. The annihilation operation $\hat{a}^-$ will always give you the lower one, until you get to the ground state when it puts its name to good use and annihilates the particle state in the sense that it reduces the state to zero, that is, no particle state at all. This is shown in Equation 13.26 and it is the proof that Equation 13.16 is indeed the ground state of the QHO:

$$\text{Annihilation of ground state:} \qquad \phi_0 = e^{-A\frac{x^2}{2}}$$

$$\hat{a}^-(\phi_0) = Ax\,\phi_0 + \frac{\partial \phi_0}{\partial x} = Ax\,e^{-A\frac{x^2}{2}} - Ax\,e^{-A\frac{x^2}{2}} = 0 \tag{13.26}$$

Some readers may want to see how Dirac's trick works. If you are so inclined, see Box 13.6.

Box 13.6   How Dirac's cunning trick works

For a QHO energy eigenstate ϕ, we know: $\frac{\partial^2 \phi}{\partial x^2} = A^2 x^2 \phi - B\phi$

$$\Psi = \hat{a}^+(\phi) = Ax\,\phi - \frac{\partial \phi}{\partial x} \qquad \textit{creation operation creates new state } \Psi$$

$$\frac{\partial \Psi}{\partial x} = Ax\,\frac{\partial \phi}{\partial x} + A\phi - \frac{\partial^2 \phi}{\partial x^2} = Ax\,\frac{\partial \phi}{\partial x} - (A^2 x^2 - A - B)\phi$$

$$\frac{\partial^2 \Psi}{\partial x^2} = Ax\,\frac{\partial^2 \phi}{\partial x^2} + A\frac{\partial \phi}{\partial x} - (A^2 x^2 - A - B)\frac{\partial \phi}{\partial x} - 2A^2 x\,\phi$$

$$= Ax\,(A^2 x^2 - B)\phi - 2A^2 x\,\phi - (A^2 x^2 - 2A - B)\frac{\partial \phi}{\partial x}$$

$$= (A^2 x^2 - 2A - B)\,Ax\,\phi - (A^2 x^2 - 2A - B)\frac{\partial \phi}{\partial x}$$

$$= (A^2 x^2 - 2A - B)\left(Ax\,\phi - \frac{\partial \phi}{\partial x}\right) = (A^2 x^2 - 2A - B)\,\Psi$$

$$\frac{\partial^2 \Psi}{\partial x^2} = A^2 x^2\,\Psi - (B + 2A)\,\Psi$$

Ψ is also a QHO energy eigenstate. For this new eigenstate, the value of B is $2A$ higher than that of the old eigenstate, that is, $B_\Psi = B_\phi + 2A$, meaning that the energy is higher by $\hbar\omega$:

$$\Delta B = 2A \quad \Longrightarrow \quad \frac{2m}{\hbar^2}\Delta E = \frac{2m\omega}{\hbar} \quad \Longrightarrow \quad \Delta E = \hbar\omega$$

13.3.3   The QHO Energy Eigenstate Ladder

The creation and annihilation operations reveal the energy eigenstates of the QHO. Let me emphasise the word *reveal*. The ladder of energy eigenstates does not come from mathematical jiggery pokery. It is baked into the Quantum Harmonic Oscillator.

Figure 13.3 shows the spatial distribution of the eigenstates. In the ground state of the QHO (see Ψ_0 in the figure), you can see that the particle will tend to be found *near* the equilibrium point. This is very different from the SHO because a pendulum spends most time at the extremes of its swing (where it is moving slowest).

Remember that each energy eigenstate is spatially stable and acts as a standing wave vibrating up and down at a rate dictated by the energy of the state. The QHO can interact only with energy values that are rungs on the ladder, separated by $\hbar\omega$ between each step so it can absorb or emit energy only in multiples of this quantum of energy. As the maths of the QHO applies generally to quantum systems close to an equilibrium point, this energy ladder with its evenly spaced rungs is widely applicable.

You should be expecting, and be familiar with a couple of other features of the ladder of energy eigenstates:

- With every step up the ladder, the eigenstate adds one more node (stationary point)
- Each eigenstate creeps into the forbidden zone where, classically, it cannot reach, then decays exponentially (quantum tunnelling)

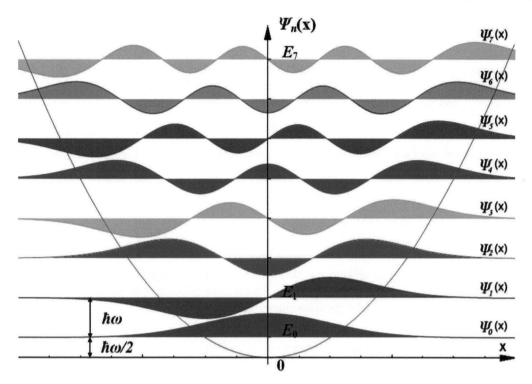

Figure 13.3 Quantum Harmonic Oscillator energy eigenstates.

How does this tally with the classical Simple Harmonic Oscillator? The $\hbar\omega$ energy steps are immeasurably small from a classical perspective so we see the ground state energy as zero and cannot perceive that the energy of the oscillator is quantised. At classical energy levels (physicists describe this as being in the *classical limit*), the allowable energy levels appear as a continuum and the quantum nature of the harmonic oscillator disappears.

13.3.4 QHO Superpositions

Never forget that quantum mechanics allows superpositions. The QHO is unlikely to be in one particular energy eigenstate. It is likely to be in a superposition of multiple energy eigenstates. If we label the QHO ground state as $|0\rangle$ and the first energy eigenstate $|1\rangle$ and so on, then the QHO can be expressed as a superposition of energy eigenstates:

$$\Psi = \alpha_0 |0\rangle + \alpha_1 |1\rangle + \alpha_2 |2\rangle + \alpha_3 |3\rangle \ldots \tag{13.27}$$

Every possible state of the QHO can be expressed in this way as a sum of energy eigenstates. The coefficients α are complex numbers indicating the relative probability of the QHO being in that particular eigenstate upon measurement. For example, the relative probability of it being in the first energy eigenstate $|1\rangle$ is $|\alpha_1|^2 = \alpha_1^* \alpha_1$ (where α_1^* is the complex conjugate of α_1). In this book, I have ignored where possible the normalisation of wave functions, but for this sort of calculation, you must use a normalised value for α_1. By normalised, I mean that the values are scaled so that adding up all the $|\alpha|^2$ totals 1.

The action of the energy operator typically called the Hamiltonian operator and labelled $\hat{H}$ on an energy eigenstate reveals the energy of that state that is, the energy of the QHO if a measurement lands it in that state. For example, for the one-dimensional QHO:

$$\hat{H}|0\rangle = E|0\rangle = \frac{1}{2}\hbar\omega |0\rangle \qquad \hat{H}|1\rangle = E|1\rangle = \frac{3}{2}\hbar\omega |1\rangle \tag{13.28}$$

Do *not* confuse the coefficients α with the energy values E. They are completely different things. The coefficients α show the amplitude of the wave function for a particular eigenstate. The value E is the observable energy value of the QHO if found in that energy eigenstate. Another thing to note is that changing the phase of the QHO does not change anything measurable as shown in Box 13.7 (this will become important later in the book).

Box 13.7   Phase change and the QHO

Changing the phase of any wave function, including a QHO, does not affect anything measurable (see Box 11.4). Superpositions, including of QHOs, are also unaffected by phase change provided that the phase change is the same for all the components of the superposition. The example below shows the result if a phase change of $e^{i\theta}$ is applied to two components ϕ_1 and ϕ_2 of a superposition Ψ. You can see that $|\Psi|^2$ is unchanged:

$$|e^{i\theta} a \phi_1 + e^{i\theta} b \phi_2|^2 = |e^{i\theta} (a \phi_1 + b \phi_2)|^2 = |e^{i\theta} \Psi|^2 = e^{-i\theta} \Psi^* e^{i\theta} \Psi = |\Psi|^2 \tag{13.29}$$

This will become relevant when we discuss the QHO in the context of quantum field theory in Chapter 19 towards the end of this book.

13.4   The QHO in Three Dimensions

The QHO is not very different in three spatial dimensions (Box 13.8). There is the same energy ladder with the same $\hbar\omega$ energy gap between rungs. The only difference is that the ground state energy is three times higher than the one-dimensional QHO and the energy eigenstates are degenerate. This is not a criticism of its moral compass. It simply means that there are multiple energy eigenstates with the same energy. Let's have a look at it.

Box 13.8   Quantum Harmonic Oscillator in three spatial dimensions

$$E \phi_E(x, y, z) = -\frac{\hbar^2}{2m} \left(\frac{\partial^2 \phi_E}{\partial x^2} + \frac{\partial^2 \phi_E}{\partial y^2} + \frac{\partial^2 \phi_E}{\partial z^2} \right) + \frac{m\omega^2 (x^2 + y^2 + z^2)}{2} \phi_E \tag{13.30}$$

I will briefly explain Equation 13.30. Don't forget that Schrödinger's equation is an energy balance. The total energy E is the total of kinetic energy and potential energy. The kinetic energy is driven by total momentum squared and that is in turn the sum of the momentum squared in each direction. The momentum squared in each direction is expressed in terms of second derivatives:

$$p_{tot}^2 = p_x^2 + p_y^2 + p_z^2 = -\frac{\hbar^2}{2m} \left(\frac{\partial^2 \phi_E}{\partial x^2} + \frac{\partial^2 \phi_E}{\partial y^2} + \frac{\partial^2 \phi_E}{\partial z^2} \right) \tag{13.31}$$

The 3D QHO has a central single point of equilibrium in a volume of space. The further the particle wanders from this point the higher the potential energy. So the potential energy depends on the square of the total distance (d^2) from the equilibrium point where ($x = y = z = 0$):

$$d^2 = x^2 + y^2 + z^2 \implies V = \frac{m\omega^2 d^2}{2} = \frac{m\omega^2 (x^2 + y^2 + z^2)}{2} \tag{13.32}$$

The fact that kinetic energy and potential energy can be pulled apart into each dimensional component means that we can rearrange the 3D QHO Equation 13.30 and express it as the addition of three independent 1D QHO oscillators, one in each dimension:

$$E\phi_E = -\frac{\hbar^2}{2m} \left(\frac{\partial^2 \phi_E}{\partial x^2} + \frac{\partial^2 \phi_E}{\partial y^2} + \frac{\partial^2 \phi_E}{\partial z^2} \right) + \frac{m\omega^2 (x^2 + y^2 + z^2)}{2} \phi_E \tag{13.33}$$

$$= -\frac{\hbar^2}{2m} \frac{\partial^2 \phi_E}{\partial x^2} + \frac{m\omega^2 x^2}{2} \phi_E - \frac{\hbar^2}{2m} \frac{\partial^2 \phi_E}{\partial y^2} + \frac{m\omega^2 y^2}{2} \phi_E - \frac{\hbar^2}{2m} \frac{\partial^2 \phi_E}{\partial z^2} + \frac{m\omega^2 z^2}{2} \phi_E$$

$$= (1D\,QHO)_X + (1D\,QHO)_Y + (1D\,QHO)_Z \tag{13.34}$$

An energy eigenstate of the 3D QHO must be a combination of an energy eigenstate in the x direction, an energy eigenstate in the y direction and an energy eigenstate in the z direction. For example the ground state of the 3D QHO is the combination of the ground state in each direction:

$$\phi_0 = e^{-\frac{m\omega}{2\hbar}x^2} \, e^{-\frac{m\omega}{2\hbar}y^2} \, e^{-\frac{m\omega}{2\hbar}z^2} = e^{-A\frac{x^2}{2}} \, e^{-A\frac{y^2}{2}} \, e^{-A\frac{z^2}{2}} \quad where \quad A = \frac{m\omega}{\hbar} \tag{13.35}$$

We know that each 1D QHO has its own energy ladder and each ground state has energy $\frac{\hbar\omega}{2}$. The 3D QHO is a combination of the three so the ground state energy is $\frac{3\hbar\omega}{2}$. And what is the first energy eigenstate above the ground state? We can use our cunning trick to generate the next energy eigenstate, but now we can do so in three different ways. For example we can go up an energy rung in the X QHO *or* up an energy rung in the Y QHO:

$$X\,QHO \quad \phi_1 = Ax\phi_0 - \frac{\partial\phi_0}{\partial x} = xe^{-A\frac{x^2}{2}} \, e^{-A\frac{y^2}{2}} \, e^{-A\frac{z^2}{2}} \qquad E = \frac{5\hbar\omega}{2} \tag{13.36}$$

$$Y\,QHO \quad \phi_1 = Ay\phi_0 - \frac{\partial\phi_0}{\partial y} = ye^{-A\frac{x^2}{2}} \, e^{-A\frac{y^2}{2}} \, e^{-A\frac{z^2}{2}} \qquad E = \frac{5\hbar\omega}{2} \tag{13.37}$$

For the 3D QHO the energy gap between allowable energy levels is still $\hbar\omega$ but there are multiple energy eigenstates that can have the same energy. Let me simplify by using Dirac notation. For example, if the 3D QHO is in the ground state in the x direction, the first energy eigenstate in the y direction and the first in the z direction, this would be: $|0\rangle_x |1\rangle_y |1\rangle_z$. Consider other ways an energy eigenstate can have energy $\frac{7\hbar\omega}{2}$ that is $2\hbar\omega$ energy above the 3D QHO ground state. There are a number of possibilities including $|0\rangle_x |0\rangle_y |2\rangle_z$ or $|1\rangle_x |0\rangle_y |1\rangle_z$ or $|0\rangle_x |2\rangle_y |0\rangle_z$... When there are multiple energy eigenstates with the same energy, they are said to be *degenerate*. The key thing to remember about the 3D QHO is that the gap between energy levels *still* is $\hbar\omega$ just as for the 1D QHO.

For a brief break from all this maths, check out Box 13.9 on the dispute between Newton and Leibniz over who invented calculus.

Box 13.9 The Calculus Wars: Newton versus Leibniz

Isaac Newton's (1643–1727) most famous dispute was not with Hooke, but with Gottfried Leibniz (1646–1716). By the way, you may see different dates for Newton's death because the calendar was changed in 1752.

Newton developed calculus in the 1660s, but did not publish. Leibniz published two papers on his version of calculus in 1684 and 1686. Over the next 40 years a bitter dispute developed as to who had priority. Finally, in 1712, the Royal Society produced a report which opined in Newton's favour. Some feel it a little unfair that Newton, as President of the Society, drafted the report. The general consensus is that Newton was indeed first, but Leibniz discovered calculus independently. It is Leibniz's notation that is used today including his integral sign $\int$, an elongated S from the Latin *summa* (total), and derivative d from *differentia* (difference).

Newton was buried in Westminster Abbey. Leibniz, out of favour and ignored, was buried in an unmarked grave. In memory of Isaac Newton, Alexander Pope's poem: *Nature and Nature's laws laid hid in night; God said, let Newton be! And all was light.*

In memory of Gottfried Leibniz, the Leibniz Butterkeks, a German biscuit brand – and no, the Fig Newton biscuit has nothing to do with Newton. I love biscuits. I think Leibniz won.

13.5 Formal Definition of the Creation and Annihilation Operators

Typically, the creation and annihilation operators are shown in terms of the position operator $\hat{x}$ and the momentum operator $\hat{p}$ and weighted, dividing by a factor of $\sqrt{2A}$. In this format i must be included to offset the phase change that the momentum operator $\hat{p}$ would introduce. Don't be confused by this. There is nothing mystical about $\hat{a}^+$ and $\hat{a}^-$. As you have seen, they are simple mathematical operations.

$$\hat{a}^+\phi = \frac{1}{\sqrt{2A}}\left(Ax\phi - \frac{\partial\phi}{\partial x}\right) \qquad Note\ that:\quad \hat{x}\phi = x\phi \quad \hat{p}\phi = -i\hbar\frac{\partial\phi}{\partial x} \tag{13.38}$$

$$= \frac{1}{\sqrt{2A}}\left(A\hat{x} - \frac{i\hat{p}}{\hbar}\right)\phi \qquad and \quad A = \frac{m\omega}{\hbar}$$

$$= \sqrt{\frac{A}{2}}\left(\hat{x} - \frac{i\hat{p}}{A\hbar}\right)\phi = \sqrt{\frac{m\omega}{2\hbar}}\left(\hat{x} - \frac{i\hat{p}}{m\omega}\right)\phi \tag{13.39}$$

Note that the multiple of ϕ in Equation 13.39 (which is the eigenvalue of $\hat{a}^+$), is *not* real – it contains an *i*. The creation and annihilation operators do not correspond to measurable quantities in the way that an operator such as $\hat{p}$ does (in technical language, $\hat{a}^+$ and $\hat{a}^-$ are not Hermitian). The operators are summarised in Box 13.10 (see also the warning on notation in Box 13.11).

Box 13.10   Creation and annihilation operators in terms of $\hat{x}$ and $\hat{p}$

$$\hat{a}^+ = \sqrt{\frac{m\omega}{2\hbar}}\left(\hat{x} - \frac{i\hat{p}}{m\omega}\right) \qquad \hat{a}^- = \sqrt{\frac{m\omega}{2\hbar}}\left(\hat{x} + \frac{i\hat{p}}{m\omega}\right)$$

There is another benefit. Our cunning creation and annihilation operations combine to form the QHO *number operator* that tells us the energy level of the QHO:

$$\hat{a}^+\hat{a}^- \phi_E = n\phi_E \qquad \textit{Energy of eigenstate: } E = \left(n + \frac{1}{2}\right)\hbar\omega \tag{13.40}$$

We can show this by using the definition of $\hat{a}^+\,\hat{a}^-$ to expand the expression for $\hat{H}$:

$$\hat{H} = \left(\hat{a}^+\,\hat{a}^- + \frac{1}{2}\right)\hbar\omega \tag{13.41}$$

$$= \frac{m\omega^2}{2}\left(\hat{x} - \frac{i\hat{p}}{m\omega}\right)\left(\hat{x} + \frac{i\hat{p}}{m\omega}\right) + \frac{1}{2}\hbar\omega$$

$$= \frac{m\omega^2\hat{x}^2}{2} + \frac{\hat{p}^2}{2m} + \frac{i\omega}{2}[\hat{x},\hat{p}] + \frac{1}{2}\hbar\omega \tag{13.42}$$

$$= \frac{m\omega^2\hat{x}^2}{2} + \frac{\hat{p}^2}{2m} + \frac{i\omega}{2}i\hbar + \frac{1}{2}\hbar\omega$$

$$= \frac{m\omega^2\hat{x}^2}{2} + \frac{\hat{p}^2}{2m} \tag{13.43}$$

Equation 13.42 uses $i\hbar$ as the value of the $[\hat{x}, \hat{p}]$ commutator (as in Equation 11.24). The result in Equation 13.43 is the equation for the energy operator of the QHO as shown in Box 13.3. The creation operator $\hat{a}^+$ the annihilation operator $\hat{a}^-$ and their combination, the number operator $\hat{a}^+\hat{a}^-$ are helpful in navigating around the QHO. The way these operations pull out the value n of the energy level will be significant when we discuss quantum field theory later. If we label the nth energy level as $|n\rangle$ and the $(n+1)^{th}$ as $|n+1\rangle$, then the impact of each is:

$$\hat{a}^+|n\rangle = \sqrt{n+1}\,|n+1\rangle \qquad \hat{a}^-|n\rangle = \sqrt{n}\,|n-1\rangle \tag{13.44}$$

These factors combine in action to form the number operator $\hat{a}^+\,\hat{a}^-$ as follows:

$$\hat{a}^+\,\hat{a}^-|n\rangle = \hat{a}^+\left(\sqrt{n}\,|n-1\rangle\right) \tag{13.45}$$

$$= \sqrt{n}\,\hat{a}^+|n-1\rangle$$

$$= \sqrt{n}\sqrt{n}\,|n\rangle$$

$$= n\,|n\rangle \tag{13.46}$$

In three dimensions, there are creation and annihilation operators for each dimension. If we take the operators that move up and down the energy ladder in the x dimension, $\hat{a}_X^+$ and $\hat{a}_X^-$, these combine to the number operator $\hat{a}_X^+\hat{a}_X^-$ that gives the energy level in the x dimension of the QHO. If you want to know the overall energy of the 3D QHO you need to combine the result of the number operators in each dimension and add that to the ground state energy of the 3D QHO:

$$E\phi(x,y,z) = \left(\hat{a}_X^+\hat{a}_X^- + \hat{a}_Y^+\hat{a}_Y^- + \hat{a}_Z^+\hat{a}_Z^- + \frac{3}{2}\right)\hbar\omega = \left(n_{\text{total}} + \frac{3}{2}\right)\hbar\omega \tag{13.47}$$

Box 13.11   Warning: Creation and annihilation operator notation

Some textbooks show the annihilation operator $\hat{a}^-$ simply as $\hat{a}$, and the creation operator $\hat{a}^+$ which is its adjoint operator (basically its complex conjugate partner) is shown as something like $\hat{a}^\dagger$ and is called *a-dagger*. I will always use $\hat{a}^+$ and $\hat{a}^-$ to keep things clear, but watch out if you are a bit messy with your lecture notes (heaven forbid!).

13.6   The Path to Quantum Field Theory (QFT)

We would be getting well ahead of ourselves if we discussed quantum field theory in detail now. I will save that for later. On the other hand, the maths that we have covered in this chapter sits at the heart of QFT so I want to flag up where it leads.

When we apply Schrödinger's equation, we build a scenario that includes a potential V. In the case of electrons in the atom, we use the Coulomb potential that varies with the distance from the nucleus. The potential varies smoothly with $\frac{1}{r}$. But where does that potential come from? It is an electromagnetic field. And what do we know about the fluctuation of an electromagnetic field? It comes in lumps that we call photons. It is *quantised*.

Do you see the problem? In advanced (very advanced) quantum mechanics, you must quantise the fields. How will those fields behave? They will fluctuate from equilibrium with different levels of excitation. What does a quantum system look like when it fluctuates around equilibrium? Step forwards QHO! That is exactly what the QHO is. It is the mathematical description of quantum equilibrium.

What happens if we model the electromagnetic field using the QHO? The excitation of the field will be a ladder of excitation states each separated by $\hbar\omega$ energy. The energy comes in lumps. Let's call each lump a photon. So zero photons is the ground state. The first energy eigenstate is one photon, the second is two photons; the difference between one photon and a hundred photons is just the level of excitation of the electromagnetic field.

In QFT, you take that model and apply it to *everything*. A photon is an excitation of the electromagnetic field. An electron is an excitation of the electron field. An up quark is an excitation of the up quark field. This also has the benefit of explaining why all electrons are the same.

I hope this helps you see why the QHO features so centrally in the study of quantum mechanics. Now, forget all of this QFT nonsense for the time being. We need to get back to Schrödinger's equation and look at what it tells us about electrons in the atom.

Chapter 14

Angular Momentum

CHAPTER MENU
14.1 A Primer on Classical Angular Momentum
14.2 Quanta of Angular Momentum
14.3 Angular Momentum's Intricate Dance
14.4 Angular Kinetic Energy and Angular Momentum
14.5 The Pattern of Angular Momentum Eigenstates
14.6 The Angular Momentum Creation Operator
14.7 Summary

Most of the models we have used so far have been one-dimensional, in the sense that we have included only one spatial dimension. It will not have escaped your notice that the universe we live in actually has not one but three spatial dimensions! The nucleus and electrons of atoms interact as a three-dimensional system. In order to understand the energy levels of atoms, we must take into account angular momentum and angular kinetic energy.

14.1 A Primer on Classical Angular Momentum

Readers with a background in physics can skim or skip this primer section as it will be well trodden territory. I include it because angular momentum L is less familiar to many than is its translation momentum counterpart, p. Angular momentum is the momentum of an object *around* a given axis. Like momentum p, it is a conserved quantity. If you are dealing with the energy or momentum of a spinning object, you must take L into account.

Figure 14.1 gives a simple scenario. Imagine a ball on a string being swung around with angular velocity ω (measured in radians per second) at a distance r from an axis. Its velocity around that axis at any moment is $v = \omega r$ and its angular momentum L is shown in Equation 14.1.

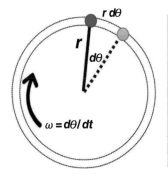

Angular Momentum

Spinning object spins around axis and travels distance $r\,d\theta$ in time dt

Velocity: $v = r\dfrac{d\theta}{dt} = r\omega$

Angular momentum: $L = mvr = mr^2\omega$

Angular kinetic energy: $\dfrac{mv^2}{2} = \dfrac{L^2}{2mr^2}$

Figure 14.1 Angular momentum overview.

Quantum Untangling: An Intuitive Approach to Quantum Mechanics from Einstein to Higgs, First Edition. Simon Sherwood.
© 2023 John Wiley & Sons Ltd. Published 2023 by John Wiley & Sons Ltd.

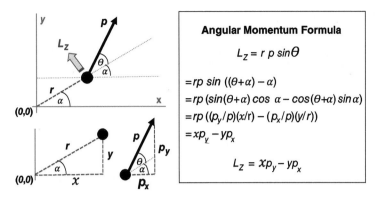

Figure 14.2 Angular momentum L_z expressed in terms of x, y, p_x and p_y.

$$L = mvr = mr^2\omega \tag{14.1}$$

This can be positive or negative. A negative value merely indicates that it is spinning in the opposite direction (clockwise/anticlockwise). Angular momentum L is a conserved quantity so if r reduces, then ω will increase. The typical example given is an ice skater who starts to spin, then pulls in his or her arms. This reduces r and he or she spins faster. The equations for angular momentum L and angular kinetic energy that I will label E_{Ang} are similar in structure to their translational counterparts:

$$\textit{Momentum} \qquad\qquad L = mr^2\omega \qquad\qquad p = mv \tag{14.2}$$

$$\textit{Energy} \qquad\qquad E_{Ang} = \frac{L^2}{2mr^2} \qquad\qquad E = \frac{p^2}{2m} \tag{14.3}$$

If an object is not spinning exactly around one particular axis, it will have components of angular momentum around more than one axis and the angular kinetic energy is calculated by combining these:

$$L_x = mr^2\omega_x \qquad L_y = mr^2\omega_y \qquad L_z = mr^2\omega_z \tag{14.4}$$

$$E_{Ang} = \frac{L_x^2 + L_y^2 + L_z^2}{2mr^2} \tag{14.5}$$

Angular momentum can be broken down into its component parts. The maths is simple, but it takes a few steps. The calculation for L_z is shown in Figure 14.2. The figure shows an object (the small black ball) moving with momentum p up and to the right. The two triangles in the lower part of the figure show the relationship of the angles so we can express L_z in terms of x, y, p_x and p_y as shown (this calculation uses $\sin(a - b) = \sin(a)\cos(b) - \cos(a)\sin(b)$). L_x and L_y break down in a similar way as shown in Equation 14.6. These relationships will be very important in the coming discussion on the quantum mechanics of angular momentum.

$$L_x = yp_z - zp_y \qquad L_y = zp_x - xp_z \qquad L_z = xp_y - yp_x \tag{14.6}$$

You will often see these relationships written as $\vec{r} \times \mathbf{p}$, the cross-product of the vector $\vec{r}$ (which has components x, y, z) and the vector $\mathbf{p}$ (which has components p_x, p_y, p_z). See Box 14.1 for a refresher on vector cross-product.

Box 14.1 Vector cross-product: $\vec{A} \times \vec{B} = \vec{C}$

The cross-product of two vectors produces another vector that is perpendicular to both. The magnitude of the resulting vector depends on the magnitude of the original vectors and the angle θ between them. If $\theta = 0$ then the cross-product is zero.

$$|\vec{C}| = |\vec{A}|\,|\vec{B}|\sin\theta \qquad \textit{where } \theta \textit{ is the angle between } \vec{A} \textit{ and } \vec{B}$$

In terms of the vector components $\vec{A} = (a_x, a_y, a_z)$, $\vec{B} = (b_x, b_y, b_z)$ and $\vec{C} = (c_x, c_y, c_z)$:

$$c_x = a_y b_z - a_z b_y \qquad c_y = a_z b_x - a_x b_z \qquad c_z = a_x b_y - a_y b_x$$

14.2 Quanta of Angular Momentum

The relationship between the particle wave function Ψ and the particle state's angular momentum L is similar to that of Ψ and p which is not surprising given their similar origin. The relationship of angular momentum L_x (around the x-axis) is with the change in the wave function rotated around that axis. Remember that θ_x refers to a rotation around the x axis and θ_y refers to a rotation around the y axis etc:

$$\frac{\partial \Psi}{\partial \theta_x} = \frac{i}{\hbar} L_x \Psi \qquad\qquad \frac{\partial \Psi}{\partial \theta_y} = \frac{i}{\hbar} L_y \Psi \qquad\qquad \frac{\partial \Psi}{\partial \theta_z} = \frac{i}{\hbar} L_z \Psi \qquad (14.7)$$

$$\frac{\partial \Psi}{\partial x} = \frac{i}{\hbar} p_x \Psi \qquad\qquad \frac{\partial \Psi}{\partial y} = \frac{i}{\hbar} p_y \Psi \qquad\qquad \frac{\partial \Psi}{\partial z} = \frac{i}{\hbar} p_z \Psi$$

There is a major consequence of this relationship. Angular momentum must come in quanta of $\hbar$. The easiest way to understand this is to look at how angular momentum must appear in the wave function. In order to generate the derivative, the wave function must include a component linking L_z and θ_z as follows:

$$\frac{\partial \Psi}{\partial \theta_z} = \frac{i}{\hbar} L_z \Psi \qquad \implies \qquad \Psi = ...e^{\frac{i}{\hbar} L_z \theta_z} \qquad (14.8)$$

What happens to the wave function if we add 2π to θ_z? That is the same as travelling 360 degrees; so full circle. We should arrive back at the same point unless there is a completely ridiculous change in our understanding of geometry (more on this later). In the example below, I consider a rotation around the z axis, but the argument holds true for any axis.

$$e^{\frac{i}{\hbar} L_z \theta_z} = e^{\frac{i}{\hbar} L_z (\theta_z + 2\pi)} = e^{\frac{i}{\hbar} L_z \theta_z} e^{\frac{i}{\hbar} L_z 2\pi} \qquad (14.9)$$

$$\implies \quad e^{i \frac{L_z}{\hbar} 2\pi} = 1 \qquad (14.10)$$

$$\implies \quad \cos\left(\frac{L_z}{\hbar} 2\pi\right) + i \sin\left(\frac{L_z}{\hbar} 2\pi\right) = 1 \qquad (14.11)$$

$$\implies \quad \frac{L_z}{\hbar} = n \qquad n \text{ must be an integer so... } L_z = n\hbar \qquad (14.12)$$

Adding 2π to θ_z makes no difference so Equation 14.10 must equal 1. This is written out in terms of the cosine and sine elements in Equation 14.11. For the cosine element to equal 1 and the sine element to equal 0, they must both be multiples of 2π. That means $\frac{L}{\hbar}$ must be an integer. This is an amazingly simple consequence of the maths of quantum mechanics. Angular momentum always comes in quanta of $\hbar$.

14.3 Angular Momentum's Intricate Dance

One thing that it is very important to note is that L_x, L_y, and L_z are *not* independent variables in the way that p_x, p_y, and p_z are. The derivatives of Ψ that drive p are working in independent dimensions (x, y and z). This is *not* true for L because θ_x, θ_y and θ_z are not independent dimensions. Consider a football. If you rotate it a half turn of θ_x followed by a half turn of θ_y you will find it is equivalent to turning it a half turn of θ_z. In contrast, no amount of x added to some y will ever give any z. As θ_x, θ_y and θ_z are related, then L_x, L_y and L_z are interlinked and it turns out that they share an uncertainty relationship.

To simplify the equations, we will define operators for each L variable. For example, $\hat{L}_x$ pulls out L_x as an eigenvalue from an L_x angular momentum eigenstate in exactly the same way as $\hat{p}$ does on a p momentum eigenstate. We also can write $\hat{L}_x$ as a composite of operations based on the relationship in Equation 14.6:

$$\hat{L}_x \Psi = -i\hbar \frac{\partial \Psi}{\partial \theta_x} \qquad L_x = y p_z - z p_y \qquad \implies \qquad \hat{L}_x \Psi = (\hat{y}\hat{p}_z - \hat{z}\hat{p}_y)\Psi \qquad (14.13)$$

We are going to need to understand the uncertainty relationships between L_x, L_y, and L_z. To do this, we calculate the effect of the commutators between them such as $[\hat{L}_x, \hat{L}_y]\Psi$. You will remember (hopefully) that if a commutator is not zero, it shows that the measurement of one variable affects the other. The value of the commutator tells us what that uncertainty relationship is (look back at Section 11.3 if you need to).

For the calculation of $[\hat{L}_x, \hat{L}_y]\Psi$ below, you need to remember that there is *no* uncertainty relationship between position and momentum in different dimensions so, the commutator is zero for such as $[\hat{x}, \hat{y}]\Psi$ or $[\hat{p}_y, \hat{p}_z]\Psi$ or $[\hat{y}, \hat{p}_x]\Psi$. There *is* an uncertainty relationship between position and momentum in the *same* dimension and the value of the commutator in each case is: $[\hat{x}, \hat{p}_x]\Psi = i\hbar\Psi$, that is, the commutator is the same as multiplying by $i\hbar$. Note that for brevity and clarity, I have not bothered to write the Ψ on every line after Equation 14.14 but do remember that it is there. These represent operations on a wave function.

$$[\hat{L}_x, \hat{L}_y]\Psi = [\hat{y}\hat{p}_z - \hat{z}\hat{p}_y, \hat{z}\hat{p}_x - \hat{x}\hat{p}_z]\Psi \tag{14.14}$$

$$= (\hat{y}\hat{p}_z - \hat{z}\hat{p}_y)(\hat{z}\hat{p}_x - \hat{x}\hat{p}_z) - (\hat{z}\hat{p}_x - \hat{x}\hat{p}_z)(\hat{y}\hat{p}_z - \hat{z}\hat{p}_y)$$

$$= \hat{y}\hat{p}_z\hat{z}\hat{p}_x - \hat{y}\hat{p}_z\hat{x}\hat{p}_z - \hat{z}\hat{p}_y\hat{z}\hat{p}_x + \hat{z}\hat{p}_y\hat{x}\hat{p}_z - \hat{z}\hat{p}_x\hat{y}\hat{p}_z + \hat{z}\hat{p}_x\hat{z}\hat{p}_y + \hat{x}\hat{p}_z\hat{y}\hat{p}_z - \hat{x}\hat{p}_z\hat{z}\hat{p}_y$$

$$= \hat{z}\hat{p}_z(\hat{x}\hat{p}_y - \hat{y}\hat{p}_x) - \hat{p}_z\hat{z}(\hat{x}\hat{p}_y - \hat{y}\hat{p}_x) \tag{14.15}$$

$$= (\hat{z}\hat{p}_z - \hat{p}_z\hat{z})(\hat{x}\hat{p}_y - \hat{y}\hat{p}_x)$$

$$= [\hat{z}, \hat{p}_z](\hat{x}\hat{p}_y - \hat{y}\hat{p}_x) \tag{14.16}$$

$$= i\hbar\hat{L}_z \qquad \Longrightarrow \qquad [\hat{L}_x, \hat{L}_y]\Psi = i\hbar\hat{L}_z\Psi \tag{14.17}$$

What a calculation! It is tedious, but it ends up so neat. A quick recap on what we have done. The key to Equation 14.15 is that everything commutes in these equations (so you can change the order) *except* $\hat{z}$ and $\hat{p}_z$, so four of the terms cancel out. After that, you need to spot in Equation 14.16 the commutator $[\hat{z}, \hat{p}_z]\Psi$ whose value we know is $i\hbar\Psi$ and to recognise that $(\hat{x}\hat{p}_y - \hat{y}\hat{p}_x)\Psi$ is $\hat{L}_z\Psi$.

The commutator tells us that there is a difference between acting first with the operator $\hat{L}_y$ on Ψ, then with $\hat{L}_x$ versus the reverse order and that the difference is the same as acting with $\hat{L}_z$ on Ψ, multiplied by $i\hbar$. Don't forget that the i that appears is just a phase change of $\frac{\pi}{2}$.

By symmetry, similar uncertainty relationships hold between $\hat{L}_y$ and $\hat{L}_z$ and between $\hat{L}_z$ and $\hat{L}_x$. This means that there is an uncertainty relationship between all three. If you know the value of the angular momentum around one axis such as L_x, you cannot know the value around either other axis. There is one and only one exception. That is when L_x, L_y and L_z are *all* zero because in this case, and only in this case, the value of all the commutators drops to zero. We can quantify the uncertainty relationship as it is always half the magnitude of the commutator (see Section 11.3):

$$[\hat{L}_x, \hat{L}_y]\Psi = i\hbar\hat{L}_z\Psi \qquad \Longrightarrow \qquad \Delta L_x \Delta L_y \geq \frac{\hbar L_z}{2} \tag{14.18}$$

To summarise, there is an uncertainty relationship between the angular momentum around each axis. The amount of uncertainty in the values around any two axes is proportional to the angular momentum around the third. For example the uncertainty $\Delta L_x \Delta L_y$ is proportional to L_z. The upshot is that you can measure the angular momentum only around one axis at a time *unless* $L_x = L_y = L_z = 0$ (i.e. you *can* measure that there is no angular momentum at all).

14.4 Angular Kinetic Energy and Angular Momentum

In order to calculate the stable energy levels of atoms (their energy eigenstates) we are going to need to know their total angular kinetic energy. This depends on a quantity that I label L_{sum}^2. In most texts, you will see this shown simply as L^2 but I think this is unhelpful because it gives the wrong impression that L_{sum}^2 is the square of an integer whereas it is the *sum* of squares as follows with, in this case, m_e being the electron's mass.

$$\textit{Angular kinetic energy:} \quad E_{\text{Ang}} = \frac{L_{\text{sum}}^2}{2m_e r^2} \qquad L_{\text{sum}}^2 = L_x^2 + L_y^2 + L_z^2 \tag{14.19}$$

The previous section on the uncertainty relationships of angular momentum raises a question. If I know the value of L_x, then the values of L_y and L_z are not precisely defined. So, what about the value of L_{sum}^2? If I know the angular momentum around one axis such as L_x, can I also know L_{sum}^2 or is that value uncertain? To assess this, we create an operator $\hat{L}_{sum}^2$ that draws out the sum of the squares of the angular momentum as follows:

$$\hat{L}_{sum}^2 \Psi = \left(\hat{L}_x^2 + \hat{L}_y^2 + \hat{L}_z^2\right) \Psi = \left(L_x^2 + L_y^2 + L_z^2\right) \Psi \tag{14.20}$$

A long calculation shows that L_{sum}^2 does *not* have an uncertainty relationship with L_x, L_y and L_z. If you want to work through it, you can find it at the end of the chapter in Box 14.6. These commutators and their related uncertainties are important because they drive the pattern of angular momentum eigenstates that in turn drives atomic structure, as we will see. They are summarised in Box 14.2.

Box 14.2 Angular momentum: commutators and uncertainty relationships

There is an uncertainty relationship between L_x, L_y and L_z unless the value of all three is zero.

$$[\hat{L}_x, \hat{L}_y]\Psi = i\hbar\hat{L}_z\Psi \qquad [\hat{L}_y, \hat{L}_z]\Psi = i\hbar\hat{L}_x\Psi \qquad [\hat{L}_z, \hat{L}_x]\Psi = i\hbar\hat{L}_y\Psi$$

$$\Delta L_x \, \Delta L_y \geq \frac{\hbar L_z}{2} \qquad \Delta L_y \, \Delta L_z \geq \frac{\hbar L_x}{2} \qquad \Delta L_z \, \Delta L_x \geq \frac{\hbar L_y}{2}$$

There is none with L_{sum}^2 because: $[\hat{L}_x, \hat{L}_{sum}^2]\Psi = [\hat{L}_y, \hat{L}_{sum}^2]\Psi = [\hat{L}_z, \hat{L}_{sum}^2]\Psi = 0$

What does this say about angular momentum? It says that you can know at any given time only one of L_x, L_y and L_z (unless they are all zero) plus you can know L_{sum}^2. That is the best you can do. If you know L_{sum}^2 and, say L_z, you can know no more. That is total quantum knowledge of the state. If you need a moment of relaxation, check out the history of one of the pioneers of angular momentum, Daniel Bernouilli, and his father in Box 14.3.

Box 14.3 Bernoulli's not so proud father

Newton introduced angular motion in *Principia* in 1687, but Daniel Bernoulli (1700–1782) took things further writing of a *moment of rotation*, now called angular momentum. He was from a family of mathematicians. His uncle, Jacob Bernoulli, discovered the constant *e* and his father, Johann, made contributions to calculus and was one of Euler's teachers.

In 1735, Daniel Bernoulli entered the Paris Academy's grand prize for maths. His father, Johann, entered the same competition. They shared the prize. It is said that his father was furious and, from that moment on, would not speak to his son or let him visit his home. Worse was to come when, in 1738, Daniel published his masterpiece *Hydrodynamica*. In a fit of pique, his father Johann plagiarised it in a book of his own called *Hydraulica* (subtle title eh?) that he falsely back-dated to 1732! In Daniel Bernoulli's own words:

I have been robbed of my entire Hydrodynamica, not one part of which I owe to my father and I have been deprived of the fruits of 10 years of work.

Fortunately, the scientific community saw through his father's ruse and Daniel received full credit – not a very touching example of paternal love.

14.5 The Pattern of Angular Momentum Eigenstates

Everything we have discussed is quite general, but things will be clearer if I tie this back to atoms. The main reason for this chapter is to help explain the energy levels of electrons. Picture in your mind the wave function of the electron in a hydrogen atom. It spreads out around the proton (the nucleus). If it is not completely spherically symmetrical, there is angular momentum simply because $\frac{\partial \Psi}{\partial \theta} \neq 0$ for at least one axis. If there is angular momentum, there is angular kinetic energy that forms a component of the total energy of the electron. That is why this is so important.

We need to figure out what the uncertainty relationships of angular momentum mean in terms of angular kinetic energy. To keep track, we label each energy level with the quantum number l so $l = 0$ indicates the ground state, $l = 1$ the next energy level up, and so forth.

14.5.1 Ground State: $l = 0$

The starting point is simple. There is *zero* angular kinetic energy when there is *zero* angular momentum: $L_x = L_y = L_z = 0$. In this case, there is no uncertainty. If you check back to Equation 14.18 you can see that if $L_z = 0$ then there is zero uncertainty in L_x and L_y. Zero angular momentum means that $\frac{\partial \Psi}{\partial \theta} = 0$ around every axis. The wave function is completely spherically symmetrical.

14.5.2 First Energy Level: $l = 1$

Things get more interesting when we add some angular energy. Imagine we start with the hydrogen electron having zero angular momentum so $L_{sum}^2 = 0$. What is the next energy level up? What is the smallest angular kinetic energy boost we can give the electron? Let's try increasing the angular momentum from $L = 0$ up to $L = \hbar$ around *one* axis (note that $-\hbar$ would simply mean the rotation is in the opposite direction).

Suppose we generate this angular momentum around the x axis. For $l = 1$ we add one quantum of angular momentum around the x axis so $L_x = \hbar$. The angular kinetic energy depends on L_{sum}^2 as shown in Equation 14.19 so, we need to calculate the minimum value of L_{sum}^2 when $L_x = \hbar$. You may be seduced into thinking that the minimum is $L_{sum}^2 = \hbar^2$. After all, if L_y and L_z are zero then $L_{sum}^2 = L_x^2 = \hbar^2$. If you think that, then please pause for a moment – it would mean that you know both L_y^2 and L_z^2 are zero, so that means you know that L_y and L_z are zero; at the same time that you know $L_x = \hbar$... ho hum... think some more!

The point is that if L_x has a value, then L_y and L_z are indeterminate because of the uncertainty principle. Neither can have a precise value. L_{sum}^2 cannot equal L_x^2. It must be higher than L_x^2 in order to reflect the uncertainty in the values of L_y^2 and L_z^2. How much higher? Equation 14.21 shows the uncertainty relationship (as in Box 14.2). It shows the calculation if we add l quanta of angular momentum to L_x. The result is that the value of $L_y^2 + L_z^2$ in L_{sum}^2 must be $l\hbar^2$ to conform with the uncertainty relationship. This is in addition to the $l^2\hbar^2$ contribution that L_x^2 makes to L_{sum}^2.

$$\Delta L_y \, \Delta L_z \geq \frac{\hbar L_x}{2} \quad \Longrightarrow \quad \Delta L_y \, \Delta L_z \geq \frac{l\hbar^2}{2} \qquad \text{excitation level } l, \text{ so } L_x = l\hbar$$

$$\Longrightarrow \quad \Delta L_y = \Delta L_z = \sqrt{\frac{l\hbar^2}{2}} \qquad \text{by symmetry} \tag{14.21}$$

$$\Longrightarrow \quad \Delta L_y^2 + \Delta L_z^2 = \frac{l\hbar^2}{2} + \frac{l\hbar^2}{2} = l\hbar^2 \tag{14.22}$$

This result is summarised in Equation 14.23. For example, in the case of the first energy level ($l = 1$), then $L_{sum}^2 = 2\hbar^2$. The upshot is that the relationship between angular kinetic energy E_{Ang} and the excitation level l depends on $l(l + 1)$ as shown in Equation 14.24.

$$L_{sum}^2 = L_x^2 + (L_y^2 + L_z^2) = l^2\hbar^2 + l\hbar^2$$

$$= l(l + 1)\,\hbar^2 \tag{14.23}$$

$$\Longrightarrow \quad \text{Angular kinetic energy: } E_{Ang} = \frac{L_{sum}^2}{2m_e r^2} = l(l + 1)\frac{\hbar^2}{2m_e r^2} \tag{14.24}$$

14.5.3 Three Distinct First Level States: $l = 1, \quad m = -1, 0, +1$

We need to return to our scenario. We have excited the electron up to $l = 1$. It has one quantum of angular momentum around the x axis meaning that $L_x = \hbar$. Imagine that we prepare more electrons in the same way so we now have a bunch of them. Then, we measure the value of L_z for each. What is the result?

The first thing to note is that this measurement can be made without disturbing the angular kinetic energy of the electron. There is no uncertainty relationship between L_z and L_{sum}^2 (see Box 14.2). Measuring L_z will not affect the angular kinetic energy. Quantum number l is unchanged.

The second thing to note is that the starting point is $L_x = \hbar$. This means the value of L_z is indeterminate. The result of measuring L_z will not be the same for all the electrons (or is unlikely to be). Angular momentum comes in quanta of $\hbar$ so the possible results for L_z are $\hbar$, 0 and $-\hbar$. The value of L^2_{sum} for $l = 1$ is $2\hbar^2$ so L_z cannot exceed $\hbar$. For example, if L_z were $2\hbar$ then L^2_{sum} would have to be in excess of its value squared which is $4\hbar^2$.

The next step is to separate the electrons according to their value of L_z. Can we tell them apart later? Absolutely! We can measure L_z again and we will get the same distinct value for each. These are three distinct angular momentum eigenstates that we label with the quantum number m where angular momentum is $m\hbar$ (to avoid confusing this with mass, I will label the electron mass as m_e for the rest of this chapter). For angular kinetic energy level $l = 1$, the value of m can be $+1, 0$ or -1.

For other energy levels, the rule is simple: m can be any integer between $-l$ and $+l$. It cannot be higher or lower because the next value would be $m = \pm(l + 1)$. This would mean L^2_{sum} could not be less than $(l + 1)^2\hbar^2$ but its actual value is only $l(l + 1)\hbar^2$... so not possible.

You might ask if there are other quantum numbers associated with the angular momentum around other axes, but no, the most you can know is the angular momentum around *one* axis (dependent on quantum number m) and the value of L^2_{sum} (dependent on quantum number l). That is total quantum knowledge of the state in terms of angular momentum. In an effort to entertain you, I offer you two more poor jokes: Box 14.4 and at the end of the chapter, Box 14.7.

Box 14.4   Just for laughs:

Why is angular momentum important?   *Because it makes the world go around.*

14.5.4   Resulting in the Pattern

This leads to the pattern of states shown in Box 14.5. The ground state $l = 0$ has no angular kinetic energy. This means that it has only one angular momentum eigenstate because $m = 0$. The first energy level $l = 1$ has 3 distinct states as we just discussed. As mentioned earlier, having more than one state with the same energy is called *degeneracy*. The next level $l = 2$ has 5 distinct states because m can be any integer from -2 up to $+2$... and so the pattern goes on.

The quantum number l tells you the level of excitation in terms of angular kinetic energy. Don't forget that the actual amount of energy depends on $l(l + 1)$ because of the uncertainty relationships. The number of distinct m states tells you how many different ways there are of spreading that angular kinetic energy around the atom. For a particular energy level l, the value of m can be any integer between $-l$ and $+l$ so there are $(2l + 1)$ distinct angular momentum states.

In Box 14.5, on the right side, you can see the link to the electron energy levels in atoms. These orbitals should be somewhat familiar, assuming you listened in your chemistry class at school. Each distinct state can accommodate two electrons which have different spin. We describe one as spin-up and the other spin-down. Spin is an intrinsic property of electrons that we examine in Chapter 17.

Box 14.5   Pattern of angular kinetic energy eigenstates

The pattern of atomic orbitals reflects multiple states with the same angular kinetic energy. We will get a fuller picture when we combine this with the Coulomb potential.

$l = 0 \implies m = 0$		*1 state, 2 electrons,* **s** *orbital*
$l = 1 \implies m = -1, 0, +1$		*3 states, 6 electrons,* **p** *orbital*
$l = 2 \implies m = -2, -1, 0, +1, +2$		*5 states, 10 electrons,* **d** *orbital*
$l = 3 \implies m = -3, -2, -1, 0, +1, +2, +3$		*7 states, 14 electrons,* **f** *orbital*

Angular kinetic energy in each case: $E_{Ang} = \dfrac{\hbar^2\, l(l + 1)}{2m_e r^2}$ *where m_e is electron mass*

The $l = 0$ level (zero angular kinetic energy, zero angular momentum) gives the spherically symmetrical **s** orbital with its two electrons. The next level of excitation $l = 1$ which accommodates 6 electrons gives the **p** orbital, and then further excitation levels give the **d** and **f** orbitals. The full picture will emerge when we tie this together with the maths behind the attraction between electron and proton in Chapter 15 and Chapter 16.

14.6 The Angular Momentum Creation Operator

This section gives more rigour to the maths behind angular momentum. For some of you, it may be more than you need (or want). If so, feel free to skip to the summary in Section 14.7. For those who want a bit more, we can employ a cunning trick similar to the one we used for the energy eigenstates of the Quantum Harmonic Oscillator (Subsection 13.3.2).

This trick uses an operator called $\hat{L}_+$ that changes the quantum number m without affecting quantum number l. The operation $\hat{L}_+ \Psi$ changes an angular momentum eigenstate with quantum numbers (l, m) to one with quantum numbers $(l, m + 1)$. There is also an annihilation operator $\hat{L}_-$ that changes the eigenstate from (l, m) to another with $(l, m - 1)$:

$$\hat{L}_+ \Psi = \hat{L}_x \Psi + i\hat{L}_y \Psi \qquad \textit{L_z creation operator} \qquad (14.25)$$

$$\hat{L}_- \Psi = \hat{L}_x \Psi - i\hat{L}_y \Psi \qquad \textit{L_z annihilation operator} \qquad (14.26)$$

These are fairly weird looking operations. Let's run through what $\hat{L}_+$ is in words. You take an L_z eigenstate with a specific value of quantum number m and then differentiate it with respect to θ_x (i.e. around the x axis). You then shift the phase of the answer by $\frac{\pi}{2}$ (i.e. multiply by $-i$). Separately, differentiate the eigenstate with respect to θ_y. Add the two derivatives together, then multiply the result by $\hbar$ and, hey presto, you have another L_z eigenstate[1]. That is quite a mouthful. Things become much clearer when we calculate the commutator of $\hat{L}_z$ with $\hat{L}_+$.

$$[\hat{L}_z, \hat{L}_+] = [\hat{L}_z, \hat{L}_x + i\hat{L}_y] = [\hat{L}_z, \hat{L}_x] + i\,[\hat{L}_z, \hat{L}_y] = i\hbar\hat{L}_y + i\,(-i\hbar\hat{L}_x)$$

$$= \hbar(\hat{L}_x + i\hat{L}_y) = \hbar\hat{L}_+$$

$$[\hat{L}_z, \hat{L}_+] = \hbar\hat{L}_+ \qquad (14.27)$$

If you ever see a result like Equation 14.27, give a yelp of joy. You have found a ladder of eigenstates! This is quite simple to demonstrate. We take an L_z eigenstate which I will label ϕ_m because it has a precise value for L_z of $m\hbar$. Then we act on it with the creation operator. First we must set the scene by using the commutator to give a new expression for $\hat{L}_+\hat{L}_z$

$$[\hat{L}_z, \hat{L}_+] = \hbar\hat{L}_+ \quad \implies \quad \hat{L}_z\hat{L}_+ - \hat{L}_+\hat{L}_z = \hbar\hat{L}_+ \quad \implies \quad \hat{L}_+\hat{L}_z = \hat{L}_z\hat{L}_+ - \hbar\hat{L}_+ \qquad (14.28)$$

We can use this new expression to show that the action $\hat{L}_+\phi_m$ creates another $\hat{L}_z$ eigenstate with the value of quantum number m increased by one:

$$\hat{L}_z\phi_m = m\hbar\phi_m \qquad \textit{for an eigenstate: } \hat{L}_z\phi_m = L_z\phi_m \qquad (14.29)$$

$$\hat{L}_+\hat{L}_z\phi_m = m\hbar\,\hat{L}_+\phi_m \qquad \textit{act on eigenstate with creation operator}$$

$$(\hat{L}_z\hat{L}_+ - \hbar\hat{L}_+)\phi_m = m\hbar\,\hat{L}_+\phi_m \qquad \textit{substitute with Equation 14.28}$$

$$\hat{L}_z\hat{L}_+\phi_m = m\hbar\,\hat{L}_+\phi_m + \hbar\,\hat{L}_+\phi_m$$

$$\hat{L}_z(\hat{L}_+\phi_m) = (m+1)\hbar(\hat{L}_+\phi_m) \qquad (14.30)$$

Compare Equation 14.30 with Equation 14.29. It has the same structure. We have produced another $\hat{L}_z$ eigenstate with quantum number $(m + 1)$. We can act again with $\hat{L}_+$ so we have a ladder of L_z eigenstates each separated by $\hbar$ angular momentum. We argued this was needed in Section 14.2 in order for the wave function to remain unchanged by a full rotation. It all fits.

We can use the creation operator to show the relationship between quantum number m and quantum number l. A particle with quantum number l is in a angular kinetic energy eigenstate with $L_{\text{sum}}^2 = \hbar^2 l(l + 1)$. We can measure its value of L_z to find m without affecting l because $\hat{L}_{\text{sum}}^2$ and $\hat{L}_z$ commute. This gives us an eigenstate with precise l and m which I will label of course ϕ_{lm}. What happens if we act on ϕ_{lm} with $\hat{L}_+$? We know it will create another eigenstate with m increased by

1, but what will it do to l? The answer is simple. It does not change l. We know this because $\hat{L}_+ = \hat{L}_x + i\hat{L}_y$ and both $\hat{L}_x$ and $\hat{L}_y$ commute with $\hat{L}_{sum}^2$ so neither will alter the value of l.

$$\hat{L}_+\phi_{l,m} \;=\; \phi_{l,m+1} \tag{14.31}$$

We can act again and again with $\hat{L}_+$ producing a ladder of eigenstates with increasing value of m, but there must be a limit. For example if l is 2, then L_{sum}^2 is $l(l+1)\hbar^2 = 6\hbar^2$. In this case m cannot be as high as 3 because, if so, $L_z^2 = 9\hbar^2$ which is in excess of the total L_{sum}^2 of the state.

There must be a maximum m_{max} and, at that stage, acting on ϕ_{lm} with $\hat{L}_+$ destroys the wave function (Equation 14.32). If we then act on this state with $\hat{L}_-$ (Equation 14.33) the result must still be zero because there is no wave function to act on; that and a little maths shows $m_{max} = l$...

$$\hat{L}_+\,\phi_{lm} \;=\; 0 \qquad\qquad\qquad \textit{when m is at } m_{max} \tag{14.32}$$

$$\hat{L}_-\hat{L}_+\,\phi_{lm} \;=\; (\hat{L}_x - i\hat{L}_y)(\hat{L}_x + i\hat{L}_y)\,\phi_{lm} \;=\; 0 \qquad \textit{when m is at } m_{max} \tag{14.33}$$

$$= (\hat{L}_x^2 + \hat{L}_y^2 + i[\hat{L}_x,\hat{L}_y])\,\phi_{lm}$$

$$= (\hat{L}_{sum}^2 - \hat{L}_z^2 + i[\hat{L}_x,\hat{L}_y])\,\phi_{lm} \qquad\qquad \hat{L}_x^2 + \hat{L}_y^2 \;=\; \hat{L}_{sum}^2 - \hat{L}_z^2$$

$$= (\hat{L}_{sum}^2 - \hat{L}_z^2 - \hbar\hat{L}_z)\,\phi_{lm} \qquad\qquad\qquad [\hat{L}_x,\hat{L}_y] \;=\; i\hbar\hat{L}_z$$

$$= (l(l+1)\hbar^2 - m^2\hbar^2 - m\hbar^2)\,\phi_{lm}$$

$$\implies \; l^2 + l - m^2 - m = 0 \quad \implies \quad m_{max} = l \tag{14.34}$$

The ladder of m eigenstates goes up to $m = +l$ and, by a similar analysis you can show it goes down to $m = -l$. This confirms the pattern of eigenstates shown in Box 14.5. I hope this section has given a warm thrill to any of you who are addicted mathmagicians.

14.7 Summary

At the start of this chapter we showed that the relationship between angular momentum L and the wave function is somewhat similar to that of translational momentum p. As a result of this relationship (summarised in Equation 14.35), angular momentum is quantised in units of $\hbar$. This is because a full rotation (adding 2π to θ) should leave the wave function unchanged.

$$\frac{\partial\Psi}{\partial\theta_x} \;=\; \frac{i}{\hbar}L_x\Psi \qquad\qquad \frac{\partial\Psi}{\partial\theta_y}\Psi \;=\; \frac{i}{\hbar}L_y\Psi \qquad\qquad \frac{\partial\Psi}{\partial\theta_z} \;=\; \frac{i}{\hbar}L_z\Psi \tag{14.35}$$

The interplay of L_x, L_y, and L_z leads to an uncertainty relationship between them. We use the quantum number l to denote the level of angular kinetic energy. The uncertainty relationship means L_{sum}^2 must be $l(l+1)\hbar^2$ so at level l, the amount of angular kinetic energy E_{Ang} depends on $l(l+1)$:

$$E_{Ang} \;=\; \frac{L_{sum}^2}{2m_e r^2} \;=\; l(l+1)\frac{\hbar^2}{2m_e r^2} \qquad\qquad \textit{r is radius, } m_e \textit{ is electron mass} \tag{14.36}$$

To understand the structure of atoms, it is important to know how many angular momentum states there are at each level of angular kinetic energy. We label these distinct states with quantum number m (not to be confused with mass). The value of m can be any integer from $-l$ to $+l$ so there are $(2l + 1)$ distinct angular momentum states at each energy level.

We will show in the next chapter how this leads to the structure of the electron orbitals in atoms. The **s** orbital ($l = 0$) has only one angular momentum state that accommodates two electrons. The **p** orbital ($l = 1$) consists of three distinct angular momentum states which equates to six electrons. The **d** orbital with 10 electrons is next ($l = 2$) followed by the **f** orbital with 14 electrons ($l = 3$).

For those needing a deeper dive into the maths, I introduced the creation and annihilation operators of angular momentum.

The next part of the story is to look at the attraction between an electron and a proton. This is called the Coulomb potential.

Box 14.6   Calculation of $\hat{L}^2_{sum}$ and $\hat{L}_x$ commutator – if you really want to see it!

In the calculations below I have not shown the wave function Ψ until the final line. This is for brevity. Please remember that these operators are being applied to a wave function.

We need to set things up using the $[\hat{L}_x, \hat{L}_y]\Psi$ and $[\hat{L}_z, \hat{L}_x]\Psi$ commutator relationships:

$$\hat{L}_x\hat{L}_y - \hat{L}_y\hat{L}_x = i\hbar\hat{L}_z \implies \hat{L}_x\hat{L}_y = i\hbar\hat{L}_z + \hat{L}_y\hat{L}_x \quad and \quad \hat{L}_y\hat{L}_x = -i\hbar\hat{L}_z + \hat{L}_x\hat{L}_y$$

$$\hat{L}_z\hat{L}_x - \hat{L}_x\hat{L}_z = i\hbar\hat{L}_y \implies \hat{L}_z\hat{L}_x = i\hbar\hat{L}_y + \hat{L}_x\hat{L}_z \quad and \quad \hat{L}_x\hat{L}_z = -i\hbar\hat{L}_y + \hat{L}_z\hat{L}_x$$

We can use the equations above to solve for the $[\hat{L}^2_{sum}, \hat{L}_x]\Psi$ commutator:

$$[\hat{L}^2_{sum}, \hat{L}_x] = (\hat{L}^2_x + \hat{L}^2_y + \hat{L}^2_z)\hat{L}_x - \hat{L}_x(\hat{L}^2_x + \hat{L}^2_y + \hat{L}^2_z)$$

$$= \hat{L}_y\hat{L}_y\hat{L}_x - \hat{L}_x\hat{L}_y\hat{L}_y + \hat{L}_z\hat{L}_z\hat{L}_x - \hat{L}_x\hat{L}_z\hat{L}_z + \hat{L}^3_x - \hat{L}^3_x$$

$$= \hat{L}_y(\hat{L}_y\hat{L}_x) - (\hat{L}_x\hat{L}_y)\hat{L}_y + \hat{L}_z(\hat{L}_z\hat{L}_x) - (\hat{L}_x\hat{L}_z)\hat{L}_z$$

$$= \hat{L}_y(-i\hbar\hat{L}_z + \hat{L}_x\hat{L}_y) - (i\hbar\hat{L}_z + \hat{L}_y\hat{L}_x)\hat{L}_y + \hat{L}_z(i\hbar\hat{L}_y + \hat{L}_x\hat{L}_z) - (-i\hbar\hat{L}_y + \hat{L}_z\hat{L}_x)\hat{L}_z$$

$$= -i\hbar\hat{L}_y\hat{L}_z - i\hbar\hat{L}_z\hat{L}_y + i\hbar\hat{L}_z\hat{L}_y + i\hbar\hat{L}_y\hat{L}_z = 0$$

And by symmetry: $[\hat{L}^2_{sum}, \hat{L}_x]\Psi = [\hat{L}^2_{sum}, \hat{L}_y]\Psi = [\hat{L}^2_{sum}, \hat{L}_z]\Psi = 0$

Box 14.7   Just for laughs:

What is a mathematician's answer to constipation?   *He works it out with a pencil!*

Note

1 If you're confused, check by calculating the expression in Equation 14.25 using the definition of $\hat{L}$ from Equation 14.13.

Chapter 15

Coulomb Potential

CHAPTER MENU
15.1 The Hydrogen Emission Spectrum 15.2 The Challenge of the Coulomb Potential 15.3 A Primitive Model 15.4 Schrödinger's Equation for Hydrogen 15.5 Discussion

In this chapter we will look at the behaviour of electrons in atoms. Specifically, we are going to use the Schrödinger equation to calculate the stable states of the electron around a single proton nucleus (hydrogen in its simplest form). The electromagnetic attraction between the electron and proton is the *Coulomb potential*. The bad news is that, outside of a few special cases such as the QHO, the Schrödinger equation is surprisingly difficult to solve, but I will try to keep the maths clear and take shortcuts where possible. The maths in this chapter is some of the most challenging in this book, so I have included the usual historical snippets to keep you entertained.

15.1 The Hydrogen Emission Spectrum

One of the first signs of quantum *stuff* came from the emission spectrum of elements, particularly hydrogen. When heated, hydrogen emits very specific grouped wavelengths of light. Back in 1885, Johann Balmer spotted a pattern in the wavelengths (see Box 15.1). The group of emissions he assessed is now called the Balmer series. Subsequently, emissions were spotted at other wavelengths, giving us what are called the Lyman and Paschen series.

Box 15.1   Life begins at 60: Johann Balmer (1825–1898)
Balmer was a maths and Latin teacher at a girls' secondary school in Basel, Switzerland. He loved number games. The story goes that he complained to a chemist of being bored. This friend gave him four numbers to puzzle over: the wavelengths of hydrogen emissions. Within days, Balmer unearthed a pattern and wrote down the formula for hydrogen that made him famous. It was Johann Balmer's first and only significant involvement in physics. He was 60 years old. Balmer was a late bloomer in other ways. He married Christine Rinck in 1868, when he was 43 years old and went on to have 6 children with her. One of his sons, Wilhelm Balmer, became a well known painter.

Johannes Rydberg, a Swedish physicist, took the pattern that Balmer had spotted and extended it into a formula that covers all three series and describes the energy levels of the electron in a hydrogen atom with considerable accuracy (Box 15.3 gives some history). The modern version of his formula is shown as Equation 15.1 and is quantified in joules and electron volts (eV) but watch out for the notation used (see Box 15.2):

Quantum Untangling: An Intuitive Approach to Quantum Mechanics from Einstein to Higgs, First Edition. Simon Sherwood.

$$E = -\frac{1}{n^2} \frac{m_e e^4}{32\pi^2 \hbar^2 \epsilon_0^2} \approx -\frac{1}{n^2}(2.2 \times 10^{-18})\, joules \approx -\frac{14}{n^2}\, eV \quad \text{for integer } n \tag{15.1}$$

$$m_e = electron\ mass \qquad e = electron\ charge \qquad \epsilon_0 = permittivity\ of\ free\ space$$

Box 15.2 Watch out for h versus $\hbar$

In quantum mechanics we mostly use $\hbar$ which is $\frac{h}{2\pi}$. In the Rydberg equation h can seem more convenient because the π factor cancels. I will always use $\hbar$, but watch out in other texts.

The electron energy is shown as negative because it represents the amount of energy that would be required to move the electron out of the clutches of the proton that is, to an infinite distance away. To illustrate, the calculation below shows that the energy required to lift the hydrogen electron from the $n = 1$ ground state up to the $n = 2$ state is 1.65×10^{-18} joules:

$$E\,(n = 1) = -2.2 \times 10^{-18}\, joules \tag{15.2}$$

$$E\,(n = 2) = -\frac{1}{4}(2.2 \times 10^{-18}) = -0.55 \times 10^{-18}\, joules \tag{15.3}$$

$$\Delta E\,(n = 1 \rightarrow 2) = +1.65 \times 10^{-18}\, joules \approx 10\, eV \tag{15.4}$$

Why does the attraction of the electron to the proton produce this $\frac{1}{n^2}$ pattern of energy levels? If we are going to explain this, we need to take Schrödinger's equation, add in the effect of the potential energy of the attraction between the electron and proton, then calculate the energy eigenstates, and if you have not guessed already, this is quite tricky; so let's start with a closer look at the Coulomb potential.

Box 15.3 Unlucky Johannes Rydberg (1854–1919)

Rydberg was not from a wealthy family. His father died when he was only 4 years old. Throughout his life he struggled to make ends meet. Even when he finally became a physics professor, he had to supplement his income with work as a bank accountant and actuary. In 1890 he published his formula for the spectral lines of elements. Thirty years later, in 1917, he finally was nominated for a Nobel Prize, but the committee decided not to award one that year. He was nominated again in 1920 by which time he had died.

15.2   The Challenge of the Coulomb Potential

The Coulomb potential (Equation 15.5) describes the potential energy from the attraction in the hydrogen atom between the electron and the proton, both of which have the same strength of charge e (for the proton positive, for the electron negative).

$$Potential\ energy\ V = -\frac{e^2}{4\pi\epsilon_0 r} \qquad for\ hydrogen \tag{15.5}$$

$$e = electron\ charge \qquad \epsilon_0 = permittivity\ of\ free\ space \qquad r = distance$$

In the QHO, V is shown as a positive number increasing from $V = 0$ at the equilibrium point $r = 0$. This approach is not possible with the Coulomb potential because the potential V is proportional to $\frac{1}{r}$ so at $r = 0$ the value of the potential V explodes to infinity. We solve this by measuring potential energy with $V = 0$ set at an infinite distance from the proton. We then count down in negative value as the electron approaches the proton. Thus, if the electron has potential energy $-A$, it would require A of energy to get out to $r = \infty$ distance from the proton.

15.3 A Primitive Model

Later in this chapter we will use the Schrödinger equation to tie together the Coulomb potential with the electron energy levels in hydrogen. However, the maths is a bit obscure. First I should try to convince you that the link between the $\frac{1}{r}$ Coulomb potential and the $\frac{1}{n^2}$ electron energy levels is not unexpected.

Back in 1913, 10 years before Heisenberg developed his uncertainty principle, Niels Bohr built a model of an atom that worked (mathematically) surprisingly well. Bohr's model treated the electron as a classical particle *except* that its energy is quantised. Let's do something similar.

If we constrain an electron close to a proton, the electron wavelength must fit in the constrained space. This means that the electron's energy is quantised. We derived the equivalent of Equation 15.6 in the one-dimensional electron-in-a-box model (Section 7.1) that relates the constrained length L to the electron's kinetic energy. If we constrain the electron within radius r of the proton, then we set $L = 2r$ to calculate the allowable levels of kinetic energy ($K.E$).

$$p = \pm\frac{2\pi\hbar}{\lambda} = \pm\frac{n\pi\hbar}{2r} \implies K.E = \frac{p^2}{2m_e} = \frac{n^2\pi^2\hbar^2}{8m_e r^2} \propto \frac{\hbar^2}{m_e}\frac{n^2}{r^2} \tag{15.6}$$

The key thing to note is the $\frac{n^2}{r^2}$ term in the energy quantisation. It turns out that, even using classical physics, when you combine this with the $\frac{1}{r}$ Coulomb potential, it always produces a $\frac{1}{n^2}$ pattern of energy states.

Classically, the lowest energy level for the electron is where the total of its potential energy ($\propto \frac{1}{r}$) and its kinetic energy ($\propto \frac{n^2}{r^2}$) is at a minimum. We don't know the precise relationship, but this is enough for a calculation. In Equation 15.7, I have labelled this proportionality as $K.E. = \frac{A}{r^2}$ and $V = \frac{B}{r}$ with A and B as constants. In Equation 15.8 setting the value of $\frac{\partial E}{\partial r}$ to zero allows us to pick out the minimum (I hope you remember this from maths at school). We feed this value of r back into the energy formula to calculate the energy at the lowest energy level.

$$E = K.E + V = \frac{A}{r^2} - \frac{B}{r} \qquad A \propto \frac{n^2\hbar^2}{m_e} \quad B \propto \frac{e^2}{\epsilon_0} \tag{15.7}$$

$$\frac{\partial E}{\partial r} = -\frac{2A}{r^3} + \frac{B}{r^2} = 0 \implies r_{min} = \frac{2A}{B} \qquad At\ minimum\ total\ energy \tag{15.8}$$

$$E_{min} = K.E + V = \frac{A}{r_{min}^2} - \frac{B}{r_{min}} = \frac{B^2}{4A} - \frac{B^2}{2A} = -\frac{B^2}{4A} \tag{15.9}$$

The final step is to substitute back the values of A and B to give Equation 15.10 for the minimum energy levels. This matches the $\frac{1}{n^2}$ pattern for the energy levels in Equation 15.1.

$$E_{min} = -\frac{B^2}{4A} \propto -\frac{1}{n^2}\frac{m_e e^4}{\hbar^2\epsilon_0^2} \propto -\frac{1}{n^2} \tag{15.10}$$

This primitive model is not meant to be rigorous. The point is simple. When a particle is constrained, one might justifiably expect its momentum to be quantised proportional to $\frac{n}{r}$. This leads to kinetic energy being quantised $\frac{n^2}{r^2}$. Put that together with a $\frac{1}{r}$ potential and the minimum energy solutions will follow a $\frac{1}{n^2}$ pattern.

The approach that Niels Bohr took in 1913 was slightly different (Box 15.4). He thought the electron orbited the proton. He calculated the orbital radius where the attractive force and the centripetal force are in balance. Mathematically, this is the same as minimising the energy.[1] Bohr proposed that angular momentum be in multiples of $\hbar$ (this is correct). He used the classical calculation to give angular kinetic energy of $\frac{n^2\hbar^2}{2m_e r^2}$ (this is wrong in a quantum system). But his calculation has the $\frac{n^2}{r^2}$ relationship for kinetic energy so, in spite of many flaws, it gives the $\frac{1}{n^2}$ pattern for energy levels. In fact, if you feed this into Equation 15.9 it produces the precise formula for Equation 15.1. This is 10 years before de Broglie suggested electrons have a wavelength. One has to conclude that Bohr was very lucky!

I hope this section has given you some intuitive comfort that the Coulomb potential's $\frac{1}{n^2}$ pattern of energy levels is not outlandish or strange. But for a proper answer, we need much more than a loose fuzzy hybrid model – we need the Schrödinger equation.

Box 15.4   Lucky Niels Bohr (1885–1962)

Bohr came up with a classical orbital model of the hydrogen atom that explained Rydberg's formula. With only a passing nod to quantum mechanics, it really has no right to work anything like as well as it does; *lucky Niels Bohr!* Carlsberg brewery funded Bohr throughout his life including giving him a house in which he lived for 30 years. One dubious story has it that the house was connected to a brewery providing him with a free supply of beer! *Lucky Niels Bohr!*

15.4   Schrödinger's Equation for Hydrogen

From the moment that Schrödinger came up with his equation, he would have known that the *big* test was to match it up versus the pattern of electron energy levels in hydrogen. This was important not only to explain the origin of the hydrogen spectra, but also to prove that quantum mechanics and his new equation were on the right track.

Constructing Schrödinger's equation for hydrogen is simple. We feed the expression for the Coulomb potential into the Schrödinger equation. We want to find the stable energy eigenstates so we use the time-independent version (as summarised in Box 11.5). Note that I label the wave function as Ψ instead of the ϕ_E I have used elsewhere for an energy eigenstate. This simply is because we are going to need the ϕ symbol when we switch to spherical coordinates:

$$E\,\Psi \;=\; -\frac{\hbar^2}{2m_e}\left(\frac{\partial^2\Psi}{\partial x^2} + \frac{\partial^2\Psi}{\partial y^2} + \frac{\partial^2\Psi}{\partial z^2}\right) - \frac{e^2}{4\pi\epsilon_0\,r}\,\Psi \tag{15.11}$$

The good news is that you are going to learn a huge amount from the solution of this equation about hydrogen and about the structure of other atoms. The bad news (you knew that was coming) is that the mathematical effort is not trivial. However, as promised, I will take it step-by-step so please bear with me. It will be worth it.

15.4.1   Spherical Harmonics – *merci* Monsieur Laplace

The first challenge with Equation 15.11 is that the potential energy varies with r but the derivative terms are in Cartesian coordinates (x, y, z). The Coulomb potential is spherically symmetrical so, it makes sense to work in spherical coordinates. Figure 15.1 shows the relationship between Cartesian coordinates (x, y, z) and spherical coordinates (r, θ, φ)

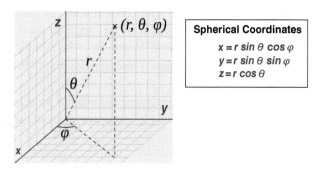

Figure 15.1   Spherical polar coordinates.

We need to switch the derivative terms to spherical coordinates. This is not a trivial task, but fortunately the pattern of derivative terms in the Schrödinger equation is well known. It is called the *Laplacian* after the amazing mathematician Pierre-Simon Laplace (1749–1827). I will use the symbol ∇^2 for it. By the end of this chapter you will be enormously grateful to Laplace, who is affectionately referred to as the French Newton. The research he did on the Laplacian back in the 1800s is going to save us a lot of work, including expressing the Laplacian in spherical coordinates as shown in Equation 15.12:

$$\nabla^2 f(x,y,z) = \frac{\partial^2 f}{\partial x^2} + \frac{\partial^2 f}{\partial y^2} + \frac{\partial^2 f}{\partial z^2}$$

$$\nabla^2 f(r,\theta,\varphi) = \frac{1}{r^2}\frac{\partial}{\partial r}\left(r^2\frac{\partial f}{\partial r}\right) + \frac{1}{r^2\sin\theta}\frac{\partial}{\partial\theta}\left(\sin\theta\frac{\partial f}{\partial\theta}\right) + \frac{1}{r^2\sin^2\theta}\frac{\partial^2 f}{\partial\varphi^2} \tag{15.12}$$

Clearly this does not simplify things! You may be reaching for matches to burn this book (yet again), but let's put ourselves in the hands of Laplace for a little longer. He wanted to know what happens when the Laplacian of a function is zero ($\nabla^2 f = 0$) and he had the brains and patience to find a solution providing that the function $f(r,\theta,\varphi)$ can be expressed as a combination of two separate functions which I will label $G(r)$, that depends only on the radial coordinate and $Y(\theta,\varphi)$ that depends only on the angular coordinates:

$$f(r,\theta,\varphi) = G(r)\,Y(\theta,\varphi) \qquad \textit{if the radial and angular variables are separable} \tag{15.13}$$

For $\nabla^2 f = 0$ in Equation 15.12 we can express the function f in terms of G and Y. I have also multiplied both sides by r^2 to simplify. This gives Equation 15.14. The next step in Equation 15.15 is to recognise that the differential operators do not affect functions that don't contain the relevant variable so, for example, $\frac{\partial(GY)}{\partial r} = Y\frac{\partial G}{\partial r}$ or, in the case of one of the angular derivatives $\frac{\partial(GY)}{\partial\theta} = G\frac{\partial Y}{\partial\theta}$.

$$0 = \frac{\partial}{\partial r}\left(r^2\frac{\partial GY}{\partial r}\right) + \frac{1}{\sin\theta}\frac{\partial}{\partial\theta}\left(\sin\theta\frac{\partial GY}{\partial\theta}\right) + \frac{1}{\sin^2\theta}\frac{\partial^2 GY}{\partial\varphi^2} \tag{15.14}$$

$$= Y\frac{\partial}{\partial r}\left(r^2\frac{\partial G}{\partial r}\right) + G\frac{1}{\sin\theta}\frac{\partial}{\partial\theta}\left(\sin\theta\frac{\partial Y}{\partial\theta}\right) + G\frac{1}{\sin^2\theta}\frac{\partial^2 Y}{\partial\varphi^2} \tag{15.15}$$

$$= \frac{Y}{G}\frac{\partial}{\partial r}\left(r^2\frac{\partial G}{\partial r}\right) + \frac{1}{\sin\theta}\frac{\partial}{\partial\theta}\left(\sin\theta\frac{\partial Y}{\partial\theta}\right) + \frac{1}{\sin^2\theta}\frac{\partial^2 Y}{\partial\varphi^2} \tag{15.16}$$

After dividing by G in Equation 15.16 we can shift the angular terms to the left side of the equation as shown in Equation 15.17. Recognising that all of the radial terms (involving G and r) are *constant* with respect to θ and φ, we finally arrive at Equation 15.18.

$$\frac{1}{\sin\theta}\frac{\partial}{\partial\theta}\left(\sin\theta\frac{\partial Y}{\partial\theta}\right) + \frac{1}{\sin^2\theta}\frac{\partial^2 Y}{\partial\varphi^2} = -Y\frac{1}{G}\frac{\partial}{\partial r}\left(r^2\frac{\partial G}{\partial r}\right) \tag{15.17}$$

$$= -C\,Y(\theta,\varphi) \qquad \textit{varying }\theta\textit{ or }\varphi\textit{ does not affect C} \tag{15.18}$$

We have separated the variables. Equation 15.18 depends only on the angular coordinates θ and φ. This may seems like a lot of work, but it is nothing compared to Laplace's effort. He solved this equation. He found that for every possible solution for Y, the value of C is always an integer of the form $l(l+1)$ where l is an integer (or zero). He also discovered that for each value of l there are $(2l+1)$ independent solutions for Y. Each of these independent solutions gives a different integer value for m in the equation $\frac{\partial^2 Y}{\partial\varphi^2} = -m^2\,Y$ with the value of m being any integer (including 0) between $-l$ and $+l$. Do not confuse this m with the mass of the electron m_e.

This should all sound eerily familiar if you were not asleep during the chapter on angular momentum. Laplace's solutions are now called the *spherical harmonics*. They are labelled Y_l^m with the values of l and m identifying each particular solution.

To make sure this is clear, let's look at one of the simplest solutions to the equation: $Y_1^0 = \cos\theta$ for which $l = 1$ and $m = 0$. One great thing about this sort of differential equation is that, while it is hard to find the solutions, it is relatively simple to check them. To confirm that it is a valid solution we substitute $Y = \cos\theta$ into Equation 15.21 and calculate l

(Equation 15.19). The value of m is easy to verify as $Y = \cos\theta$ is not dependent on variable φ, so $m = 0$ as shown in Equation 15.20.

$$-l(l+1)\,Y = \frac{1}{\sin\theta}\frac{\partial}{\partial\theta}\left(\sin\theta\,\frac{\partial\cos\theta}{\partial\theta}\right) + \frac{1}{\sin^2\theta}\frac{\partial^2\cos\theta}{\partial\varphi^2}$$

$$= \frac{1}{\sin\theta}\frac{\partial}{\partial\theta}\left(-\sin^2\theta\right) + 0$$

$$= -\frac{2\sin\theta\cos\theta}{\sin\theta} = -2\cos\theta = -2\,Y \implies l = 1 \tag{15.19}$$

$$\textit{Rule:}\quad \frac{\partial^2 Y}{\partial\varphi^2} = -m^2\,Y \qquad \frac{\partial^2(\cos\theta)}{\partial\varphi^2} = 0 \implies m = 0 \tag{15.20}$$

Another of the solutions is $Y = \sin\theta\,e^{i\varphi}$. Can *you* prove it's a solution and work out l and m? The answer is shown later in Box 15.8, but don't peek until you have tried. Solving some examples will help you understand the spherical harmonics equation (shown as Equation 15.21 in Box 15.5). There is more to say about these harmonics, but first we need to tie things back to Schrödinger's equation.

Box 15.5 Spherical harmonics

$$\frac{1}{\sin\theta}\frac{\partial}{\partial\theta}\left(\sin\theta\,\frac{\partial Y_l^m}{\partial\theta}\right) + \frac{1}{\sin^2\theta}\frac{\partial^2 Y_l^m}{\partial\varphi^2} = -l(l+1)\,Y_l^m \tag{15.21}$$

15.4.2 The Angular Equation

You may not know it, but we have at our mercy what is called the *angular equation* for hydrogen. In this section, we are going to express Ψ as a combination of separable radial and angular parts: $\Psi = R(r)f(\theta,\varphi)$, and show that the angular part $f(\theta,\varphi)$ is the equation for Y_l^m of the spherical harmonics. Don't be scared by the number of terms in the equations. All we are going to do is to move bits about and then use Laplace's result.

$$E\,\Psi = -\frac{\hbar^2}{2m_e}\nabla^2\Psi + V\,\Psi \qquad \textit{where}\ \ V = -\frac{e^2}{4\pi\epsilon_0 r} \tag{15.22}$$

Equation 15.22 is the Schrödinger equation for the Coulomb potential. It is the same as Equation 15.11 but I have used ∇^2 for the Laplacian and V for the Coulomb potential. The equation is an energy balance. In words it is: *Total energy = Kinetic energy + Potential energy*.

We can shift the $V\Psi$ across and open up the Laplacian in spherical coordinates (using the expression from Equation 15.12) to give Equation 15.24, which I will come back to later.

$$(E-V)\Psi = -\frac{\hbar^2}{2m_e}\left(\frac{1}{r^2}\frac{\partial}{\partial r}\left(r^2\frac{\partial\Psi}{\partial r}\right) + \frac{1}{r^2\sin\theta}\frac{\partial}{\partial\theta}\left(\sin\theta\frac{\partial\Psi}{\partial\theta}\right) + \frac{1}{r^2\sin^2\theta}\frac{\partial^2\Psi}{\partial\varphi^2}\right) \tag{15.23}$$

$$= -\frac{\hbar^2}{2m_e r^2}\frac{\partial}{\partial r}\left(r^2\frac{\partial\Psi}{\partial r}\right) - \frac{\hbar^2}{2m_e r^2}\left(\frac{1}{\sin\theta}\frac{\partial}{\partial\theta}\left(\sin\theta\frac{\partial\Psi}{\partial\theta}\right) + \frac{1}{\sin^2\theta}\frac{\partial^2\Psi}{\partial\varphi^2}\right) \tag{15.24}$$

Returning to Equation 15.23, we want to separate it into radial (r dependent) and angular (θ and φ dependent) terms. To prepare for this we multiply everything by $-\frac{2m_e r^2}{\hbar^2}$ and then shift the angular θ and φ terms over to the left side, and the radial r terms to the right, giving Equation 15.25

$$\frac{1}{\sin\theta}\frac{\partial}{\partial\theta}\left(\sin\theta\frac{\partial\Psi}{\partial\theta}\right) + \frac{1}{\sin^2\theta}\frac{\partial^2\Psi}{\partial\varphi^2} = -\frac{\partial}{\partial r}\left(r^2\frac{\partial\Psi}{\partial r}\right) - \frac{2m_e r^2}{\hbar^2}(E-V)\Psi \tag{15.25}$$

Notice that all the angular terms are on the left and all the radial r terms (including the Coulomb potential) are on the right. Now comes the big step. We *assume* that Ψ can be separated into two components, one radial and the other angular such that: $\Psi = R(r)f(\theta, \varphi)$. This may seem a bold assumption, but it turns out to be correct. You can take this sort of risk when searching for answers because it is much easier to check solutions than to find them. The radial derivative components affect only $R(r)$ and the angular derivative terms affect only $f(\theta, \varphi)$. For example:

$$\frac{\partial \Psi}{\partial r} = \frac{\partial R(r)}{\partial r} f(\theta, \varphi) \qquad and\ similarly \qquad \frac{\partial \Psi}{\partial \theta} = \frac{\partial f(\theta, \varphi)}{\partial \theta} R(r) \qquad (15.26)$$

We then express Equation 15.25 with the Ψ terms as $R(r)f(\theta, \varphi)$. The next step is to divide both sides by $R(r)$ giving Equation 15.27. The final step (phew!) is to recognise that all the terms on the right side of this equation are constant with respect to f, taking us at last to Equation 15.28.

$$\frac{1}{\sin \theta} \frac{\partial}{\partial \theta} \left(\sin \theta \frac{\partial f}{\partial \theta} \right) R + \frac{1}{\sin^2 \theta} \frac{\partial^2 f}{\partial \varphi^2} R = -\frac{\partial}{\partial r} \left(r^2 \frac{\partial R}{\partial r} \right) f - \frac{2m_e r^2}{\hbar^2} (E - V) R f$$

$$\frac{1}{\sin \theta} \frac{\partial}{\partial \theta} \left(\sin \theta \frac{\partial f}{\partial \theta} \right) + \frac{1}{\sin^2 \theta} \frac{\partial^2 f}{\partial \varphi^2} = -\left(\frac{1}{R} \frac{\partial}{\partial r} \left(r^2 \frac{\partial R}{\partial r} \right) + \frac{2m_e r^2}{\hbar^2} (E - V) \right) f \qquad (15.27)$$

$$= -C f(\theta, \varphi) \qquad C\ is\ a\ constant \qquad (15.28)$$

We have finally arrived. Equation 15.28 is the angular equation for hydrogen and it is none other than the spherical harmonics' equation (Equation 15.21). The angular term that I labelled $f(\theta, \varphi)$ is actually Y so, $\Psi = R(r)Y(\theta, \varphi)$. Thanks to Laplace we know the solutions for Y. We also know the value of the constant C so Equation 15.28 becomes:

$$\frac{1}{\sin \theta} \frac{\partial}{\partial \theta} \left(\sin \theta \frac{\partial Y}{\partial \theta} \right) + \frac{1}{\sin^2 \theta} \frac{\partial^2 Y}{\partial \varphi^2} = -l(l+1) Y \qquad (15.29)$$

You can multiply Y by any constant (by which I mean anything that does *not* vary with θ or φ) and it does not change the result of Equation 15.29 so it is also valid for $\Psi = R(r)Y(\theta, \varphi)$. We can use this to express the Schrödinger in a simpler form. Starting from Equation 15.24:

$$(E - V) \Psi = -\frac{\hbar^2}{2m_e r^2} \frac{\partial}{\partial r} \left(r^2 \frac{\partial \Psi}{\partial r} \right) - \frac{\hbar^2}{2m_e r^2} \left(\frac{1}{\sin \theta} \frac{\partial}{\partial \theta} \left(\sin \theta \frac{\partial \Psi}{\partial \theta} \right) + \frac{1}{\sin^2 \theta} \frac{\partial^2 \Psi}{\partial \varphi^2} \right)$$

$$= -\frac{\hbar^2}{2m_e r^2} \frac{\partial}{\partial r} \left(r^2 \frac{\partial \Psi}{\partial r} \right) - \frac{\hbar^2}{2m_e r^2} (-l(l+1) \Psi)$$

$$(E - V) \Psi = -\frac{\hbar^2}{2m_e r^2} \frac{\partial}{\partial r} \left(r^2 \frac{\partial \Psi}{\partial r} \right) + \frac{l(l+1) \hbar^2}{2m_e r^2} \Psi \qquad (15.30)$$

Let's review the energy balance in Equation 15.30. The left side $(E - V)$ is energy less potential energy so, it is the total *kinetic* energy. On the far right you should recognise the angular kinetic energy (see Equation 14.36). The central term is the radial kinetic energy that we will get to in a moment. In words, Equation 15.30 can be summarised as:

Total kinetic energy = Radial kinetic energy + Angular kinetic energy

15.4.3 The Shape of the Atomic Orbitals

Equation 15.29 reveals the shape of the electron orbitals of hydrogen. If there is no angular kinetic energy ($l = 0$) then the wave function does not vary with θ and φ. If you examine the surface of a sphere radius r away from the proton, the value of the wave function is identical at every point. The probability of finding the electron varies with distance r from the proton but is the same at all angles θ and φ. The atomic orbital for ($l = 0$) is spherically symmetrical.

The spherical symmetry disappears if the electron has any angular kinetic energy. The orbital shape is then determined by the relevant spherical harmonic Y_l^m which depends on l (which reflects the amount of angular kinetic energy) and m (which reflects how that energy is distributed). In this case, if you examine the surface of a sphere radius r away from the proton, the relative value of the wave function is the value of Y_l^m which can be derived from the values of θ and φ at each point.

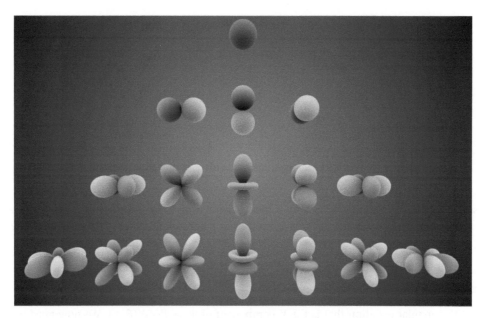

Figure 15.2   Spherical harmonics (Inigo.quilez/Wikimedia Commons/CC BY-SA 3.0).

Figure 15.2 shows the shape of the spherical harmonics. The blue portions indicate Y is positive and the yellow that it is negative. However, it is the absolute value that matters because the probability of the electron being at any location is determined by $|\Psi|^2$. Uppermost is ($l = 0$) which is spherically symmetrical, has one state ($m = 0$) and is called the **s** shell. The next level down in the figure is the **p** shell ($l = 1$) that has 3 states ($m = -1, 0, +1$). Lower down in the figure is the **d** shell ($l = 2$) with 5 states ($m = -2, -1, 0, +1, +2$) and the **f** shell ($l = 3$) with 7 states ($m = -3, -2, -1, 0, +1, +2, +3$). We discussed these distinct angular momentum states back in Box 14.5. Thanks to Monsieur Laplace, you now know their orbital shapes.[2]

15.4.4   Radial Kinetic Energy

Figure 15.3 illustrates the difference between radial and angular excitation. The figure is in only *two spatial dimensions*. Imagine that the electron wave function is a spherical cloud around the nucleus and we cut a slice across the centre. The grey circle shows a distance r away from the nucleus.

An example of *radial* excitation is shown on the left. The wave function is constrained by the attraction of the proton. The vertical height represents the numerical amplitude of the wave function at that point (it is not a physical wave in the z direction) and that value is identical at all points that are distance (r) from the proton. Switching to three dimensions, the value is the same for all points on the surface of a sphere at a distance r from the proton. The change in the value of the wave function with radius is $\frac{\partial\Psi}{\partial r}$ and determines the radial momentum (and therefore radial kinetic energy) of the electron.

I must remind you that the electron is *not* actually moving. The probability of finding it at a particular point remains stable (in the language of quantum mechanics, the *probability density function* does not change over time). In classical terms you might describe the electron as being in a superposition of moving towards and away from the nucleus.

For comparison, an example of *angular* excitation is shown on the right of Figure 15.3. The wave function is not radially symmetrical. Its value varies with the angle around the circle. If we call that angle θ then the value of $\frac{\partial\Psi}{\partial\theta}$ determines the angular momentum around that particular axis. If there is angular momentum, there is angular kinetic energy. Again, let me stress that the electron is *not* moving in the classical sense. If it were actually rotating around the proton, it would radiate energy and the state would be unstable.

There is an important difference between radial and angular excitation. The *radial* excitation *cannot* be zero. Think back to the electron in a box. For the wave function to exist, it must oscillate in the potential. The ground state is not zero.

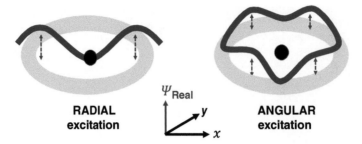

Figure 15.3 Two-dimensional illustration of radial and angular excitation.

In sharp contrast the *angular* excitation *can* be zero. If the wave function is perfectly spherically symmetrical, then $\frac{\partial \Psi}{\partial \theta}$ is zero around every axis and there is no angular momentum or angular kinetic energy. This is true of the **s** shell we just discussed.

15.4.5 The Radial Equation

We have looked at the angular equation that is the Y component of $\Psi = R(r)Y(\theta, \varphi)$. We now need to solve the radial equation that tells us about $R(r)$. We can start from Equation 15.27. We know that f in that equation is Y and so $-C = -l(l+1)$.

$$-\left(\frac{1}{R} \frac{\partial}{\partial r} \left(r^2 \frac{\partial R}{\partial r} \right) + \frac{2m_e r^2}{\hbar^2} (E - V) \right) Y = -l(l+1)Y \tag{15.31}$$

We can generate an equation that contains only radial terms by dividing both sides by Y and multiplying by $-R$ to get Equation 15.32. We need to keep track of the $\frac{1}{r}$ dependency of the Coulomb potential so I have written it as $V = -\frac{q^2}{r}$ where $q^2 = \frac{e^2}{4\pi\epsilon_0}$, simply to avoid writing all those terms again and again.

$$\frac{\partial}{\partial r} \left(r^2 \frac{\partial R}{\partial r} \right) + \frac{2m_e r^2}{\hbar^2} \left(E + \frac{q^2}{r} \right) R - l(l+1)R = 0 \qquad Note:\ q^2 = \frac{e^2}{4\pi\epsilon_0} \tag{15.32}$$

This radial equation for hydrogen is a one-dimensional challenge. That is encouraging. However, it is not trivial (as a mathematician would say with understatement). Even a brilliant physicist like Schrödinger had to get a smart mathematician to help him solve it; but would you choose someone who is sleeping with your wife (see Box 15.7)? I will run fairly quickly through the solution. If you find this too much, jump to the discussion at Section 15.5.

When you have an equation such as Equation 15.32, a good approach is to find something similar with known solutions and then massage your equation into that form. That is what we are going to do. Equation 15.33 in Box 15.6 is a variation of the *Associated Laguerre equation*. This equation has solutions if j and k (which are just numbers) are both positive *integers* or zero. Equation 15.34 shows the solutions. They are somewhat complicated involving binomials, but you can look up what you need in a table. For example, for $j = 2$ and $k = 1$, you find the binomial is $L_2^1 = 3x^2 - 18x + 18$ and with that, you can write down the solution.

Box 15.6   Our target equation: an Associated Laguerre variation

$$\frac{\partial^2 y}{\partial x^2} + \left(-\frac{1}{4} + \frac{2j + k + 1}{2x} - \frac{k^2 - 1}{4x^2} \right) y = 0 \tag{15.33}$$

which has the solutions: $y(x) = e^{-\frac{x}{2}} x^{\frac{k+1}{2}} L_j^k(x)$ \hfill (15.34)

The first step is the trickiest. We need to get rid of the rather nasty looking first term in Equation 15.32. We achieve this by substituting $R(r)$ with another function, which I will label $y(r)$, such that $y = rR$. This means that $R = \frac{y}{r}$ and so simplifies the first term as follows:

$$\frac{\partial}{\partial r} r^2 \frac{\partial R}{\partial r} = \frac{\partial}{\partial r} r^2 \frac{\partial}{\partial r} \left(\frac{y(r)}{r} \right) = \frac{\partial}{\partial r} r^2 \left(-\frac{y}{r^2} + \frac{1}{r} \frac{\partial y}{\partial r} \right)$$

$$= \frac{\partial}{\partial r} \left(-y + r \frac{\partial y}{\partial r} \right) = -\frac{\partial y}{\partial r} + \frac{\partial y}{\partial r} + r \frac{\partial^2 y}{\partial r^2} = r \frac{\partial^2 y}{\partial r^2} \tag{15.35}$$

Box 15.7   A shared endeavour: Schrödinger and Weyl

Even Schrödinger needed help to solve his own equation. He called on the assistance of an old friend, the mathematician Hermann Weyl, who lived nearby. At the time, Weyl was sleeping with Schrödinger's wife, but that seems fair enough as Schrödinger was sleeping with pretty much everyone else, and Weyl's wife was sleeping with another physicist, Paul Scherrer. Fortunately all these amorous liaisons did not discourage Schrödinger and Weyl from rolling up their sleeves and producing a solution between them.

We substitute R with y into Equation 15.32, using the result of Equation 15.35 for the first term, using $\frac{y}{r}$ for the other R terms and then dividing the lot by r to simplify. This gives Equation 15.36 which is neater, and the factors of r now match with those of x in the target Equation 15.33.

$$\frac{\partial^2 y}{\partial r^2} + \left(\frac{2m_e E}{\hbar^2} + \frac{2m_e q^2}{\hbar^2 r} - \frac{l(l+1)}{r^2} \right) y = 0 \tag{15.36}$$

To match with Equation 15.33, we need the second term of 15.36 to become $-\frac{1}{4}$. To simplify the notation, we introduce a new symbol β such that $\beta^2 = -\frac{8m_e E}{\hbar^2}$.

$$\frac{\partial^2 y}{\partial r^2} + \left(-\frac{\beta^2}{4} + \frac{2m_e q^2}{\hbar^2 r} - \frac{l(l+1)}{r^2} \right) y = 0 \qquad \text{Note: } \beta^2 = \frac{-8m_e E}{\hbar^2} \tag{15.37}$$

We then change coordinates, substituting r with what I will label x such that $x = \beta r$ and so $r = \frac{x}{\beta}$. This x has nothing to do with the x axis (I chose it simply to match the target equation). This means that the first term in Equation 15.37 becomes:

$$\frac{\partial^2 y}{\partial r^2} = \frac{\partial}{\partial r} \left(\frac{\partial y}{\partial x} \frac{\partial x}{\partial r} \right) = \frac{\partial}{\partial r} \left(\beta \frac{\partial y}{\partial x} \right) = \frac{\partial}{\partial x} \left(\beta \frac{\partial y}{\partial x} \right) \frac{\partial x}{\partial r} = \beta^2 \frac{\partial^2 y}{\partial x^2} \tag{15.38}$$

Using the answer from Equation 15.38 and substituting elsewhere in Equation 15.37 with $r = \frac{x}{\beta}$, gives Equation 15.39 which we can simplify to Equation 15.40, which is now in a form that matches the target Equation 15.33.

$$\beta^2 \frac{\partial^2 y}{\partial x^2} + \left(-\frac{\beta^2}{4} + \frac{2m_e q^2 \beta}{\hbar^2 x} - \frac{l(l+1)\beta^2}{x^2} \right) y = 0 \tag{15.39}$$

$$\frac{\partial^2 y}{\partial x^2} + \left(-\frac{1}{4} + \frac{2m_e q^2}{\hbar^2 \beta x} - \frac{l(l+1)}{x^2} \right) y = 0 \tag{15.40}$$

The hard work is done! We can now match up the radial equation for hydrogen with the known solutions of the Associated Laguerre equation. You can sit back, rest your brain cells and sip a glass of champagne while we bask in all that this tells us. Let's compare the two equations:

$$\frac{\partial^2 y}{\partial x^2} + \left(-\frac{1}{4} + \frac{2j+k+1}{2x} - \frac{k^2-1}{4x^2}\right) y = 0 \qquad \text{\textit{Associated Laguerre}} \qquad (15.41)$$

$$\frac{\partial^2 y}{\partial x^2} + \left(-\frac{1}{4} + \frac{2m_e q^2}{\hbar^2 \beta\, x} - \frac{l(l+1)}{x^2}\right) y = 0 \qquad \text{\textit{Hydrogen Radial}} \qquad (15.42)$$

Let's start with the $\frac{1}{x^2}$ term. In order for the two equations to match, $k = 2l + 1$ as shown in Equation 15.43. This is very encouraging. We know from the angular equation that l is zero or a positive integer (Laplace proved this). That means for any l, there is a value of k that is a positive integer. Therefore we can use the Associated Laguerre equation to find solutions for any value of l.

$$\frac{k^2-1}{4} = l(l+1) \quad \implies \quad k^2 = 4l^2 + 4l + 1 = (2l+1)^2 \qquad (15.43)$$

Now let's look at the second term. We know that $k = 2l + 1$ so in the Associated Laguerre equation the factor in the $\frac{1}{x}$ term becomes $l + j + 1$ as shown in Equation 15.44. For all the solutions of the Associated Laguerre equation, both k and j are zero or positive integers. We know that l is also zero or a positive integer. Taken together, this means that $l + j + 1$ must be a positive integer of one or higher. Typically this is labelled n.

$$\frac{2j+k+1}{2} = \frac{2j+2l+2}{2} = l+j+1 = n \qquad (15.44)$$

This must also be the case for the solutions to the hydrogen radial equation. The factor in the $\frac{1}{x}$ term of the radial equation must also be a positive integer n that is 1 or higher. We call this n the *principle quantum number*. We can substitute back for the values of q^2 and β to give us an expression for the energy of the solutions as shown in Equation 15.45.

$$n = \frac{2m_e q^2}{\hbar^2 \beta} = \frac{2m_e}{\hbar^2}\frac{e^2}{4\pi\epsilon_0}\left(\frac{-\hbar}{\sqrt{8m_e E}}\right) = -\frac{m_e e^2}{2\pi\hbar\epsilon_0\sqrt{8m_e E}}$$

$$8m_e E = -\frac{m_e^2 e^4}{4\pi^2\hbar^2\epsilon_0^2 n^2} \quad \implies \quad E = -\frac{1}{n^2}\frac{m_e e^4}{32\pi^2\hbar^2\epsilon_0^2} \qquad (15.45)$$

The solutions to the equation have energy levels with exactly the pattern that Rydberg had spotted (check back to Equation 15.1). This was a tremendous victory for Schrödinger and a stunning vindication of his equation.

In case you are interested, Equation 15.46 shows examples of the radial wave functions for $R_{1,0}$ which is the ground state ($n=1$, $l=0$) and $R_{2,0}$ ($n=2$, $l=0$). The symbol a is the Bohr radius which is shorthand for $a = \frac{4\pi\epsilon_0\hbar^2}{m_e e^2}$ and the symbol $\mathcal{N}$ is a normalisation factor which needs to be calculated in order that the total probability of the electron being *somewhere* is 100%.

$$R_{1,0} = \mathcal{N} a^{-\frac{3}{2}} e^{-\frac{r}{a}} \qquad\qquad R_{2,0} = \mathcal{N} a^{-\frac{3}{2}}\left(1 - \frac{r}{2a}\right) e^{-\frac{r}{2a}} \qquad (15.46)$$

15.5 Discussion

I fear I may have lost a few readers with the heavy maths in the last few pages so let me take a moment to recap in words what we have done. The aim was to find the energy levels for the electron in hydrogen. We inserted the Coulomb potential into the time-independent version of the Schrödinger equation and switched to spherical coordinates.

We assumed that the wave function Ψ can be separated into an angular component $Y(\theta, \varphi)$ and a radial component $R(r)$. The angular equation turns out to be that of the spherical harmonics Y_l^m. These define the shape of the various atomic orbitals. Solving the radial equation was tougher. We massaged it into the same form as a variation on the Associated Laguerre equation which has known solutions. For any solution, j and k in the equation both must be positive integers or zero. Applying this to the hydrogen radial equation showed that the energy levels of its solutions accurately match Rydberg's $\frac{1}{n^2}$ pattern – a famous victory for Schrödinger and his equation.

The energy level depends on the quantum number n where $n = l + j + 1$. Both l and j are any positive integer or zero. The value of l is the level of *angular* excitation which can be zero. The value of $n - l$ (which is $j + 1$) is the *radial* excitation level

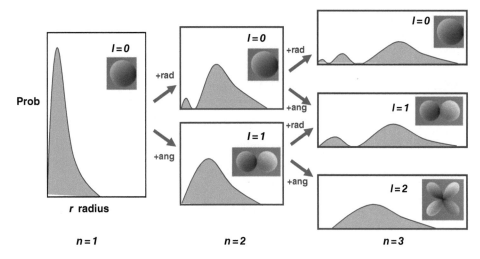

Figure 15.4 Shape of hydrogen energy levels.

which cannot be zero because there is a minimum ground state energy. Increasing either radial or angular excitation raises n. Thus, both the state $n = 2$, $l = 0$ (radial excitation 2, angular excitation 0) and the state $n = 2$, $l = 1$ (radial excitation 1, angular excitation 1) are at the same energy level $n = 2$.

Figure 15.4 illustrates some different hydrogen electron wave functions. The box on the left of the figure is the ground state, $n = 1$. The blue curve shows the relative probability of the electron being at different distances from the proton which depends on $R(r)$ and is proportional to $r^2|R|^2$. The inset picture shows the angular variation of the wave function (its shape). This depends on $Y(\theta, \varphi)$. In the ground state $l = 0$, which means the wave function is spherically symmetrical.

In this ground state, there is no angular momentum so all the energy and momentum is radial. If you run the numbers, the electron is moving at about 2,000 kilometres per second. What is happening to the electron? Using classical terminology, it is in a superposition of momentum towards the nucleus and away from it! This is yet another example of how hard it is to describe quantum effects in classical terms.

The electron can be excited to the $n = 2$ level *either* by raising its radial excitation (shown as +*rad* in the figure) or by raising its angular excitation (+*ang*). Both are shown in the centre of Figure 15.4. Raising its *radial* excitation (to $n = 2$, $l = 0$) adds a stationary zero-point node to the radial curve. Another step up in radial excitation (to $n = 3$, $l = 0$) adds a further node in just the same way as each step up in the energy level of the electron in a box adds a node (Figure 7.1 in Chapter 7). Raising its *angular* excitation (to $n = 2$, $l = 1$) alters the *angular* variation in the wave function changing the shape to that shown in the inset. Excite the electron with even more angular kinetic energy and this shape will change again (see $n = 3$, $l = 2$ in the figure).

It is somewhat surprising that raising the radial excitation and angular excitation appear to result in the same change in energy level. The term *degeneracy* is used when more than one state has the same energy so, this is called the *accidental degeneracy* of hydrogen. It is caused by a symmetry in the Schrödinger equation rather than something that is fundamental in nature. In reality there is a small difference between the energy required for radial and angular excitation, but you have to dig down to show it. The difference appears when you improve the approximation for kinetic energy to include the next term in the Taylor expansion (see Box 3.3):

$$Kinetic\ energy \approx \frac{p^2}{2m_e} - \frac{3p^4}{8m_e^3 c^2} \cdots \tag{15.47}$$

This reveals that raising n requires very slightly more energy if it is done with an increase in angular excitation rather than radial excitation. Therefore, at any given energy level n, the $l = 0$ states are slightly lower energy.

At the risk of boring you, I want to end this chapter with yet another reminder that the electron is not moving in the classical sense. These electron energy eigenstates are spatially stable. Don't forget!

In the next chapter, we will discuss how these energy levels shape the chemical behaviour of elements which, in turn, explains the structure of the periodic table.

Box 15.8   Solution to problem in Subsection 15.4.1: $Y = \sin\theta\, e^{i\varphi}$ $l = 1$ $m = 1$

$$-l(l+1)\,Y = \frac{1}{\sin\theta}\frac{\partial}{\partial\theta}\left(\sin\theta\,\frac{\partial(\sin\theta\, e^{i\varphi})}{\partial\theta}\right) + \frac{1}{\sin^2\theta}\frac{\partial^2(\sin\theta\, e^{i\varphi})}{\partial\varphi^2}$$

$$= \frac{1}{\sin\theta}\frac{\partial}{\partial\theta}(\sin\theta\,\cos\theta\, e^{i\varphi}) + \frac{1}{\sin^2\theta}(-\sin\theta\, e^{i\varphi})$$

$$= \frac{1}{\sin\theta}(\cos^2\theta - \sin^2\theta)\,e^{i\varphi} - \frac{1}{\sin\theta}\,e^{i\varphi} = \frac{1}{\sin\theta}(\cos^2\theta - 1 - \sin^2\theta)\,e^{i\varphi}$$

$$= -\frac{2\sin^2\theta}{\sin\theta}\,e^{i\varphi} = -2Y \implies l = 1$$

Rule: $\dfrac{\partial^2 Y}{\partial\varphi^2} = -m^2 Y$ $\dfrac{\partial^2(\sin\theta\, e^{i\varphi})}{\partial\varphi^2} = -\sin\theta\, e^{i\varphi} \implies m = 1$

Notes

1 Moving in either direction from the minimum takes energy so there must be opposing balanced forces at that radius.

2 Note that the *only* thing we have assumed in the angular equation is that the potential is spherically symmetrical. Therefore, the spherical harmonics are the angular solution for *any* spherically symmetrical potential, providing the angular and radial components are separable.

Chapter 16

The Periodic Table

CHAPTER MENU
16.1 Introduction 16.2 Adding More Protons 16.3 The Periodic Table 16.4 Molecular Bonds 16.5 Bonds in the Nucleus 16.6 Virtual Particles 16.7 Fusion and Fission 16.8 Module Summary 16.9 Module Memory Jogger

16.1 Introduction

We are reaching the end of the module on *Complex Quantum Mechanics*. The grand finale is to explain the structure of the periodic table. I assume you know a fair amount about the periodic table of elements. If you do not, then hang your head in shame because you must have slept through a lot of classes at school.

Box 16.1   The periodic table: it came in a dream
Dmitri Ivanovich Mendeleev (1834–1907) was the son of a Russian orthodox priest. Clearly his father took his reproductive responsibilities extremely seriously as Mendeleev was the youngest of 14 children (some say only 13 but that is still impressive). In 1865 he received his doctorate for his thesis, *On the combinations of water with alcohol* – in my opinion a very short-sighted concept versus the obvious alternative. In 1869 Mendeleev published his structure of the periodic table that brought him fame. Reputedly, he claimed to have seen it in a dream: *In a dream I saw a table where all the elements fell into place as required. Awakening, I immediately wrote it down on a piece of paper.* He predicted the existence of gallium, scandium and germanium, all of which were discovered in the following twenty years.

Mendeleev found a pattern (Box 16.1). If you lay the elements out in a specific way (the periodic table) then elements with similar chemical properties sit together neatly. We now know that the chemical properties of an element are largely determined by the number of electrons in the higher energy levels. This is called its *outer shell* and the electrons in it are called *valence* electrons. What is it that determines how many electrons are in this shell? The calculation of the energy levels of hydrogen goes a long way towards explaining the pattern.

Quantum Untangling: An Intuitive Approach to Quantum Mechanics from Einstein to Higgs, First Edition. Simon Sherwood.
© 2023 John Wiley & Sons Ltd. Published 2023 by John Wiley & Sons Ltd.

16.2 Adding More Protons

What happens to the pattern of energy levels if there are more protons in the nucleus? The Coulomb potential between a nucleus with Z protons ($+Ze$) and a single electron ($-e$) is shown in Equation 16.1 (the formula is from Equation 15.5). The number of protons in the nucleus is called the *atomic number* of the element. The factor Z flows through the calculation and changes each of the electron energy levels of the multi-proton nucleus by a factor of Z^2 versus the equivalent energy level for hydrogen as shown in Equation 16.2.

$$V = -\frac{Ze^2}{4\pi\epsilon_0 r} \qquad \textit{for a nucleus with Z protons} \qquad (16.1)$$

$$E_{hydrogen} = -\frac{1}{n^2}\frac{m_e e^4}{32\pi^2\hbar^2\epsilon_0^2} \implies E_Z = -\frac{Z^2}{n^2}\frac{m_e e^4}{32\pi^2\hbar^2\epsilon_0^2} \qquad (16.2)$$

As an example, consider the nucleus of helium. It would take four times as much energy to strip off a lone electron from a helium nucleus (two protons) than a hydrogen nucleus (one proton). Bereft of electrons, helium is called an alpha particle. It should come as no surprise that alpha particles are strongly *ionising*, meaning that they can rip electrons from many atoms. Indeed, some smoke detectors contain a source of alpha particle radiation and detect the subsequent ionisation of smoke particles.

Importantly, adding more protons to the nucleus doesn't change the angular equation (spherical harmonics) nor the *pattern* of energy levels predicted by the Schrödinger equation. These are listed below by energy level, low to high (note that at any n, the lower l-value states are very slightly lower energy).

Only two electrons (with opposite spin) can co-exist in each energy state. This is a consequence of the Pauli exclusion principle which we will discuss along with spin in the next chapter. This means two electrons in the first shell ($n = 1$) and eight electrons in the second shell ($n = 2$). The right side of the table below shows how the **s**, **p**, **d** and **f** atomic orbitals would fill for larger and larger Z based on this pattern of electron energy levels.

$n = 1$	$\implies$	$n, l, (m) = 1, 0, (0)$	*1 state, 2 electrons,* **1s** *orbital*
$n = 2$	$\implies$	$n, l, (m) = 2, 0, (0)$	*1 state, 2 electrons,* **2s** *orbital*
	$\implies$	$n, l, (m) = 2, 1, (-1, 0, +1)$	*3 states, 6 electrons,* **2p** *orbital*
$n = 3$	$\implies$	$n, l, (m) = 3, 0, (0)$	*1 state, 2 electrons,* **3s** *orbital*
	$\implies$	$n, l, (m) = 3, 1, (-1, 0, +1)$	*3 states, 6 electrons,* **3p** *orbital*
	$\implies$	$n, l, (m) = 3, 2, (-2, -1, 0, +1, +2)$	*5 states, 10 electrons,* **3d** *orbital*
$n = 4$	$\implies$	$n, l, (m) = 4, 0, (0)$	*1 state, 2 electrons,* **4s** *orbital*
	$\implies$	$n, l, (m) = 4, 1, (-1, 0, +1)$	*3 states, 6 electrons,* **4p** *orbital*
	$\implies$	$n, l, (m) = 4, 2, (-2, -1, 0, +1, +2)$	*5 states, 10 electrons,* **4d** *orbital*
	$\implies$	$n, l, (m) = 4, 3, (-3 \text{ to } + 3)$	*7 states, 14 electrons,* **4f** *orbital*

As an example, silicon (Si, atomic number: $Z = 14$) would have its 14 electrons distributed with two in ($n = 1$), eight in ($n = 2$) and four in ($n = 3$). This means that silicon has four electrons in its outer shell (*valence* electrons). The number of valence electrons is the key to explaining the pattern that Mendeleev discovered.

16.3 The Periodic Table

Figure 16.1 shows the periodic table. The elements are ordered by atomic number (the number of protons in the nucleus). The right side shows how the shells actually fill as we consider progressively larger atoms. Up to the third shell (**3s**, **3p**), the order matches the energy levels listed in Section 16.2. This includes silicon which we have just discussed. However, things get a bit more complicated when we deal with heavier elements because the electrons filling the lower energy levels

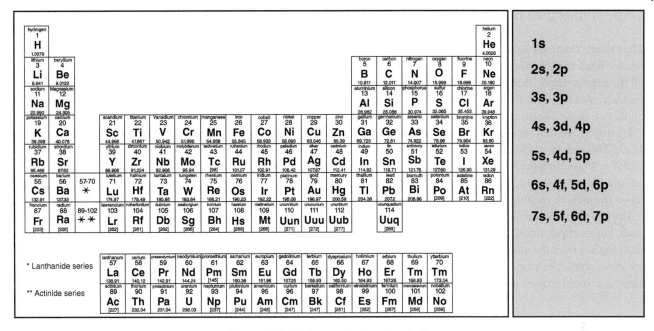

Figure 16.1 Periodic table showing shells.

affect the relative energy of the higher energy levels. This alters the pattern. For example, after filling up the **3p** orbitals, it requires less energy to fill the **4s** orbital than the **3d** orbitals. As a result, once the **3p** orbitals are full, the fourth shell starts to fill (**4s** followed by **3d** and then **4p**).

Generally the electron configuration of an element is shown either as a straight list of the number of electrons in each orbital, or it is summarised to show only the valence electrons. For example, the electron configuration of silicon (Si) is $1s^2 2s^2 2p^6 3s^2 3p^2$ or, in summary form, $[Ne]\, 3s^2 3p^2$ where the inner shells are shown as [Ne] because they have the same structure as the noble gas neon. To give you a couple more examples, nitrogen (N) is $[He]\, 2s^2 2p^3$ where [He] stands for the electron configuration of helium while potassium (K) is $[Ar]\, 4s^1$ where [Ar] stands for the electron configuration of argon.[1]

When there are multiple available states such as within the **p** orbitals, one electron will go into each available state. This is called the Hund rule. This gives the maximum spatial separation between electrons, as you might expect based on their mutual repulsion. I should mention that the lone electrons in these orbitals have the same spin. We distinguish the different possible spin states of an electron by labelling one up (↑) and the other down (↓) as we will discuss in detail in the next chapter. Thus, in the case of nitrogen (N), its three **2p** electrons will fill the three available states [↑][↑][↑]. If there is a fourth electron such as with oxygen (O), the configuration across the three available states of **2p** is [↑↓][↑][↑].

I want to pause a moment to address one common misconception. A novice might expect that the *size* of the atom increases as the atomic number increases. If you think carefully about the analysis of the last chapter, you should be suspicious about this – and you would be right.

Consider the left-most column in the periodic table. As you read *down* the table the atoms are indeed progressively *larger*. The single valence electron of the first element, hydrogen, is in **1s**. That of the next, lithium, is in **2s**. The next is in **3s** and so forth. The atomic radius of hydrogen (H) is 37 picometres (pm) which is 37×10^{-12} metres, lithium (Li) is 152pm, sodium (Na) is 227pm and potassium (K) is 280pm.

However, when you read *across* the table, the atoms get *smaller*. The nucleus has progressively more protons, but the additional valence electrons are accommodated without a step-up to a higher energy shell. As the number of protons increases, the electron shells are held more and more tightly in to the nucleus. Reading across, the atomic radius of lithium (Li) is 152pm, beryllium (Be) is 112pm, boron (B) is 85pm, carbon (C) is 70pm, nitrogen (N) is 65pm, oxygen (O) is 60pm, fluorine (F) is 42pm and finally neon (Ne) is 38pm. Neon has 10 electrons but an atomic radius barely larger than that of hydrogen.

16.4 Molecular Bonds

The valence electrons (those in the outer shell) are the ones that are involved in molecular bonds. You can get a quantitative feel for what is happening by looking at the *ionisation* energies of elements. This is the energy required to strip electrons off from the nucleus. Table 16.1 shows a summary for the first 11 elements in the periodic table. The outermost electron of each element is shown furthest to the right in the table. For example, for carbon it requires 11 eV (electron-volts) to strip off its outermost electron, 24 eV to strip off another, 48 eV to strip off a third... and so forth.

Hydrogen has only one electron which sits in the **1s** orbital. The energy required to remove this electron is just under 14 eV. This can be calculated from the solution for the Coulomb potential using the Schrödinger equation setting $n = 1$ and $Z = 1$ (see Equation 16.2). As you can see from Equation 16.2, the energy level depends on Z squared. Therefore, for helium ($Z = 2$) it is four times greater and takes 54 eV to strip the last innermost electron from the nucleus, and so forth down the table to sodium ($Z = 11$) which is 121 times greater than for hydrogen.

The key point to note in Table 16.1 is the change in ionisation energy of an element when its electrons start to fill a new shell. For example, lithium ([He] **2s**^{1}) has only one valence electron which sits in the **2s** orbital. It takes only 5 eV to strip this electron from the atom. To strip another electron from lithium would take 76 eV because the second electron is not in the outer shell. The story is similar for sodium ([Ne] **3s**^{1}). The outermost electron is easily ionised, but removing further electrons would require much more energy.

16.4.1 Ionic Bonds

The valence electron of lithium and sodium is bound so weakly that it will happily go walk-about and jump ship. These two elements are in the left-most column of the periodic table, called *Group 1* (see Figure 16.1). They readily lose the valence electron to form the positive ions Li$^+$ and Na$^+$. In a similar fashion, those elements in the second column, *Group 2*, can lose both their valence electrons to form ions such as Be^{2+} and Mg^{2+}.

Towards the far right side of the periodic table there are some groups of elements which readily accept the released electrons. Fluorine and chlorine are in *Group 7*, the halogens. These each take one electron to fill their **p** orbital, forming F$^-$ and Cl$^-$ ions. Achieving the same for *Group 6* elements such as oxygen and sulphur forms O^{2-} and S^{2-} ions.

The electrostatic attraction between the resulting positive and negative ions draws them together in molecules with what are called *ionic* bonds. For example, if you mix sodium ([Ne] **3s**^{1}) and chlorine ([Ne] **3s**^{2}, **3p**^{5}), the valence electron of sodium will migrate to the chlorine atom to form Na$^+$ and Cl$^-$ ions which, in turn, form the molecule NaCl. This is of course the familiar salt you will have had on your fish and chips, if you are a Brit. Another example is the mix of magnesium and chlorine. In this case the ions are Mg^{2+} and Cl$^-$ and bond 1:2 which forms the molecule MgCl$_2$. This is the magnesium chloride some of you may take as a dietary supplement.

The configuration of the valence electrons affects many aspects of the behaviour of an element such as its reactivity, its pattern of molecular bonding and its conductivity (which depends on the mobility of the valence electrons). This should

Table 16.1 Ionisation energy to nearest electron-volt.

Electron	1s^1	1s^2	2s^1	2s^2	2p^1	2p^2	2p^3	2p^4	2p^5	2p^6	3s^1
Hydrogen (H)	14										
Helium (He)	54	25									
Lithium (Li)	122	76	5								
Beryllium (Be)	218	154	18	9							
Boron (B)	340	259	38	25	8						
Carbon (C)	500	392	64	48	24	11					
Nitrogen (N)	667	552	98	77	47	30	15				
Oxygen (O)	871	739	138	114	77	55	35	14			
Fluorine (F)	1103	954	185	157	114	87	63	35	17		
Neon (Ne)	1362	1196	239	207	158	126	97	63	41	22	
Sodium (Na)	1649	1465	300	264	209	172	138	99	72	47	5

all be boringly familiar to you from your high school chemistry lessons. The outcome is that elements with a similar configuration of valence electrons react in similar ways. This explains the patterns that Mendeleev discovered in the periodic table (for more on Mendeleev, see Box 16.2).

Box 16.2   The other side of Mendeleev

Dmitri Ivanovich Mendeleev (1834–1907), creator of the periodic table, was married to Feozva Nikitichna Leshcheva, 6 years his senior, in 1862. About 18 years into the marriage, at the age of 46, he became obsessed with his niece's best friend, Anna Ivanova Popova, a 19 year old some 32 years younger than his wife. He proposed to Anna threatening suicide if she refused to marry him.

Mendeleev and Anna had a daughter Lyubov, born in December 1881. They married in April 1882 in spite of opposition from the Russian Orthodox Church. His second marriage caused scandal and uproar in Russia. He was 48 years old. Anna was 22. Some sources say it was bigamous as his divorce to Feozva was not finalised for another month.

16.4.2   Covalent Bonds

Your chemistry teacher also will have told you about *covalent* bonds. The term was coined by Gilbert Lewis (flick back to Box 4.7 if you are interested in his tragic story). A covalent bond typically is described as occurring when a molecule is formed by atoms *sharing* pairs of electrons between them. All things considered, this is quite a fair description.

Covalent bonding will be very familiar to biology students or anyone who has studied organic chemistry. Carbon ($\mathbf{1s^2 2s^2 2p^2}$) has four valence electrons in its ($n = 2$) outer shell. It would require about 150 eV to strip all four electrons (see Table 16.1) so it almost never forms free-standing C^{4+} ions. Instead it can share each of its four valence electrons with valence electrons from other atoms. For example they can pair up with the lone valence electrons of four hydrogen atoms to create CH_4 which is methane. This pairing gives each hydrogen atom an extra electron filling its $n = 1$ shell, and gives the carbon four extra electrons filling its $n = 2$ shell. Importantly carbon also bonds with other carbon atoms. This creates the backbone to the larger complex molecules that are essential to life.

If you want to experience directly a few simple examples of covalent bonding, just take a deep breath. The N_2, O_2 and CO_2 in the air around you all involve covalent bonds. One of the simplest examples of covalent bonding is the hydrogen molecule H_2. Each hydrogen atom consists of a proton and electron. The atoms merge together as shown schematically on the left side of Figure 16.2. Each black dot represents a proton initially surrounded by a larger blue circle which represents its electron in the spherical **1s** orbital. The result of the merger is an H_2 molecule.

This raises an obvious question. Each of the hydrogen atoms is electrically neutral so what is it that attracts them to each other? *Why* do they approach each other? *Why* do they bond?

Let's start with a simple and intuitive explanation of what is happening in terms of energy. I have labelled the two protons A and B as shown on the right side of Figure 16.2. Each has an electron. The figure shows a schematic of their wave functions. Initially, electron 1 surrounds proton A and electron 2 surrounds proton B. If the protons approach each other the two electrons can form a shared state as shown on the far right of the figure (they can share the state providing they have opposite spins). The result is that each electron is less spatially constrained. The lower the constraint, the less the energy of each electron. As a result, the protons are drawn together until this energy reduction is offset by the energy of the protons' mutual repulsion.

That is a reasonable description, but let's dig a bit deeper to get a slightly different perspective on how the interaction occurs. Think about the scenario labelled *Separate* in the figure. Electron 1 surrounds proton A and electron 2 surrounds

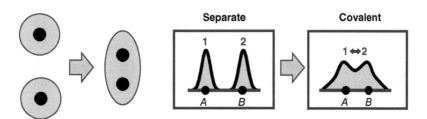

Figure 16.2   Covalent bonding: Feynman's flip-flop of electrons in a shared state.

proton *B but* remember *quantum tunnelling* (see Figure 12.3 if you need a reminder). There is always a small probability that electron 1 will be around proton *B* and electron 2 around proton *A*. Don't forget that the two electrons are distinguishable because they must have different spin or it would be impossible for either to enter the state of the other. As the protons get closer, the chance grows of 1 and 2 flipping. The easing of the spatial constraint on each electron is a result of their ability to flip between protons $A \leftrightarrow B$. Therefore, we can describe the attraction as being *mediated* by the constant interchange of the electrons.

You can think of ionic bonds in a similar way except that the electrons are not shared equally between the atoms, which results in the atoms having net ionic charges. In the words of the famous Richard Feynman, *chemical binding usually involves this flip-flop game played by two electrons.* Later in this book, we will discuss quantum field theory and Feynman diagrams which quantify all attraction and repulsion in terms of particle exchange. Some students find it challenging to visualise how particle exchange can lead to an *attractive* force. Hopefully, the example of covalent bonding helps.

16.5 Bonds in the Nucleus

So far, we have been discussing the electromagnetic binding of the negatively charged electrons to the positively charged atomic nucleus. As shown in Table 16.1, this is about 5 eV for the outermost electron of lithium. What about the binding energy holding the nucleus together? For lithium, the average binding energy of the protons and neutrons (collectively called *nucleons*) in the nucleus is about 5 *million* eV.

The attraction between nucleons is a result of the *strong* force. This is covered in detail in Chapter 23. I must thank my friend, Gareth Williams, for his recommendation that I introduce it now. There are two good reasons to do so. The first is that the force between nucleons works in a similar way to the covalent bonding we have just discussed. The second is because some of you may take a while to reach Chapter 23 – so I'd better give you the basics now.

Each nucleon is made up of three *quarks*. The proton is two *up quarks* (charge $+\frac{2}{3}$) and one *down quark* (charge $-\frac{1}{3}$) giving it a net electric charge of +1. The neutron is two down quarks and one up quark so it has no net electric charge. The quarks are held together by the *strong* force which, as its name suggests is very strong. While electromagnetic charge is simply positive or negative, the charge of the strong force comes in three types (red, blue and green). The attraction between these three so-called *colour* charges is so powerful that you *never* find a standalone object with net colour charge. Hence, a nucleon always consists of one red, one blue and one green quark. Each proton and neutron is neutral overall in terms of its colour charge.

This leads us to exactly the same question that arose in the discussion on covalent bonding. If the protons and neutrons in the nucleus have no net colour charge, then why do they attract each other and bond? The answer is similar.

The quarks in a nucleon are very tightly spatially constrained. This means that they have high momentum and energy. Tunnelling into a nearby nucleon would ease this constraint and result in a lower energy configuration. This is exactly analogous to the electron tunnelling from atom to atom in a covalent bond, but in the case of quarks, there is a major hitch. No quark can exist as a standalone entity. Its wave function would have a net colour charge. The strong force is too powerful to allow this. However, there are short-lived particles called *mesons*. Each meson is the combination of a quark and an antiquark. We will cover the physics of antiparticles later. For now, the important thing is that the antiquark carries the *negative* of the quark's colour charge. If the meson's quark is blue, its antiquark is anti-blue. This means that mesons are colour neutral. The meson *can* tunnel across between nucleons.

The result is that the protons and neutrons in the nucleus attract and bond. The attraction is mediated by the ability of mesons to flip-flop between nucleons which eases the spatial constraints imposed by the strong force. This is analogous with how a covalent bond is mediated by the ability of electrons to flip-flop between atoms which eases the spatial constraints imposed by the electromagnetic force.

16.6 Virtual Particles

One word of warning. I hedged my words when discussing the interchange of electrons in covalent bonds and mesons in nuclear bonds. I said the attraction is due to their *ability* to flip-flop. They do not *physically* jump to and fro in the normal sense of the word.

Think back to the Double-slit experiment (Section 4.4) which I have referred to many times. There are two pathways available for the particles to travel to the screen. This leads to an interference pattern that persists even if you send particles through one at a time. You could say this is due to the *ability* of each particle to travel along two different pathways, but obviously the particle does not *physically* travel through both slits.

You need to think the same way about the interchange of particles in interactions. For example, how might the scenario develop in a bond between nucleons? One possibility is that a meson flips over from say nucleon *A* to nucleon *B*. Alternatively, a meson might flop from *B* to *A*. I like to call these different *development pathways*. Every legitimate pathway will affect the outcome, but they do not *physically* happen.

So what is the outcome? We touched on this when we discussed the Feynman path integral and particle interactions back in Section 6.8. The outcome depends on the sum of all the development pathways. The meson might flip. It might flop. All the possibilities matter.

What if you were somehow able to devise an experiment that could detect if the meson actually moves from nucleon *A* to *B*? In doing so, you would change the whole interaction. It is analogous in the Double-slit experiment to detecting which slit the particle goes through. The detector changes the available pathways. It destroys the superposition and the interference pattern disappears.

We call mediating particles such as these mesons *virtual particles*. If you want to understand interactions and bonds, you have to take them into account. However, be careful not to think of these virtual particles as physically bouncing to and fro. We will discuss virtual particles in much more detail in later chapters.

16.7 Fusion and Fission

Let's return to the subject of bonding in the nucleus. Figure 16.3 shows the binding energy per nucleon compared with the total number of nucleons in the nucleus (each MeV is a million electron-volts). At the bottom left, with binding energy zero, is the most abundant isotope[2] of hydrogen which has only one nucleon, a proton. Introducing further nucleons rapidly increases the binding energy per nucleon, reaching a maximum at iron (26 protons, 30 neutrons).

The aim of nuclear *fusion* is to harness the energy released by this tighter binding, typically by fusing hydrogen to helium as occurs in the sun. The attraction of the strong force between nucleons is extremely short range. Neutrons are stable only when they are alongside protons[3] so, for fusion between nucleons, protons must be very close to each other (see the discussion on this in Box 12.1). The challenge is to overcome the electrostatic repulsion between protons and squeeze them close enough together for fusion. I proudly note that, at the time of writing, my eldest daughter is in the design team for a tokamak fusion reactor. Fingers crossed that they make progress. The result would be an almost limitless source of energy.

As you can see in Figure 16.3, the binding energy starts to decrease when the size of the nucleus exceeds that of iron. This includes many familiar elements such as copper, zinc, silver, platinum and gold. If you were to try to make these elements by fusing together smaller nuclei, it would *require* energy. Where do they come from? They were forged by the tremendous energy of supernova eruptions. So, the next time you see someone wearing a wedding ring, you can joke that it all started with a massive explosion!

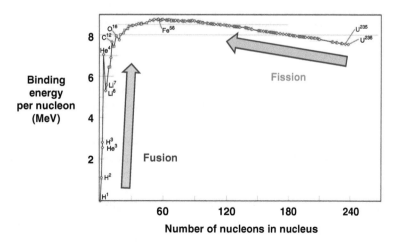

Figure 16.3 Binding energy in the nucleus.

Beyond lead (82 protons, 126 neutrons), all the elements are unstable. Nuclear *fission* takes advantage of the instability of very large nuclei which release energy when they decay. Most commonly, uranium U^{235} (92 protons) is bombarded with neutrons. This starts a process that leads it to decay into barium (56 protons), krypton (36 protons) and *three more neutrons*. Those neutrons destabilise more of the uranium leading to a chain reaction. The job of the nuclear fission reactor is to control the reaction (hopefully!) and to generate electricity from the energy released. For some history related to fission and chain reactions, see Box 16.3.

Box 16.3   Leo Szilard: the ideas man

Leo Szilard (1898–1964) was a brilliant Hungarian physicist with a penchant for new ideas. He held the first patents for the linear accelerator, the cyclotron and the electron microscope although he didn't develop practical models of them and thus missed out on a Nobel Prize. In 1933, he had what was perhaps his biggest idea. He conceived the idea that stimulating uranium with neutrons might result in a fission reaction producing more neutrons, leading to a *neutron-induced nuclear chain reaction* – the key concept in a nuclear fission reactor.

It was the same year that Adolf Hitler became Chancellor. Szilard had fled Germany a year earlier and was aware that his idea might be used for a highly destructive weapon so, he patented it and assigned it to the British Admiralty for secrecy. In 1938, he moved to New York where, one evening, his experiments proved the chain reaction worked. He realised the race to build an atomic bomb was under way. In his words: *That night, I knew the world was headed for trouble.*

A year later, with the help of Einstein, he convinced President Roosevelt of its importance and the Manhattan project was born. As the project progressed, Szilard worried about the destructive power of an atomic bomb and campaigned against its use in Japan – unsuccessfully.

16.8   Module Summary

Congratulations. You have reached the end of this module on *Complex Quantum Mechanics*. We have covered a lot of ground. Chapter 9 gave an overview of how and why we use complex vector notation. It allows us to track the phase of the wave function and eliminates special points with peaks and troughs that conflict with the requirements of special relativity. In Chapter 10 we studied how Fourier transforms take us from spacetime coordinates to energy-momentum coordinates and what this means for the quantum footprint.

In Chapters 11 and 12, we worked with the Schrödinger equation including the time-independent version which is so helpful in identifying energy eigenstates. These are stationary states in the sense that they are spatially stable. We looked at quantum tunnelling into classically forbidden zones, the different way that quantum particles respond to barriers and how localised wave packets develop over time.

In Chapter 13 we covered the Quantum Harmonic Oscillator and its ladder of energy levels with evenly spaced $\hbar\omega$ energy steps. We discussed the QHO creation and annihilation operators, cunning tricks that allow us to jump between energy eigenstates. Don't forget that the QHO is important because it tells us a lot about the behaviour of all quantum systems near equilibrium.

Chapter 14 covered angular momentum. The first step was to show why you would expect angular momentum (l) to be quantised in units of $\hbar$. We then looked at the uncertainty relationship in the angular momentum around different axes with the result that total angular kinetic energy is proportional to $l(l+1)\hbar^2$ but is not affected by how the angular momentum is distributed (which is indicated using quantum number m). All this helps explain the different atomic electron orbitals: **s**, **p**, **d** and **f**.

That brings us to Chapter 15 on the Coulomb potential. I used a primitive model to give comfort that the $\frac{1}{n^2}$ pattern of energy levels is not too weird. We dug into the Schrödinger equation with plenty of maths (perhaps too much for some). This led to the angular equation and the spherical harmonics that define the shape of the atomic orbitals. Then it was on to the radial equation to give the full explanation for the $\frac{1}{n^2}$ pattern of the atomic energy levels of hydrogen.

In this chapter, we linked the results to the periodic table showing how the difference in the binding energies of electrons leads to molecular bonding. Also, we discussed how quantum tunnelling can lead to attraction between neutrally charged particles. In the case of covalent bonds this is mediated by what Feynman described as the flip-flop interchange of electrons between atoms. In the case of nuclear bonding, it is the interchange of mesons between neighbouring nucleons.

To finish this module, I give you Box 16.4 which jokingly summarises some of the weird stuff of quantum mechanics. Enjoy!

Box 16.4   Just for laughs: Schrödinger and Heisenberg are in a car –

Heisenberg is driving like a maniac because they are late for a conference. A police officer pulls them over and asks: *Do you know how fast you were going?*

Heisenberg replies: *No, but we know exactly where we are.*

The officer looks confused and says: *You were going 150 kilometres per hour.*

Heisenberg throws up his arms and cries: *Great! Now, we're lost!*

The officer asks Schrödinger if there is anything in the trunk of the car.

Schrödinger answers: *A cat*

The officer opens the trunk and yells: *This cat is dead.*

Schrödinger angrily replies: *Well, it is now!*

Now it is time to move on to the module on *Relativistic Quantum Mechanics* with chapters on spin, the Dirac equation, quantum field theory, gauge invariance, the strong force, the weak force and the Higgs mechanism, together called the *Standard Model*. Our starting point in the next chapter will be to address an important question which remains unanswered. Why does each distinct state that we have studied (energy eigenstate) fill with *two* electrons? This involves the Pauli exclusion principle and spin.

I would describe spin and the associated Dirac equation as *next level* quantum mechanics. Spin involves a change in our understanding of geometry. The Dirac equation involves 4x4 matrices. If that is too much for you, do not despair. Many introductory texts say little about spin and most do not cover the Dirac equation. If you decide to drop out at this stage, you do so with honour, but if you want to continue, then gird your loins, prepare for your head to spin, and let's move on.

16.9   Module Memory Jogger

There is a memory jogger below and some suggestions for further study in Box 16.5. Note that all the formulas include only one spatial dimension for simplicity:

- *Complex notation for wave function of a free particle:* $e^{\frac{i}{\hbar}(\mathbf{p}x - \mathbf{E}t)}$
- *Schrödinger's equation:* $E = \frac{p^2}{2m} + V \implies i\hbar \frac{\partial \phi}{\partial t} = -\frac{\hbar^2}{2m} \frac{\partial^2 \phi}{\partial x^2} + V\phi$
- *Time-independent version* $E\phi_E(x) = -\frac{\hbar^2}{2m} \frac{\partial^2 \phi_E(x)}{\partial x^2} + V\phi_E(x)$
- *Energy eigenstates are stationary (standing waves)*
- *Quantum tunnelling, even if classically disallowed (wave function decays exponentially)*
- *QHO energy eigenstates separated by $\hbar\omega$:* $E = \left(n + \frac{1}{2}\right)\hbar\omega$
- *Angular momentum quantised in units of $\hbar$*
- *Angular eigenstate quantum numbers l (zero or any positive integer) and m ($-l$ to $+l$)*
- *Angular momentum is proportional to l, angular kinetic energy is proportional to $l(l + 1)$*
- *The Coulomb potential:* $V = -\frac{Z^2 e^2}{4\pi\epsilon_0 r}$
- *Angular solution for any spherically symmetrical potential is the spherical harmonics*
- *Radial solution for Coulomb potential gives $\frac{1}{n^2}$ pattern:* $E_Z = -\frac{Z^2}{n^2} \frac{m_e e^4}{32\pi^2\hbar^2\epsilon_0^2}$
- *Periodic table: configuration of valence electrons is key to chemical behaviour*
- *Covalent bonds: attraction mediated by ability of electrons to flip-flop between atoms*
- *Similar mechanism for bonding between nucleons, mediated by meson interchange*
- *Bonding in the nucleus as an energy source: fission and (potentially) fusion*

Box 16.5   Other Resources: Complex Quantum Mechanics

Massachusetts Institute of Technology (MIT) has a fabulous free online course. Allan Adams has boundless energy and brings a spark of fun to the subject. By the way, do not be put off by lecture 8 – things suddenly jump to a bedazzling level of maths, but it is just for that one lecture, so skip it.

https://ocw.mit.edu/courses/physics/8-04-quantum-physics-i-spring-2013/lecture-videos/ (as at December 2022).

Stanford has some excellent free lectures by Leonard Susskind including a series on quantum mechanics in his Theoretical Minimum course. The pace is less demanding than that of the MIT course (slowed by Professor Susskind's love of biscuits) – but the teaching is superb.

https://theoreticalminimum.com/courses (as at December 2022).

Notes

1 There are some small anomalies to the order in which shells fill. For example chromium (Cr) is [Ar] **4s^{1}3d^5** because it is more energetically favourable to half-fill the **3d** states ahead of filling the **4s** states.

2 Deuterium H^2 (one proton, one neutron) is stable, but H^1 is over 99.98% of the natural abundance.

3 Protons are slightly less massive than neutrons. Standalone neutrons would decay into protons in about 10 minutes.

Module IV

Relativistic Quantum Mechanics

Chapter 17

Spin

CHAPTER MENU

17.1 Intrinsic Angular Momentum: Spin
17.2 Spin-half Particles and the Pauli Exclusion Principle
17.3 Integer-spin: The Photon
17.4 Bell's Inequality and the Aspect Experiment
17.5 Summary

17.1 Intrinsic Angular Momentum: Spin

I am going to start with a confession. Most people find the *concept* of intrinsic spin weird, but I don't. It has bizarre consequences – just wait until we get to spin-half particles! But, let me repeat, I do not find the underlying *concept* of intrinsic spin weird, at least not much weirder than the quantum version of angular momentum that we dealt with in the last few chapters, or indeed the quantum version of linear momentum that we have dealt with through most of this book.

Let's start by comparing the angular momentum of the electron with the classical angular momentum of the earth going round the sun. I should warn you that this sort of comparison is dangerous if taken too far. Indeed it is considered a heinous crime by many, but it is important to address it if only to highlight the problems; so, having allowed a moment for physicists around the world to choke on their gin and tonics, let's move on.

The total angular momentum of the earth's movement has two components that must be summed: the first is the *orbital* angular momentum from the orbit of the earth around the sun, and the second is the *spin* angular momentum from the earth spinning on its axis (I am ignoring the moon). To calculate the angular momentum and angular kinetic energy of the earth you need to sum both the orbital and spin components of angular momentum.

It turns out that there are also two components to the angular momentum of elementary particles such as the electron. In addition to the angular momentum L that we have discussed in earlier chapters, the electron has $\frac{\hbar}{2}$ of intrinsic angular momentum that is labelled S. The overall angular momentum of the electron is the sum of the two and is labelled J so, $J = L + S$. When the intrinsic component S was discovered, physicists leapt to a comparison with the classical orbital model and called it *electron spin*, but this comparison fails on all levels. From a purely physical perspective, if the electron is a point particle then what exactly is spinning? How can something with zero radius have angular momentum? On the other hand, if you assume the electron does have a finite radius, then calculations show that, if the electron were really spinning classically, it would have to do so at faster than light speed to generate the required angular momentum.

The comparison is no better from the perspective of quantum mechanics and the wave function. In the classical orbital model of the earth, we add the orbital angular momentum around the sun (let's call the angle around that axis θ) plus the spin angular momentum around the earth's axis (let's call that Φ). In contrast, the electron wave function in the atom is *centred* on the nucleus. If you need to, check back to the spherical harmonics in Figure 15.2 and the excitation diagram in Figure 15.3. You cannot distinguish a spin axis that is independent of the orbital axis. The idea of θ versus Φ makes no sense.

Quantum Untangling: An Intuitive Approach to Quantum Mechanics from Einstein to Higgs, First Edition. Simon Sherwood.
© 2023 John Wiley & Sons Ltd. Published 2023 by John Wiley & Sons Ltd.

Instead of worrying about classical preconceptions, we must embrace the quantum definition of angular momentum as the angular *variation* in the wave function. S has no more to do with the electron spinning than L has with the electron orbiting the nucleus. What is called *intrinsic spin* is an underlying angular variation in the wave function that exists for almost all elementary particles. In Figure 15.3, we compared angular excitation with radial excitation for the electron in hydrogen. Using the Schrödinger equation we derived its ground state ($n = 1$, $l = 0$, $m = 0$) as having radial excitation but no angular excitation. This is not the full story. The electron's intrinsic spin means that in the ground state it actually has $\frac{\hbar}{2}$ of angular momentum.

The angular momentum can be clockwise or anticlockwise around the axis so there are two distinct ground states that we label ($n = 1$, $l = 0$, $m = 0$, $s = +\frac{1}{2}$) and ($n = 1$, $l = 0$, $m = 0$, $s = -\frac{1}{2}$). We distinguish between the electron spin states using the names *spin-up* and *spin-down*. This means that for each energy-momentum eigenstate of hydrogen (listed in the table in Section 16.2) there are two distinct electron states, one spin-up and one spin-down. Hence each one can accommodate two electrons.

Why does the electron have this intrinsic angular momentum S? You will learn in the next chapter that Schrödinger's equation is a good approximation, but not quite right for the electron. The mass-energy-momentum balance that sits behind the correct equation (the Dirac equation) is subtly different and *requires* that this intrinsic angular momentum S exist. It is perhaps unfortunate that physicists called it spin which is misleading, but the name has stuck.

17.2 Spin-half Particles and the Pauli Exclusion Principle

It is time to dig into the weirder side of spin. Yes, there is a weird side as promised. I would love to start by showing you Dirac's equation and how spin-half particles fall out of it. This would be the most elegant approach, but there is a lot of maths in the Dirac equation so perhaps it is better to start with the answer, the punch line, and address the maths of the Dirac equation later.

17.2.1 The Stern-Gerlach Experiment

One of the most famous experiments in quantum physics was the Stern-Gerlach experiment in 1922 that demonstrated electron spin. Devised by Stern and conducted by Gerlach (see Box 17.1 for some history), it involved projecting silver atoms through an inhomogeneous magnetic field. Why silver? Because silver has one lonely electron sitting in its outer **5s** shell. The intrinsic angular momentum of this outer electron creates a magnetic moment that deflects the silver atom in the magnetic field. The result of the experiment is that the beam of silver atoms splits into two depending on the direction of the electron's intrinsic angular momentum (its spin) relative to the magnetic field (see Figure 17.1).

Think about this experiment from a classical perspective. If the electron had some sort of spin, you would expect it to align at random with the experimental apparatus. In some cases the outer electron of the silver atom would have intrinsic angular momentum anticlockwise or clockwise around the z axis. In others it would be around the y axis and in others around the x axis. But no, the result shows only two results for the intrinsic spin angular momentum: either $+\frac{\hbar}{2}$ or $-\frac{\hbar}{2}$.

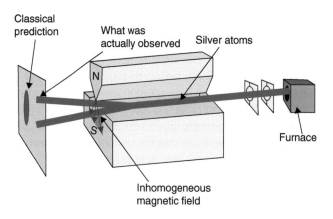

Figure 17.1 The Stern-Gerlach Experiment.

Zero is not an option, nor is any other value. If an elementary particle is spin-half, then its intrinsic angular momentum is *always* $\pm \frac{\hbar}{2}$ whatever the axis you measure around.

Box 17.1 Stern-Gerlach: diverging paths

Otto Stern (1888–1969), a Jew, fled Germany in 1933 after the Nazis seized power, moving first to Pittsburgh and then on to California where he spent the rest of his life. Walter Gerlach (1889–1979) followed a different path, working with the Nazis and becoming head of the nuclear physics section of the Reich Research Council.

Gerlach was arrested after the war and interned with nine other German physicists at a farm near Cambridge for six months. Called *Operation Epsilon*, the farmhouse was bugged in an effort to eavesdrop and determine how close Germany had been to building an atomic bomb. Transcripts show Gerlach was dismayed at having failed Hitler. Among the other physicists were Heisenberg and Diebner, who were recorded having the following conversation:

Diebner: *I wonder whether there are microphones installed here?*

Heisenberg (laughing): *Microphones installed? Oh no, they're not as cute as all that. I don't think they know the real Gestapo methods; they're a bit old fashioned in that respect.*

...which goes to show that brilliant physicists are not always street smart.

17.2.2 Spin-half and Spinors

You might be forgiven if you wonder why there is a whole chapter in this book about spin. Why does it matter that the electron has this intrinsic spin? Why all this fuss over an itsy bitsy $\frac{\hbar}{2}$ of angular momentum in the electron ground state? Well, it changes everything – make yourself a nice cup of tea, prepare your brain for fun, things are going to get freaky.

In Chapter 14 on angular momentum, we showed with creation and annihilation operators that angular momentum excitation comes in steps of $\hbar$ (check back to Section 14.6 if you need a reminder). Therefore, if we excite the electron, its total angular momentum must rise from $\frac{\hbar}{2}$ to $\frac{3\hbar}{2}$ to $\frac{5\hbar}{2}$ and so on. Putting this another way, the total angular momentum of the electron J is the sum of the intrinsic angular momentum $S = \frac{\hbar}{2}$ plus the angular excitation L, which comes in units of $\hbar$ so, $J = (N + \frac{1}{2})\hbar$ where N is an integer (I am using capital N to avoid any confusion with the hydrogen energy levels). The important point to note is that, however much angular excitation is added, the total angular momentum J will *not* be an integer multiple of $\hbar$. There is always the extra $\frac{\hbar}{2}$ component from the intrinsic spin of the electron in its ground state.

But in that same chapter, I dismissed the possibility that total angular momentum is a non-integer multiple of $\hbar$ as *ridiculous*, and with very good reason. So let's step back through the argument we covered in Section 14.2 and review the logic. The key point is shown again in Equation 17.1 separating out the angular dependence of Ψ on θ from other variables such as t, and showing total angular momentum as J. Given the derivative relationship with angular momentum, the wave function must contain a component that is structured with respect to J and θ as shown on the right of Equation 17.1. How should the value of $\Psi(\theta)$ compare with that of $\Psi(\theta + 2\pi)$? If we travel 2π (360 degrees) around the wave function then we will arrive back at the same spot so surely the value of the wave function must be the same.

Adding 2π to θ is the equivalent of multiplying Ψ by $e^{\frac{i}{\hbar}J2\pi}$ as shown in Equation 17.2. For this not to change Ψ it must be the same as multiplying by 1. For this to be true, the value of J should be an integer multiple of $\hbar$ but we know it isn't! Let me repeat that it *must* be an integer *if* we require that travelling a full 2π circle gets you back to the same value of Ψ. Surely this has to be true for any wave function?

$$\frac{\partial \Psi}{\partial \theta} = \frac{i}{\hbar} J \Psi \quad \implies \quad \Psi = \ldots e^{\frac{i}{\hbar} J\theta} \tag{17.1}$$

$$e^{\frac{i}{\hbar} J\theta} = e^{\frac{i}{\hbar} J(\theta + 2\pi)} = \Psi e^{\frac{i}{\hbar} J 2\pi} \quad \implies \quad e^{\frac{i}{\hbar} J 2\pi} = 1 \quad \implies \quad \frac{J}{\hbar} = \textit{integer???} \tag{17.2}$$

The universe begs to differ, not only for the wave function of the electron but also for that of the quarks that make up protons and neutrons. These are spin-half particles. Let's apply the analysis in Equation 17.1 to a spin-half particle with total angular momentum $J = (N + \frac{1}{2})\hbar$.

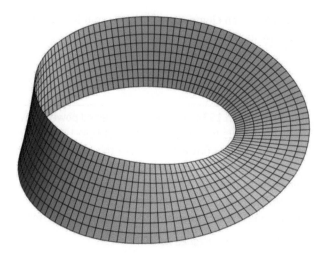

Figure 17.2   The Mobius Strip.

$$\frac{\partial \Psi}{\partial \theta} \; = \; \frac{i}{\hbar}(N + \frac{1}{2})\hbar\,\Psi \qquad \Longrightarrow \qquad \Psi \; = \; ...\, e^{\frac{i}{\hbar}\left(N+\frac{1}{2}\right)\hbar\theta}$$

$$\Psi(\theta + 2\pi) \; = \; ...\, e^{\frac{i}{\hbar}\left(N+\frac{1}{2}\right)\hbar(\theta+2\pi)} \; = \; \Psi e^{i\left(N+\frac{1}{2}\right)2\pi} \; = \; \Psi\, e^{i(N)\,2\pi}\,e^{i\pi} \; = \; -\Psi \tag{17.3}$$

$$\Psi(\theta + 4\pi) \; = \; ...\, e^{\frac{i}{\hbar}\left(N+\frac{1}{2}\right)\hbar(\theta+4\pi)} \; = \; \Psi e^{i\left(N+\frac{1}{2}\right)4\pi} \; = \; \Psi\, e^{i(2N)\,2\pi}\,e^{i2\pi} \; = \; \Psi \tag{17.4}$$

The difference between the results of Equation 17.3 and 17.4 is that while $e^{iN\,2\pi}$ is +1 (for any N), $e^{i\pi}$ is −1. What this tells us is that if we follow the electron wave function Ψ full circle (2π or 360 degrees) we do not end up back at the same value of Ψ. We end up with $-\Psi$. We have to go round twice (4π or 720 degrees) to get back to the same value of Ψ. This sort of mathematical object is called a *spinor*. In your study of quantum mechanics so far, you have seen somewhat unfamiliar mathematical representations of the wave function. For example, we use $e^{i\theta}$ to track phase and keep things compatible with special relativity. These spinors are another gargantuan mental leap away from anything that is physically familiar.

How can you get your head around the idea of a mathematical object which you have to circle twice in order to return to the same spot? There is no easy analogy, but the example of the Mobius strip may help (see Figure 17.2 and for a laugh Box 17.2). If you have never made a Mobius strip then now is the time. Cut a strip of paper, give it a twist, and then stick it together. If you run your finger around it, touching the surface all the way, you will see that you have to go round twice to reach the same spot. Don't overthink this. There is no deep lesson. It is just to give you a warm feeling inside that this sort of topology makes sense when, in truth, the mathematics is ... well, weird.

A smart student might say: *So, Simon, what you are telling us is that elementary particles have an intrinsic angular momentum that we call spin although it has nothing to do with spinning. And there is a spin-half variety where the mathematics of the wave function follows a different geometry from anything imaginable. If you circle round the wave function once, you arrive at its negative. You need to do two full circles to get back to where you started. Seriously, is that what you are telling us?*

And my reply is: *Exactly. Nice summary.*

Box 17.2   Just for laughs:

Why did the chicken cross the Mobius strip?   *To get to the other ... errr ... um ... wait ...*

17.2.3   The Pauli Exclusion Principle

Some of the elementary particles such as electrons and quarks are spin-half. Others are integer-spin meaning that the intrinsic angular momentum is an integer multiple of $\hbar$, For example, photons and gluons are spin-1 so they have $\hbar$ intrinsic spin. The spin-half particles (spinors) behave very differently from the integer-spin particles. That is hardly surprising given that the mathematics of their wave function obeys a completely different geometry. A fundamental difference between integer-spin and spin-half particles is that spin-half particles cannot share the same quantum state. This is called the Pauli exclusion principle after Wolfgang Pauli who proposed it in 1925.

As electrons are spin-half it explains why all the electrons in the atomic shell don't collapse together to the lowest allowable energy level, the **1s** orbital. It also explains why exactly two and only two electrons can occupy each of the quantum states that we identified in Section 16.2, one spin-up and the other spin-down. If you try to add a third, it must be either spin-up or spin-down. That means it would match the quantum state of one of the two resident electrons. Not allowed!

Spin-half particles are categorised together as *fermions* while integer-spin particles are categorised as *bosons*. The most familiar fermions are quarks (that make up the protons and neutrons in the atomic nucleus) and electrons that sit in shells away from the nucleus and so define the size and behaviour of atoms. The most familiar boson is the photon. While fermions such as the electron keep their distance (in the sense of quantum states), bosons such as the photon are happy to bunch together: the more in any quantum state, the merrier.

What explains this different behaviour? The usual argument involves two spin-half particles, say a and b, in an identical quantum state Ψ. The particles are identical so we can describe this as $\Psi(a, b)$ or as $\Psi(b, a)$. In the case of bosons $\Psi(a, b) = \Psi(b, a)$ and there is no conflict. In the case of fermions the wave function is antisymmetric so $\Psi(a, b) = -\Psi(b, a)$ and the two alternatives cancel each other out leaving no possible wave function for two fermions in the same state. You may like the argument that exchanging a and b is like a 2π rotation of the wave function and this explains the antisymmetry. If you want to get more complicated than this author can cope with, you can turn to the spin-statistics theorem and group theory. At the end of the day I suspect you will come away with the same feeling I have, that these arguments are all somewhat circular and pre-assume the result.

Neither Pauli nor Feynman believed we have a true theoretical explanation for Pauli's exclusion principle, but it clearly works! And surely nobody can be surprised. The mathematics of a fermion's wave function obeys a completely different geometry from that of a boson so we should not be shocked that they behave differently.

We will dig further into the difference between fermions and bosons when we address quantum field theory in a later chapter. The picture will emerge of spin-half fermions (such as electrons and the quarks that make up protons and neutrons) interacting via force-carrying integer-spin bosons (such as photons for the electromagnetic force, W/Z bosons for the weak force and gluons for the strong force).

Note that there are occasions when two fermions can combine together such that their $\frac{\hbar}{2}$ intrinsic spins sum up to integer-spin. In some metals, at very low temperatures, this happens with electrons forming what are called Cooper pairs. The result is a shift in behaviour to that of a boson often with startling results, such as, in the case of Cooper pairs, superconductivity.

17.2.4 The Pauli Matrices

Pauli (see Box 17.3 if you want some history) came up with a rather nifty way of describing spin-half angular momentum using matrices. This is more descriptive than prescriptive, but I want to step through it because it is a useful launching pad before we deep dive into the Dirac equation.

The measurement of spin around any axis can always be labelled *up* (↑) or *down* (↓). If an x measurement is x ↑ then subsequent measurements around that axis are x ↑ *unless* you measure spin around another axis first. This is because there is an uncertainty relationship between spin around each axis. If you know an electron is x ↑, it means that the y and z spins are undefined in the same way that an electron with an exact value of momentum has no defined position.

Figure 17.3 outlines the results of a couple of experiments (shown in separate boxes A and B). The blue blocks are devices that measure spin around the x and y axis as labelled (the equipment would be similar to that shown in Figure 17.1). The first step in both experiments is to create a beam of x ↑ electrons. In experiment A the beam of x ↑ electrons is measured for y spin. The result is 50%–50% y ↑ and y ↓. The next step in the A experiment is to take the y ↑ beam and the y ↓ beams and measure (separately) their x spin. Each beam splits 50%–50% into x ↑ and x ↓ because the y spin measurement has changed the wave function to y ↑, a state that has undefined x spin. The previous x ↑ nature of the electron is lost.

Experiment B is the same experiment *except* the y ↑ and y ↓ beams are recombined before the last x spin measurement. As shown, the result is 100% x ↑ electrons. The mixing of the outputs means there is no way to determine the y spin of an electron going into the final x spin measurement. The x ↑ state which is a superposition of half y ↑ and half y ↓ is undisturbed (if this seems oddly familiar, it is essentially the same as the Double-slit experiment with polarisers in Figure 7.2).

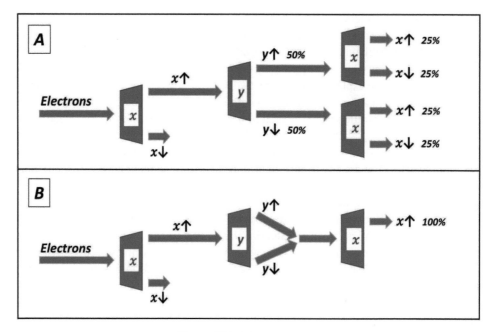

Figure 17.3 Measuring spin.

Pauli described these relationships mathematically. We will use $z \uparrow$ and $z \downarrow$ spin as our starting point (this is just convention). If a particle is in a $z \uparrow$ eigenstate, we label it $|z \uparrow>$. We are dealing with angular momentum so, similar to the angular momentum operator (Equation 14.13), Pauli proposed a spin operator such that $\hat{S}_z |z \uparrow> = +\frac{\hbar}{2}|z \uparrow>$. If it is spin-down then $\hat{S}_z |z \downarrow> = -\frac{\hbar}{2}|z \downarrow>$. Also (critically) the maths of the spin operator must respect the angular momentum uncertainty relationship between S_x, S_y and S_z that can be expressed using the commutators as shown in Equation 17.5 (these are the same as the angular momentum commutators in Box 14.2):

$$[\hat{S}_x, \hat{S}_y] = i\hbar \hat{S}_z \qquad\qquad [\hat{S}_y, \hat{S}_z] = i\hbar \hat{S}_x \qquad\qquad [\hat{S}_z, \hat{S}_x] = i\hbar \hat{S}_y \qquad (17.5)$$

The commutators have non-zero value so we know that $\hat{S}_x \hat{S}_y \Psi \neq \hat{S}_y \hat{S}_x \Psi$. Pauli realised that this kind of non-commutative maths calls for matrices so he set about developing a group of matrices that gives the pattern required. The first step is to label $z \uparrow$ and $z \downarrow$ using a simple two-component matrix as shown below along with an example of an electron in a superposition of $z \uparrow$ and $z \downarrow$:

Spin-up: $|z \uparrow> = \begin{bmatrix} 1 \\ 0 \end{bmatrix}$ Spin-down: $|z \downarrow> = \begin{bmatrix} 0 \\ 1 \end{bmatrix}$

Example superposition: $\Psi = \frac{1}{\sqrt{2}} \begin{bmatrix} 1 \\ 1 \end{bmatrix}$ $\frac{1}{\sqrt{2}}$ *normalises total probability* $|\Psi|^2$ *to 1*

Pauli's defined the $\hat{S}_z$ operator in the form of a 2x2 matrix so that the operator $\hat{S}_z$ produces the necessary $+\frac{\hbar}{2}$ result when applied to the $|z \uparrow>$ eigenstate and $-\frac{\hbar}{2}$ when applied to $|z \downarrow>$. The structure of the required matrix is shown in Equation 17.6. Equation 17.7 shows an example of it operating on the $|z \uparrow>$ matrix.

$$\hat{S}_z = \frac{\hbar}{2} \begin{bmatrix} 1 & 0 \\ 0 & -1 \end{bmatrix} \qquad (17.6)$$

$$\hat{S}_z |z \uparrow> = \frac{\hbar}{2} \begin{bmatrix} 1 & 0 \\ 0 & -1 \end{bmatrix} \begin{bmatrix} 1 \\ 0 \end{bmatrix} = \frac{\hbar}{2} \begin{bmatrix} 1 \\ 0 \end{bmatrix} = \frac{\hbar}{2}|z \uparrow> \qquad (17.7)$$

Box 17.3   Wolfgang Pauli (1900–1958): troubled genius

Admired by Einstein for his *sureness of mathematical deduction* and *profound physical insight*, Pauli's crowning achievement was the exclusion principle for which he won the 1945 Nobel Prize. He also predicted the existence of the neutrino. He was famed in scientific circles for not only his physics but also his ability to destroy laboratory equipment. When he was nearby, things would break in what became dubbed the *Pauli Effect*.

Pauli was a night owl, talking and drinking until late into the evening, but he suffered from periods of deep depression triggered initially by the suicide of his mother and a short very unhappy first marriage. Known for his forthright views he frequently criticised papers as: *Utterly wrong*. But his ultimate put-down was for a paper that he found unclear: *It is not even wrong*.

What have we done so far? We have described the spin-state with a 1x2 matrix. We have described the spin operator $\hat{S}_z$ with a 2x2 matrix that acts on the 1x2 spin-state matrix. Pauli's challenge was to find similar 1x2 matrices for x and y spin eigenstates and to find 2x2 matrices for the operators $\hat{S}_x$ and $\hat{S}_y$. Pauli's solution is shown below. The factors of $\frac{1}{\sqrt{2}}$ are normalisation factors so that the combined probability of finding the particle in one state or the other is 1. The 2x2 Pauli matrices for the operators $\hat{S}_x, \hat{S}_y$ and $\hat{S}_z$ are also shown.

$$|x \uparrow> \; = \; \frac{1}{\sqrt{2}} \begin{bmatrix} 1 \\ 1 \end{bmatrix} \qquad\qquad |y \uparrow> \; = \; \frac{1}{\sqrt{2}} \begin{bmatrix} 1 \\ i \end{bmatrix} \qquad\qquad |z \uparrow> \; = \; \begin{bmatrix} 1 \\ 0 \end{bmatrix}$$

$$|x \downarrow> \; = \; \frac{1}{\sqrt{2}} \begin{bmatrix} 1 \\ -1 \end{bmatrix} \qquad\qquad |y \downarrow> \; = \; \frac{1}{\sqrt{2}} \begin{bmatrix} 1 \\ -i \end{bmatrix} \qquad\qquad |z \downarrow> \; = \; \begin{bmatrix} 0 \\ 1 \end{bmatrix}$$

$$\hat{S}_x \; = \; \frac{\hbar}{2} \begin{bmatrix} 0 & 1 \\ 1 & 0 \end{bmatrix} \qquad\qquad \hat{S}_y \; = \; \frac{\hbar}{2} \begin{bmatrix} 0 & -i \\ i & 0 \end{bmatrix} \qquad\qquad \hat{S}_z \; = \; \frac{\hbar}{2} \begin{bmatrix} 1 & 0 \\ 0 & -1 \end{bmatrix}$$

Let's check that everything works. As an example below, I show $|y \downarrow>$ is an eigenstate of $\hat{S}_y$ and that it does throw off the value $-\frac{\hbar}{2}$ as required. You may want to try a few others.

$$\hat{S}_y \; |y \downarrow> \; = \; \frac{\hbar}{2} \begin{bmatrix} 0 & -i \\ i & 0 \end{bmatrix} \frac{1}{\sqrt{2}} \begin{bmatrix} 1 \\ -i \end{bmatrix} \; = \; \frac{\hbar}{2} \frac{1}{\sqrt{2}} \begin{bmatrix} -1 \\ i \end{bmatrix} \; = \; -\frac{\hbar}{2} \; |y \downarrow> \tag{17.8}$$

We can also check that the commutator relationships hold. This is crucial because the commutators contain the uncertainty relationship between different spin directions. The example below shows the commutator result: $[\hat{S}_x, \hat{S}_y] = i\hbar\hat{S}_z$

$$[\hat{S}_x, \hat{S}_y] \; = \; \frac{\hbar^2}{4} \begin{bmatrix} 0 & 1 \\ 1 & 0 \end{bmatrix} \begin{bmatrix} 0 & -i \\ i & 0 \end{bmatrix} - \frac{\hbar^2}{4} \begin{bmatrix} 0 & -i \\ i & 0 \end{bmatrix} \begin{bmatrix} 0 & 1 \\ 1 & 0 \end{bmatrix}$$

$$= \; \frac{\hbar^2}{4} \left(\begin{bmatrix} i & 0 \\ 0 & -i \end{bmatrix} - \begin{bmatrix} -i & 0 \\ 0 & i \end{bmatrix} \right) \; = \; \frac{\hbar^2}{4} \begin{bmatrix} 2i & 0 \\ 0 & -2i \end{bmatrix} \tag{17.9}$$

$$= \; i\hbar \frac{\hbar}{2} \begin{bmatrix} 1 & 0 \\ 0 & -1 \end{bmatrix} \; = \; i\hbar\hat{S}_z \tag{17.10}$$

This looks obscure, but bear with me. It reveals the relationship between spins that is hidden in the commutators. Consider $|x \uparrow>$. The matrix representation tells you that when a particle is spin-up in the x direction, it is in a superposition of being spin-up and spin-down in the z direction and therefore there is a 50:50 chance of measuring $+\frac{\hbar}{2}$ versus $-\frac{\hbar}{2}$ in that z direction.

$$|x \uparrow> \; = \; \frac{1}{\sqrt{2}} \begin{bmatrix} 1 \\ 1 \end{bmatrix} \; = \; \frac{1}{\sqrt{2}} |z \uparrow> + \frac{1}{\sqrt{2}} |z \downarrow> \tag{17.11}$$

Similarly, we can show that $|y \downarrow>$ can be expressed as a 50:50 superposition of x spin-states or z spin-states. Any spin-state around one axis can be written as a superposition of spin-states around another (don't forget that to calculate the probability of measuring a state you multiply the amplitude by its complex conjugate):

$$|y \downarrow> = \frac{1}{\sqrt{2}} \begin{bmatrix} 1 \\ -i \end{bmatrix} = \frac{1}{\sqrt{2}} |z \uparrow> - \frac{i}{\sqrt{2}} |z \downarrow> \tag{17.12}$$

$$|y \downarrow> = \frac{1}{\sqrt{2}} \begin{bmatrix} 1 \\ -i \end{bmatrix} = \frac{1}{\sqrt{2}} \left(\frac{(1-i)}{2} \begin{bmatrix} 1 \\ 1 \end{bmatrix} + \frac{(1+i)}{2} \begin{bmatrix} 1 \\ -1 \end{bmatrix} \right)$$

$$= \frac{(1-i)}{2} |x \uparrow> + \frac{(1+i)}{2} |x \downarrow> \tag{17.13}$$

Back in Box 7.1, I introduced the idea of *quantum knowledge* as knowing everything you possibly can about a quantum state. What information is needed for full quantum knowledge of the intrinsic spin of a spin-half particle? If you know the quantum state of intrinsic spin around *one* axis, you have full quantum knowledge. Let me phrase that differently. If you know the spin-state around one axis you know the probability of spin measurements around *any* axis.

17.3 Integer-spin: The Photon

The main topic of this chapter has been spin-half particles because spin-half is so strange. These are the fermions such as the quarks, the electron and the elusive neutrino, all of which are spin-half. However, we should spare a moment to discuss the intrinsic angular momentum of the photon.

The photon is a massless spin-1 boson. What intrinsic angular momentum states can the photon have? Photons move at the speed of light. If a photon is moving in the z direction, let's consider the variation in the wave function relative to this axis of motion. It is a spin-1 particle so the maximum variation around any axis is $\hbar$. Based on the ladder of states separated by $\hbar$, you might expect there to be three spin-states: $S_z = \hbar$, $S_z = 0$ and $S_z = -\hbar$. The first and last have intrinsic angular momentum anticlockwise and clockwise around the z axis. The $S_z = 0$ means that there is no variation in the wave function around the z axis, so the photon's $\hbar$ of intrinsic angular momentum would be split around the x or y axis in an undetermined way (the uncertainty rule must be respected). You might well expect this, but you would be wrong.

Photons can have intrinsic angular momentum (spin) *only* around the *axis of motion*. This is true for all massless particles because they travel at the speed of light. Why? Imagine a particle travelling at light speed. You try to measure the variation in its wave function in a plane such as that shown in Box A of Figure 17.4. If you think back to the section on special relativity, you will recall that when an object is observed passing at speed, its length shortens in the direction of motion (see Section 1.8). It shortens to $\frac{1}{\gamma}$ the static length. At the speed of light γ is ∞ so $d = 0$. The plane shown in Box A does not geometrically exist for any observer. Therefore, the wave function can vary only around the axis of motion as shown in Box B.

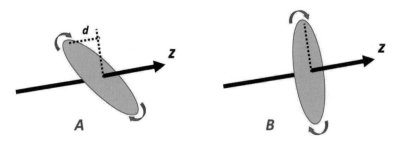

Figure 17.4 Massless particles are restricted to spin states around the direction of motion. With foreshortening, $d = 0$, so only the plane in scenario B exists.

The photon is spin-1 and all its $\hbar$ of intrinsic angular momentum must be around the axis of motion. It can be anticlockwise $|z \uparrow>$, clockwise $|z \downarrow>$ or in a superposition of the two, $a|z \uparrow> + b|z \downarrow>$, where a and b are complex numbers. You are probably more familiar with this than you think, in the form of light polarisation and polarised sunglasses.

17.3.1 Photon Polarisation

Polarised light consists of transverse waves moving along a distinct plane. Light reflecting from surfaces such as water tends to be disproportionately horizontally polarised. Thin dark vertical lines on the surface of sunglasses can block much of this horizontal light and reduce glare.

Polarised light can be expressed as a superposition of $|z \uparrow>$ and $|z \downarrow>$ states as shown in Figure 17.5. The beam of light is travelling in the z direction into the page. The arrows represent the direction of the peak in the electric field (I could equally have chosen the magnetic field). Diagram A shows the $|z \uparrow>$ and $|z \downarrow>$ components of the superposition. Let's discuss first the $|z \uparrow>$ component. The angular variation in the wave function means that the undulations of the wave function peak at different times for different angles. The result, in effect, is that the peak in the electric field appears to rotate anticlockwise. Let me stress that the photon is *not spinning*. The same is true for the $|z \downarrow>$ component, but clockwise.

Diagram B shows the superposition of the two states. The undulations come into sync. If you imagine the electric field peaks rotating, they meet at the same point every rotation so the peaks and troughs in the superposition's electric field are in one plane (illustrated in diagram B). This is vertically polarised light which I will label $|V>$.

Diagram C shows horizontally polarised light $|H>$. In terms of the rotating electric field peak, we take the black arrow of the $|z \uparrow>$ component of the superposition and wind it back against its direction of motion, by $\frac{\pi}{2}$ radians (90 degrees) which is the same as multiplying by i (see Equation 9.3). The $|z \downarrow>$ component rotates in the other direction so needs to be wound backwards by $\frac{3\pi}{2}$ radians (270 degrees), which is equivalent to multiplying by $i^3 = -i$. The equations below show vertical and horizontal light expressed in terms of spin around the z axis (the $\sqrt{2}$ is a normalisation factor). Let me emphasise again that all of this is variation in the value of the wave function. The photon is *not* rotating.

$$|V> = \frac{1}{\sqrt{2}}(|z \uparrow> + |z \downarrow>) \qquad\qquad |H> = \frac{1}{\sqrt{2}}i(|z \uparrow> - |z \downarrow>) \qquad\qquad (17.14)$$

The neat thing is that you can equally express $|z \uparrow>$ and $|z \downarrow>$ as superpositions of $|V>$ and $|H>$. Either *basis* is valid, but note that both have only two orthogonal components. This is because there are only two underlying spin states.

Polarisation can be used in the classroom for a simple display of quantum spin. Light that has passed through a vertical polariser is in state $|V>$ and will be completely blocked by a horizontal polariser. What about a polariser at a different angle? If it is at 45 degrees to vertical, then 50% of the $|V>$ photons will pass. The percentage of photons that pass depends on the relative angle θ between the light polarisation and the polariser as follows:

$$\textit{Passing amplitude } = \cos\theta \qquad\qquad\qquad \textit{Passing probability } = \cos^2\theta \qquad\qquad (17.15)$$

$$\textit{Blocked amplitude } = \sin\theta \qquad\qquad\qquad \textit{Blocked probability } = \sin^2\theta \qquad\qquad (17.16)$$

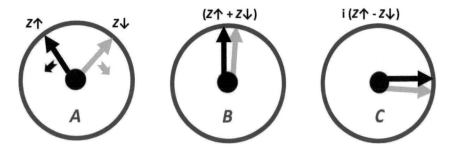

Figure 17.5 Spin and the polarisation of light. The photon is moving at speed of light into the page. The arrows illustrate the direction of the electric field for $|z \uparrow>$ and $|z \downarrow>$.

Here then is the experiment. Take two sheets of polarised plastic or two pairs of polarised glasses and cross them. The result will be black. No light will pass through because θ is $\frac{\pi}{2}$ radians (90 degrees) and $\cos^2(\frac{\pi}{2}) = 0$. Then, slip a third polariser between the two crossed polarisers but at an angle of $\frac{\pi}{4}$ radians (45 degrees) to both. What happens? The angle between the first and middle polariser is $\frac{\pi}{4}$ (45 degrees) and so is the angle between the middle and the last polarisers. As $\cos^2(\frac{\pi}{4}) = 0.5$, the result is that 50% of the light passes in each case. Thus 25% of the light makes it the whole way through all three polarisers. When you slip in that third extra blocking polariser *more* light gets through! You can see the mathematics of spin at work – worth a try.

17.4 Bell's Inequality and the Aspect Experiment

The discussion of photon polarisation allows me to address a different topic. Einstein had concerns about the probabilistic nature of quantum mechanics. Along with colleagues (Podolsky and Rosen), he came up with the EPR paradox. Produce two particles in a reaction such that they separate with related position and momentum to a great distance apart. You can measure the position of the first and you know exactly where the second is. Similarly, you might measure the momentum of the first and know the momentum of the second. The point is that you can determine the position or momentum of the second particle without making any physical measurement on it. Therefore, Einstein argued, position and momentum must be tangible values and not probabilistic.

As Einstein put it: *If, without in any way disturbing a system, we can predict with certainty the value of a physical quantity, then there exists an element of physical reality corresponding to that quantity.* Einstein admitted one loophole. His argument breaks down if the measurement of the first particle affects the second. He dismissed this possibility as *spooky action at a distance* (Section 7.7).

The argument was theoretical until John Bell came up with an ingenious experiment to test it using spin measurements (usually in the form of photon polarisation measurements) as outlined in Figure 17.6:

- Separate two entangled photons with the same polarisation (such pairs are usually produced with *parametric downconversion* to convert high energy photons into pairs).
- Pass both photons through polarisers but at slightly different angles. This difference in angle is varied during the experiment (labelled in the diagram as vertical V, angle θ and 2θ).
- Collect results for whether each photon passes through its polariser. Compare the results.

Let's keep things clear with the following notation. If a photon passes its polariser we will show this as ($\uparrow$). If not, we will use ($\downarrow$). For example if the V polariser is used on one photon and the 2θ polariser on the other and both pass, then the result would be ($V \uparrow, 2\theta \uparrow$). The very clever bit is the maths behind the experiment, which assumes nothing about the nature of quantum mechanics. If Einstein is right, then the measurements of the polarisation of the photons will have no effect on each other. The result of every possible polarisation measurement must be identically precoded in the two photons by some hidden variable. Otherwise, they could not show identical results when their polarisation is measured at the same angle (entangled photons always do). Bell argued the following. If every polarisation measurement is predetermined, then for each pair of entangled photons there is the same preset list of probability values of measuring $V \uparrow$, $\theta \uparrow$ and $2\theta \uparrow$. This gives the following inequality where P stands for the probability of that combination of results for the photons.

$$\text{Bell's inequality: } P(V \uparrow, \theta \downarrow) + P(\theta \uparrow, 2\theta \downarrow) \geq P(V \uparrow, 2\theta \downarrow) \tag{17.17}$$

This can be proved using simple logic:

$P(V \uparrow, \theta \downarrow, 2\theta \uparrow)$	$+$	$P(V \uparrow, \theta \uparrow, 2\theta \downarrow)$	$\geq$	$P(V \uparrow, \theta \uparrow, 2\theta \downarrow)$	(17.18)
$P(V \uparrow, \theta \downarrow, 2\theta \downarrow)$	$+$	$P(V \downarrow, \theta \uparrow, 2\theta \downarrow)$	$\geq$	$P(V \uparrow, \theta \downarrow, 2\theta \downarrow)$	(17.19)
$P(V \uparrow, \theta \downarrow)$	$+$	$P(\theta \uparrow, 2\theta \downarrow)$	$\geq$	$P(V \uparrow, 2\theta \downarrow)$	(17.20)

Consider Equation 17.18. The second and third terms are exactly the same so the left side of the equation *must* be greater or equal to the right. This is also true for Equation 17.19 because the first and third terms are the same. The next step is to

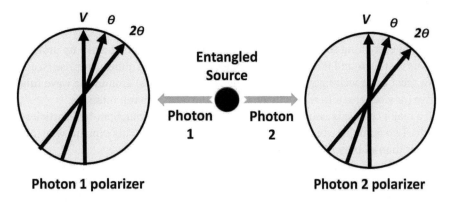

Figure 17.6 The polarisation of entangled photons is measured at slightly different angles.

add Equations 17.18 and 17.19 vertically. If the probabilities are preset then the first term of Equation 17.20 must equal the sum of the first terms of Equations 17.18 and 17.19 because $P(2\theta \uparrow) + P(2\theta \downarrow) = 1$. The same is true for the other terms in Equation 17.20 so the inequality must be correct. Equation 17.20 is Bell's inequality.

But what if Einstein was wrong? What if the measurements do affect each other? If we use a small angle for θ, the outcomes on the left side of Bell's inequality become very unlikely. When one photon passes through a polariser, the entangled pair will *both* be in that polarised eigenstate with a very low chance of the other being blocked by a polariser set at a small angle θ away. In contrast, the outcome on the right side of Bell's inequality has a larger angle difference of 2θ and so has a greater probability of occurring. We know the probability of a polarised photon passing an angled polariser (Equation 17.16) and can use this to show that entanglement breaches Bell's inequality for small θ. The calculation below uses θ of 10 degrees.

$$P(V \uparrow, \theta \downarrow) + P(\theta \uparrow, 2\theta \downarrow) = \sin^2\theta + \sin^2\theta = 2\sin^2(10°) = 0.06$$

$$P(V \uparrow, 2\theta \downarrow) = \sin^2(2\theta) = \sin^2(20°) = 0.12$$

$$\implies \quad P(V \uparrow, \theta \downarrow) + P(\theta \uparrow, 2\theta \downarrow) \leq P(V \uparrow, 2\theta \downarrow) \qquad \textit{Bell breach} \tag{17.21}$$

As you can see, at small angles not only is Bell's inequality breached, but it is significantly breached. Bell published his inequality in 1964. Experimental evidence for quantum breaches began to appear in the 1970s, but the most famous is the Aspect experiment in the 1980s because the photons were far enough apart and the polarisation measurements fast enough that any information passed between the photons would have had to travel faster than light speed. This and subsequent experiments confirm the breach. To Bell's surprise, Einstein was wrong (see Box 17.4). Entangled particles share a single quantum state. A measurement of one of the entangled particles affects the quantum state of both.

Box 17.4   A surprise for Mr Bell

John Stewart Bell (1928–1990) fully expected his inequality would vindicate Einstein, but accepted the overwhelming evidence: *For me, it is so reasonable to assume that the photons in those experiments carry with them programs, which have been correlated in advance, telling them how to behave. This is so rational that I think that when Einstein saw that – he was the rational man. So for me, it is a pity that Einstein's idea doesn't work. The reasonable thing just doesn't work.*

Pause and think on this. Imagine a photon of light produced in an entangled event just after the Big Bang. Some of that light is still visible (it is called cosmic microwave background radiation). We measure its polarisation and that measurement has implications for the quantum state of a photon on the other side of the universe. But don't forget that nothing in this allows information to be sent faster than the speed of light (refer back to Section 7.7 for more on this).

17.5 Summary

Spin-half particles make up everything solid that we see around us: the quarks that make up protons and neutrons in the nucleus and the electrons that surround it. The discovery of electron spin was quite a surprise (see Box 17.5). To describe their wave function we must use a weird geometry. Rotating 2π (360 degrees) around the wave function turns the value from positive to negative (or vice versa). To return to the same value takes two full rotations.

Unsurprisingly such a major change is associated with very different behaviour. Spin-half particles (fermions) follow the Pauli exclusion principle. Two fermions *cannot* share the same quantum state. This explains why the electrons around the nucleus sit in shells rather than all bunching together in the ground state **1s** shell.

We discussed the way Pauli used matrices to represent spin. His matrices show that we need to adopt *multi-component* wave functions to cater for spin. Pauli used a two-component representation. You will discover in the next chapter that even two components are not enough! It is time to study the equation that sits behind spin-half particles – it is time for Dirac.

Box 17.5 Samuel Goudsmit (1902–1978) – a spin of the dice

Goudsmit is credited, along with a colleague Uhlenbeck, as the first to discover electron spin in 1925. He charmed an audience in 1971 with a witty description of the discovery. Here are some excerpts that have been tweaked in translation. I hope you find them instructive:

It is said of someone who does something in physics 'yes, a really clever person'. Admittedly, there are cases such as Heisenberg, Dirac and Einstein. There are some exceptions but for most of us luck plays a very important role and that should not be forgotten...

Uhlenbeck said to me: 'But don't you see what this implies? It means that there is a fourth degree of freedom for the electron. It means that the electron has a spin, that it rotates'... then I asked him: 'What is a degree of freedom?'... and that was it: spin was discovered.

We wrote a short article for our professor Ehrenfest to send to 'Naturwissenschaften'. Uhlenbeck got frightened, went to Ehrenfest and said: 'Don't send it off, because it probably is wrong; it is impossible for an electron to rotate at such high speed'. And Ehrenfest replied: 'It is too late, I have sent it off already... anyway, you don't yet have a reputation, so you have nothing to lose'.

I received a letter from Heisenberg. In it he writes a formula. I did not understand a bit of it. And then he says somewhere: 'What have you done with the factor 2?' Which factor? I had not the slightest notion ... but there was a young man, Thomas, who worked out that Heisenberg's factor of two was a relativistic correction and everything was in order!

One aspect stands out that is of particular importance for young people – you need not be a genius to make an important contribution; one should not always aspire to tackle what is most important, but just try to have fun working in physics.

A physicist with the humility to admit luck played a part: what a wonderful man.

Chapter 18

The Dirac Equation

CHAPTER MENU

18.1 Yet Another Equation?
18.2 Bi-spinors and Four-component Wave Functions
18.3 The Dirac Equation
18.4 Spin-half Is Built in
18.5 Interpreting the Dirac Equation
18.6 The Dirac Equation and Hydrogen
18.7 Dirac Equation: Modern Formulation
18.8 The Aftermath: Physics Falls Apart Again

Box 18.1 Paul Dirac...

Once upon a time, a weird guy called Paul Dirac decided to try something very weird. It was so weird that he had to trawl through mathematics to do it. He came across some weird matrices. When he used them he got a really weird looking calculation that made no sense at all, not even weird sense. But Dirac was weird enough to continue with this weird tortuous piece of maths and out popped something really really weird... but true.

The Dirac equation is one of my favourite pieces of physics. It is an extraordinary mathematical leap as you can tell from my comments in Box 18.1. How did Dirac dare to put effort into something that seems so half-witted? I remember first learning of it. I scratched my head as the mathematics unfolded. It seemed nonsensical. And then that moment when it all comes together and it screams: *Spin-half, here I am!* And it is obvious that the result cannot be a coincidence. It as if the universe speaks to you through Dirac. OK, OK... this may be a bit over the top!

The downside is that it is going to take some mathematical effort to follow things. And if I rush through, it will only be worse. I am not going to abbreviate the formulas (most texts do) because that makes things harder, not simpler. Sadly, that means lots of matrices and a few extra pages. If you find the maths too much, I suggest you skim through this chapter as best you can and/or you jump to the aftermath Section 18.8. So here goes, and I hope you get as much joy from the Dirac equation as I do.

18.1 Yet Another Equation?

How on earth can we justify another equation? Don't we have enough already? We had the Klein-Gordon equation that looked great but does not work for electrons (the predicted hydrogen energy levels are wrong), so we switched to the Schrödinger equation, albeit a classical approximation. I mean, how many equations do we need? There are two steps in answering this very fair question. The first is to review the weaknesses of the Schrödinger equation. The second is to remind ourselves what exactly this book is all about. Let's start with the first.

Quantum Untangling: An Intuitive Approach to Quantum Mechanics from Einstein to Higgs, First Edition. Simon Sherwood.
© 2023 John Wiley & Sons Ltd. Published 2023 by John Wiley & Sons Ltd.

Let's be honest. The Schrödinger equation was never going to give us what we need. In this book we have built quantum mechanics on the back of special relativity, and then resorted to an equation that shows no respect for relativity! In the Schrödinger equation, time is treated on a completely different footing from space. We take a first derivative with respect to time and match it with second derivatives of space. Given that space and time move in tandem in special relativity, it is obvious that this is just plain wrong. Clearly, we need equations that are relativistic.

The second point is more general. The Schrödinger and Klein-Gordon equations build on the relationship of the derivatives of the wave function with energy and momentum (such as $\frac{\partial \Psi}{\partial x}$ with p_x and $\frac{\partial \Psi}{\partial t}$ with E). The equations combine these derivatives in ways that reflect the relationship between energy and momentum. As such, they respect the maths of Minkowski spacetime. There may be other ways, other equations, that achieve the same thing (indeed there are). You might expect these different solutions to result in different types of particles.

Putting it the other way round, there is a wide variety of particles ranging from electrons to photons to quarks, so it should not surprise you that the mathematical equations that describe them might be different. The maths of spin-half particles may seem (*will* seem) much more complicated, but that does not matter. What matters is that spin-half particles respect the same rules of Minkowski spacetime. The universe does not care if you and I find the maths a bit harder. In coming chapters of this book you will learn that the difference between these equations reflects differences in the structure of underlying quantum fields; but we will leave that for later.

Returning to our topic, the mathematical description of the spin-half wave function is an unexpected surprise right down to the level of basic geometry, so it should not be a shock that it involves a different approach to the relationship of energy and momentum; that is the Dirac equation. Box 18.2 gives some history of this remarkable man.

Box 18.2   Paul Dirac (1902–1984): a balance of genius and madness

Such was Einstein's opinion and in regard to Paul Dirac, everyone agrees on one word: *genius*. In his own words, Dirac had no childhood. His strict father spoke to him only in French. His elder brother committed suicide. But amidst this mayhem an extraordinary brain developed. Dirac was Bristol (UK) born, Bristol schooled and then supported by Bristol University when others would not give him a scholarship. The city should be very proud of their amazing son.

Originally an engineer, he abandoned the field because he could not find a job. When he switched to theoretical physics, Cambridge grabbed him, and the rest is history. Described as an Edwardian geek, he was so taciturn that colleagues defined a *dirac* as one word per hour. He was reputed to offer only three answers: *Yes, No* and *I don't know*. Unsurprisingly he eschewed publicity. He refused a knighthood and accepted the 1933 Nobel Prize only because he was warned that refusing it would create a hue and cry in the papers.

Dirac's theories were built on mathematics. He believed that mathematical beauty trumps all and lights the way. I am not a mathematician, but I admire him for that. He created the Dirac equation, quantum electrodynamics and quantum field theory. One should listen to success.

18.2   Bi-spinors and Four-component Wave Functions

I think it best to warn you of what is coming before it hits you like a high-speed train. The upshot of Dirac's work is that the mathematical description of the state of a spin-half particle requires four components for each wave function. That's right, even a single particle wave function is a combination of four separate, but related equations, called a *bi-spinor*.

Stop! I am guessing that some of you are about to burn this book, give up studying and settle for a hermit's life in a distant cave far away from any physics equations. But you have already absorbed the idea of a particle described by a complex wave function which, for precise momentum, would stretch across the infinity of space. You have learned that particles can be in a superposition of states, places and velocities, that they can tunnel through impossible barriers, and that experimental results are defined by probability rather than causal certainty. Is it really that much more of a leap to go from one bizarre component to four?

Again, stop! Earlier in the book (Section 5.8), I compared the wave function with π. You can get your brain in a whirl thinking about π being irrational and transcendental, but it is just a mathematical description of the relationship that *has* to exist between the diameter and circumference of a circle in flat space. In the same way, I like to think of the bi-spinor as being a mathematical description of the relationship that *has* to exist for a spin-half particle to be consistent with Minkowski spacetime. Perhaps this is a cop-out, but it is the way I stop my brain from over-heating.

Anyway, I should warn you that I am not a philosopher. My brother Charles is, so if you want to think more on this, go ask him. Just be grateful that I have given you more than Dirac would have. He never bothered himself with such things. Dirac wrote in one of his papers: *The interpretation of quantum mechanics has been dealt with by many authors, and I do not want to discuss it here. I want to deal with more fundamental things.*

18.3 The Dirac Equation

The next sections that describe the Dirac equation are intentionally slow and steady. This equation is not something that springs out as obvious. It is about as counter-intuitive as anything can be.

18.3.1 The Ingredients

Let's put ourselves inside Dirac's mind as he muses over the problem. The building blocks are the derivatives of the wave function for a particle with definite energy-momentum as shown in Equation 18.1 (check back to Equation 11.2 if needed). Dirac's equation is relativistic so I have switched to bold for energy and momentum.

$$\mathbf{E}\phi \;=\; i\hbar\frac{\partial\phi}{\partial t} \qquad \mathbf{E}^2\phi \;=\; -\hbar^2\frac{\partial^2\phi}{\partial t^2} \qquad \mathbf{p}_x\phi \;=\; -i\hbar\frac{\partial\phi}{\partial x} \qquad \mathbf{p}_x^2\phi \;=\; -\hbar^2\frac{\partial^2\phi}{\partial x^2} \tag{18.1}$$

The relationship between these derivatives needs to be in line with the relativistic energy balance that Einstein unveiled (Equation 18.2). Dirac *cannot* use the classical approximation that Schrödinger used (Total energy = Kinetic energy + Potential energy) because he specifically wants to find a *relativistic* equation.

$$m^2c^4 \;=\; \mathbf{E}^2 - \mathbf{p}^2 c^2 \quad \implies \quad \mathbf{E}^2 = (\mathbf{p}_x^2 + \mathbf{p}_y^2 + \mathbf{p}_z^2)c^2 + m^2c^4 \tag{18.2}$$

Dirac will have been familiar with what I will describe as the obvious solution: the Klein-Gordon equation, shown again for good measure in Equation 18.5. This can be derived from Equation 18.2 by substituting in the derivatives. If it works for a particle with precise energy-momentum, an energy-momentum eigenstate ϕ, it also works for any superposition Ψ because it is *linear*. Specifically it uses second derivatives so it avoids any powers (see Box 8.3).

$$\mathbf{E}^2 \;=\; (\mathbf{p}_x^2 + \mathbf{p}_y^2 + \mathbf{p}_z^2)c^2 + m^2c^4 \tag{18.3}$$

$$\frac{\partial^2\phi}{\partial t^2} \;=\; \left(\frac{\partial^2\phi}{\partial x^2} + \frac{\partial^2\phi}{\partial y^2} + \frac{\partial^2\phi}{\partial z^2}\right)c^2 + \frac{m^2c^4}{\hbar^2}\phi \qquad \textit{Energy-momentum eigenstate} \tag{18.4}$$

$$\frac{\partial^2\Psi}{\partial t^2} \;=\; \left(\frac{\partial^2\Psi}{\partial x^2} + \frac{\partial^2\Psi}{\partial y^2} + \frac{\partial^2\Psi}{\partial z^2}\right)c^2 + \frac{m^2c^4}{\hbar^2}\Psi \qquad \textit{Klein-Gordon} \tag{18.5}$$

As mentioned earlier, the Klein-Gordon equation presents some technical difficulties that come with second derivatives (such as negative probabilities) but worst of all, it produces the wrong results for the electron energy levels of hydrogen. So while it is useful (indeed, we will use it later in analysing the Higgs boson), Dirac knows it is not the equation that applies to electrons.

What does Dirac *not want* in his equation? He wants to avoid using second derivatives and he cannot use squares of the first derivatives because that would not work for superpositions.

What does Dirac *want*? Ideally he would like to use the square root of the energy relationship as in Equation 18.6. He can then replace the term on the left side with the first derivative with respect to time of the wave function. That leaves the right side. He needs to convert the momentum terms from squares of momentum into single powers so that he can then substitute them for first derivatives with respect to space. How do you do that? How do you convert the right side to single powers? Obviously you cannot, unless you are a stubborn dreamer like Dirac – and a genius too.

$$\mathbf{E} = \sqrt{(\mathbf{p}_x^2 + \mathbf{p}_y^2 + \mathbf{p}_z^2)c^2 + m^2c^4} \quad \Longrightarrow \quad i\hbar\frac{\partial \Psi}{\partial t} = ???? \tag{18.6}$$

18.3.2 Dirac's Crazy Insight

Dirac's insight was that it *can* be done if you are crazy enough and throw all intuitive rationale out of the window. To explain things as clearly as possible, I am going to switch to natural units where $c = 1$. We can add the c factor back in later. This simplifies the energy equation that Dirac was trying to solve to the following:

$$\mathbf{E} = \sqrt{\mathbf{p}_x^2 + \mathbf{p}_y^2 + \mathbf{p}_z^2 + m^2} \tag{18.7}$$

Dirac's approach was to assume, against all the obvious objections, that there *is* a solution. Let's follow in his footsteps and assume that there is some combination that works and see what happens. Suppose we weight each term with some factor: $\mathbf{p}_x$ with α_x, $\mathbf{p}_y$ with α_y, $\mathbf{p}_z$ with α_z and mass m with factor β as shown in Equation 18.8.

$$\mathbf{E} = \sqrt{\mathbf{p}_x^2 + \mathbf{p}_y^2 + \mathbf{p}_z^2 + m^2} = \alpha_x\mathbf{p}_x + \alpha_y\mathbf{p}_y + \alpha_z\mathbf{p}_z + \beta m \tag{18.8}$$

While we have said nothing about what the factors α_x, α_y, α_z and β are, they by definition must square to the necessary total as shown in Equation 18.9.

$$(\alpha_x\mathbf{p}_x + \alpha_y\mathbf{p}_y + \alpha_z\mathbf{p}_z + \beta m)^2 = \mathbf{p}_x^2 + \mathbf{p}_y^2 + \mathbf{p}_z^2 + m^2 \tag{18.9}$$

We can expand the square on the left side of Equation 18.9 and this tells us a lot about the factors we have introduced. Some of you will be thinking that this is all a bit stupid, but remember that we are treading in the footsteps of a genius, so bear with me. The expansion of the square on the left side of Equation 18.9 is shown below. Note that 18.10, 18.11 and 18.12 are all *one* long equation. This generates the $\mathbf{p}_x^2$, $\mathbf{p}_y^2$, $\mathbf{p}_z^2$ and m^2 terms shown on 18.10, plus there are a large number of cross-terms. They include cross-terms such as factors of $\mathbf{p}_x\mathbf{p}_y$ and $\mathbf{p}_y\mathbf{p}_z$. An important point to remember is that there is no uncertainty relationship between momenta in different dimensions so it does not matter if the order is $\mathbf{p}_x\mathbf{p}_y$ or $\mathbf{p}_y\mathbf{p}_x$. Therefore, we can group terms together as shown on 18.11 and 18.12

$$(\alpha_x\mathbf{p}_x + \alpha_y\mathbf{p}_y + \alpha_z\mathbf{p}_z + \beta m)^2 = \alpha_x^2\mathbf{p}_x^2 + \alpha_y^2\mathbf{p}_y^2 + \alpha_z^2\mathbf{p}_z^2 + \beta^2 m^2 \ldots \tag{18.10}$$

$$\ldots + (\alpha_x\alpha_y + \alpha_y\alpha_x)\mathbf{p}_x\mathbf{p}_y + (\alpha_x\alpha_z + \alpha_z\alpha_x)\mathbf{p}_x\mathbf{p}_z + (\alpha_y\alpha_z + \alpha_z\alpha_y)\mathbf{p}_y\mathbf{p}_z \ldots \tag{18.11}$$

$$\ldots + (\alpha_x\beta + \beta\alpha_x)\mathbf{p}_x m + (\alpha_y\beta + \beta\alpha_y)\mathbf{p}_y m + (\alpha_z\beta + \beta\alpha_z)\mathbf{p}_z m \tag{18.12}$$

If you are a normal human being, you throw your hands up in horror and head for a glass of wine – but not if you are Dirac. He looked long and hard at it and realised that this huge expression *can* equal the total on the right side of Equation 18.9 *providing* the following things are true. First:

$$\alpha_x^2 = \alpha_y^2 = \alpha_z^2 = \beta^2 = 1 \tag{18.13}$$

Look back at 18.10 and you will see that if the square of each factor is 1, it gives the correct total. However, there is second requirement. In order to match the total on the right side of Equation 18.9, we need to get rid of all those pesky cross-terms. For that to happen we need:

$$\alpha_x\alpha_y = -\alpha_y\alpha_x \qquad\qquad \alpha_x\alpha_z = -\alpha_z\alpha_x \qquad\qquad \alpha_y\alpha_z = -\alpha_z\alpha_y \tag{18.14}$$
$$\alpha_x\beta = -\beta\alpha_x \qquad\qquad\quad \alpha_y\beta = -\beta\alpha_y \qquad\qquad\quad \alpha_z\beta = -\beta\alpha_z$$

If you look at 18.11 and 18.12, you will see that these relationships reduce all the cross-terms to zero as required. If you take the combination of any two *different* factors such as $\alpha_x\alpha_y$, then adding it to the same factors reversed cancels out the term.

At this stage even a crazed mathematician should conclude that this is going nowhere. She or he should drop the wine glass and head for a bottle of whisky, but not Dirac – he realised that this is non-commutative mathematics and the factors α_x, α_y, α_z and β have to be matrices. And not simple matrices. For this to work, you need 4x4 matrices so Box 18.3 gives a quick refresher on how this sort of matrix works, plus an example.

Box 18.3 4x4 matrix

A 4x4 matrix operates on a 1x4 matrix in the following way to generate another 1x4 matrix:

$$\begin{bmatrix} a & b & c & d \\ e & f & g & h \\ i & j & k & l \\ m & n & o & p \end{bmatrix} \begin{bmatrix} q \\ r \\ s \\ t \end{bmatrix} = \begin{bmatrix} aq+br+cs+dt \\ eq+fr+gs+ht \\ iq+jq+ks+lt \\ mq+nr+os+pt \end{bmatrix} \quad e.g. \quad \begin{bmatrix} 1 & 0 & 2 & 1 \\ 2 & 0 & 0 & 3 \\ 2 & 1 & 2 & 1 \\ 0 & 2 & 1 & 0 \end{bmatrix} \begin{bmatrix} 1 \\ 2 \\ 0 \\ 1 \end{bmatrix} = \begin{bmatrix} 2 \\ 5 \\ 5 \\ 4 \end{bmatrix}$$

This is heavy stuff. We are about to introduce 4x4 matrices into the equation. Things may seem to be getting crazy and unrealistic so let's pause for a moment to reflect. I want to repeat that Dirac was searching for something that is viable in Minkowski spacetime. It does not have to be simple, just viable. And Dirac was the person for the job. He loved to play with maths. So as we start to work with some obscure-looking matrices, I ask you to suspend disbelief and to follow Dirac's own words: *If you are receptive and humble, mathematics will lead you by the hand.*

To give you a slight break for air before you suffer further matrix-torture, you may want to browse some other comments attributed to Dirac shown in Box 18.4. Enjoy!

Box 18.4 Dirac: more comments from this man of few words

On maths and physics: *A theory with mathematical beauty is more likely to be correct than an ugly one that fits some experimental data.*

On silence: *There are always more people willing to speak than there are to listen.*

On reflection: *I was taught at school never to start a sentence without knowing the end of it.*

As an avowed atheist: *Religion is a jumble of false assertions, with no basis in reality... the very idea of God is a product of the human imagination.*

On poetry: *The aim of science is to make difficult things understandable in a simpler way; the aim of poetry is to state simple things in an incomprehensible way. The two are incompatible.*

Popularly attributed to him: *Pick a flower on Earth and you move the farthest star.*

18.3.3 Dirac's Matrices

To recap, we need four matrices α_x, α_y, α_z and β with the following characteristics:

$$\alpha_x^2 = \alpha_y^2 = \alpha_z^2 = \beta^2 = 1 \quad \text{(matrices square to 1)} \tag{18.15}$$

$$\alpha_x\alpha_y = -\alpha_y\alpha_x \quad \alpha_x\alpha_z = -\alpha_z\alpha_x \quad \alpha_x\beta = -\beta\alpha_x \text{ e.t.c. (cross-terms cancel)} \tag{18.16}$$

Before we address Dirac's matrices it is worth noting that the 2x2 Pauli's matrices (Subsection 17.2.4) have exactly these properties. This is the first whiff of a link between Dirac's approach and spin. The Pauli matrices are shown in Equation 18.17 along with one example calculation showing them operating on a 1x2 matrix in the way needed, that is, $\sigma_x^2 = 1$... and $\sigma_x\sigma_y = -\sigma_y\sigma_x$:

$$\sigma_x = \begin{bmatrix} 0 & 1 \\ 1 & 0 \end{bmatrix} \quad \sigma_y = \begin{bmatrix} 0 & -i \\ i & 0 \end{bmatrix} \quad \sigma_z = \begin{bmatrix} 1 & 0 \\ 0 & -1 \end{bmatrix} \tag{18.17}$$

$$\sigma_y^2 \begin{bmatrix} a \\ b \end{bmatrix} = \begin{bmatrix} 0 & -i \\ i & 0 \end{bmatrix}\begin{bmatrix} 0 & -i \\ i & 0 \end{bmatrix}\begin{bmatrix} a \\ b \end{bmatrix} = \begin{bmatrix} 1 & 0 \\ 0 & 1 \end{bmatrix}\begin{bmatrix} a \\ b \end{bmatrix} = \begin{bmatrix} a \\ b \end{bmatrix} \tag{18.18}$$

$$\sigma_y\sigma_z \begin{bmatrix} a \\ b \end{bmatrix} = \begin{bmatrix} 0 & -i \\ i & 0 \end{bmatrix}\begin{bmatrix} 1 & 0 \\ 0 & -1 \end{bmatrix}\begin{bmatrix} a \\ b \end{bmatrix} = \begin{bmatrix} 0 & i \\ i & 0 \end{bmatrix}\begin{bmatrix} a \\ b \end{bmatrix} = \begin{bmatrix} ib \\ ia \end{bmatrix} \tag{18.19}$$

$$\sigma_z\sigma_y \begin{bmatrix} a \\ b \end{bmatrix} = \begin{bmatrix} 1 & 0 \\ 0 & -1 \end{bmatrix}\begin{bmatrix} 0 & -i \\ i & 0 \end{bmatrix}\begin{bmatrix} a \\ b \end{bmatrix} = \begin{bmatrix} 0 & -i \\ -i & 0 \end{bmatrix}\begin{bmatrix} a \\ b \end{bmatrix} = -\begin{bmatrix} ib \\ ia \end{bmatrix}$$

Yahoo! I hear you rejoice that we can do this with 2x2 matrices. Sorry. Not possible. To handle the requirements of Equation 18.15, we need *four* factors so we need *four* of these matrices. There are only *three* Pauli matrices and no more exist. To get what we need, we must use 4x4 matrices.

Here are the Dirac 4x4 matrices in all their glory. By now you should be relaxed about the appearance of i as you know it just signals a phase change (multiplying a wave function by i shifts the phase of each sinusoid within it by $\frac{\pi}{2}$). Please remember that the only point of these matrices is that they conform by design to the requirements shown in Equations 18.15 and 18.16. And why does Dirac want that? It is so Equation 18.21 holds. And why does he want that? So he can create the first order relationship between energy and momentum shown in Equation 18.22

$$\alpha_x = \begin{bmatrix} 0 & 0 & 0 & 1 \\ 0 & 0 & 1 & 0 \\ 0 & 1 & 0 & 0 \\ 1 & 0 & 0 & 0 \end{bmatrix} \qquad \alpha_y = \begin{bmatrix} 0 & 0 & 0 & -i \\ 0 & 0 & i & 0 \\ 0 & -i & 0 & 0 \\ i & 0 & 0 & 0 \end{bmatrix} \qquad (18.20)$$

$$\alpha_z = \begin{bmatrix} 0 & 0 & 1 & 0 \\ 0 & 0 & 0 & -1 \\ 1 & 0 & 0 & 0 \\ 0 & -1 & 0 & 0 \end{bmatrix} \qquad \beta = \begin{bmatrix} 1 & 0 & 0 & 0 \\ 0 & 1 & 0 & 0 \\ 0 & 0 & -1 & 0 \\ 0 & 0 & 0 & -1 \end{bmatrix}$$

$$\mathbf{E}^2 = \mathbf{p}_x^2 + \mathbf{p}_y^2 + \mathbf{p}_z^2 + m^2 = (\alpha_x \mathbf{p}_x + \alpha_y \mathbf{p}_y + \alpha_z \mathbf{p}_z + \beta m)^2 \qquad (18.21)$$

$$\implies \mathbf{E} = \alpha_x \mathbf{p}_x + \alpha_y \mathbf{p}_y + \alpha_z \mathbf{p}_z + \beta m \qquad (18.22)$$

You might want to try a few and check that $\alpha_x^2 = 1$, that $\alpha_x \alpha_z = -\alpha_z \alpha_x$ and so forth. I include Box 18.5 with an example for reference below. Remember that these 4x4 matrices must act on something like a 1x4 matrix, that is, there must be at least four components. Or you may just prefer to believe and move on...

Box 18.5 Dirac matrices in action

Example: $\alpha_z^2 = 1$

$$\alpha_z \alpha_z \begin{bmatrix} a \\ b \\ c \\ d \end{bmatrix} = \begin{bmatrix} 0 & 0 & 1 & 0 \\ 0 & 0 & 0 & -1 \\ 1 & 0 & 0 & 0 \\ 0 & -1 & 0 & 0 \end{bmatrix} \begin{bmatrix} 0 & 0 & 1 & 0 \\ 0 & 0 & 0 & -1 \\ 1 & 0 & 0 & 0 \\ 0 & -1 & 0 & 0 \end{bmatrix} \begin{bmatrix} a \\ b \\ c \\ d \end{bmatrix} = \begin{bmatrix} 1 & 0 & 0 & 0 \\ 0 & 1 & 0 & 0 \\ 0 & 0 & 1 & 0 \\ 0 & 0 & 0 & 1 \end{bmatrix} \begin{bmatrix} a \\ b \\ c \\ d \end{bmatrix} = \begin{bmatrix} a \\ b \\ c \\ d \end{bmatrix}$$

Example: $\alpha_y \alpha_z = -\alpha_z \alpha_y$

$$\alpha_y \alpha_z \begin{bmatrix} a \\ b \\ c \\ d \end{bmatrix} = \begin{bmatrix} 0 & 0 & 0 & -i \\ 0 & 0 & i & 0 \\ 0 & -i & 0 & 0 \\ i & 0 & 0 & 0 \end{bmatrix} \begin{bmatrix} 0 & 0 & 1 & 0 \\ 0 & 0 & 0 & -1 \\ 1 & 0 & 0 & 0 \\ 0 & -1 & 0 & 0 \end{bmatrix} \begin{bmatrix} a \\ b \\ c \\ d \end{bmatrix} = \begin{bmatrix} 0 & i & 0 & 0 \\ i & 0 & 0 & 0 \\ 0 & 0 & 0 & i \\ 0 & 0 & i & 0 \end{bmatrix} \begin{bmatrix} a \\ b \\ c \\ d \end{bmatrix} = \begin{bmatrix} ib \\ ia \\ id \\ ic \end{bmatrix}$$

$$\alpha_z \alpha_y \begin{bmatrix} a \\ b \\ c \\ d \end{bmatrix} = \begin{bmatrix} 0 & 0 & 1 & 0 \\ 0 & 0 & 0 & -1 \\ 1 & 0 & 0 & 0 \\ 0 & -1 & 0 & 0 \end{bmatrix} \begin{bmatrix} 0 & 0 & 0 & -i \\ 0 & 0 & i & 0 \\ 0 & -i & 0 & 0 \\ i & 0 & 0 & 0 \end{bmatrix} \begin{bmatrix} a \\ b \\ c \\ d \end{bmatrix} = \begin{bmatrix} 0 & -i & 0 & 0 \\ -i & 0 & 0 & 0 \\ 0 & 0 & 0 & -i \\ 0 & 0 & -i & 0 \end{bmatrix} \begin{bmatrix} a \\ b \\ c \\ d \end{bmatrix} = - \begin{bmatrix} ib \\ ia \\ id \\ ic \end{bmatrix}$$

Some of you, if you are sharp, may be wondering if Equation 18.22 should not be ± (positive or negative) given that it is a square root. We don't have to worry about this because the negative option is actually contained in the positive solution. This sounds odd, but all will become clear.

18.3.4 We Are Finally There: Dirac's Equation

What Dirac established is another permissible relationship between energy, momentum and mass in Minkowski spacetime. The second order relationship $\mathbf{E}^2 = \mathbf{p}_x^2 + \mathbf{p}_y^2 + \mathbf{p}_z^2 + m^2$ was already well known and leads straight to the Klein-Gordon equation based on second derivatives. Dirac showed that a *first* order relationship $\mathbf{E} = \alpha_x \mathbf{p}_x + \alpha_y \mathbf{p}_y + \alpha_z \mathbf{p}_z + \beta m$ can exist consistent with Minkowski spacetime *if* the wave function has four components that are related in a very specific way.

This may strike you as a convoluted and complex structure, but the crucial point is that it *conforms* to Minkowski spacetime. Therefore in theory at least it *can* exist. The next (and relatively easy) step is to use this weird but permitted relationship to build a relativistic equation by substituting first order derivatives for the $\mathbf{E}$ and $\mathbf{p}$ terms.

Substituting in the derivatives:

$$\hat{\mathbf{E}}\Psi = i\hbar \frac{\partial \Psi}{\partial t} \qquad \hat{\mathbf{p}}_x \Psi = -i\hbar \frac{\partial \Psi}{\partial x} \tag{18.23}$$

This takes us to the Dirac equation for a free particle. I am going to continue with natural units so the speed of light $c = 1$ and also $\hbar = 1$. This makes things clearer. I will add them back later:

$$\mathbf{E} = \alpha_x \mathbf{p}_x + \alpha_y \mathbf{p}_y + \alpha_z \mathbf{p}_z + \beta m \tag{18.24}$$

$$\hat{\mathbf{E}}\Psi = \left(\alpha_x \hat{\mathbf{p}}_x + \alpha_y \hat{\mathbf{p}}_y + \alpha_z \hat{\mathbf{p}}_z + \beta m \right) \Psi \tag{18.25}$$

$$i\frac{\partial \Psi}{\partial t} = -i\left(\alpha_x \frac{\partial \Psi}{\partial x} + \alpha_y \frac{\partial \Psi}{\partial y} + \alpha_z \frac{\partial \Psi}{\partial z} \right) + \beta m \, \Psi \tag{18.26}$$

Equation 18.26 is the Dirac equation in a form similar to that of the Schrödinger equation. Let's look at how this works. We must acknowledge the major difference that Ψ must consist of four components that we can label Ψ_1, Ψ_2, Ψ_3 and Ψ_4. The action of the factors α_x, α_y, α_z and β mixes the order of these components around and/or changes some of the values. For example shown below is the action of α_y on $-i\frac{\partial \Psi}{\partial y}$. It mixes the components and changes their phase by multiplying them with positive or negative i.

$$\alpha_y \left(-i\frac{\partial \Psi}{\partial y} \right) = -i \begin{bmatrix} 0 & 0 & 0 & -i \\ 0 & 0 & i & 0 \\ 0 & -i & 0 & 0 \\ i & 0 & 0 & 0 \end{bmatrix} \begin{bmatrix} \partial \Psi_1/\partial y \\ \partial \Psi_2/\partial y \\ \partial \Psi_3/\partial y \\ \partial \Psi_4/\partial y \end{bmatrix} = \begin{bmatrix} -\partial \Psi_4/\partial y \\ \partial \Psi_3/\partial y \\ -\partial \Psi_2/\partial y \\ \partial \Psi_1/\partial y \end{bmatrix} \tag{18.27}$$

If you run through these calculations for all of the elements of the Dirac Equation 18.26, then you can show how the components inside line up:

$$i\frac{\partial \Psi}{\partial t} = -i\left(\alpha_x \frac{\partial \Psi}{\partial x} + \alpha_y \frac{\partial \Psi}{\partial y} + \alpha_z \frac{\partial \Psi}{\partial z} \right) + \beta m \, \Psi \tag{18.28}$$

$$i \begin{bmatrix} \partial \Psi_1/\partial t \\ \partial \Psi_2/\partial t \\ \partial \Psi_3/\partial t \\ \partial \Psi_4/\partial t \end{bmatrix} = -i \begin{bmatrix} \partial \Psi_4/\partial x \\ \partial \Psi_3/\partial x \\ \partial \Psi_2/\partial x \\ \partial \Psi_1/\partial x \end{bmatrix} + \begin{bmatrix} -\partial \Psi_4/\partial y \\ \partial \Psi_3/\partial y \\ -\partial \Psi_2/\partial y \\ \partial \Psi_1/\partial y \end{bmatrix} - i \begin{bmatrix} \partial \Psi_3/\partial z \\ -\partial \Psi_4/\partial z \\ \partial \Psi_1/\partial z \\ -\partial \Psi_2/\partial z \end{bmatrix} + m \begin{bmatrix} \Psi_1 \\ \Psi_2 \\ -\Psi_3 \\ -\Psi_4 \end{bmatrix} \tag{18.29}$$

This looks somewhat daunting. It takes some thought, but it is not as bad as it seems at first. What it means is that the four components of Ψ must be related in a very specific way in order to work in Minkowski spacetime. There are four individual equations linking the components together in a merry dance:

$$i\frac{\partial \Psi_1}{\partial t} = -i\frac{\partial \Psi_4}{\partial x} - \frac{\partial \Psi_4}{\partial y} - i\frac{\partial \Psi_3}{\partial z} + m\,\Psi_1 \tag{18.30}$$

$$i\frac{\partial \Psi_2}{\partial t} = -i\frac{\partial \Psi_3}{\partial x} + \frac{\partial \Psi_3}{\partial y} + i\frac{\partial \Psi_4}{\partial z} + m\,\Psi_2$$

$$i\frac{\partial \Psi_3}{\partial t} = -i\frac{\partial \Psi_2}{\partial x} - \frac{\partial \Psi_2}{\partial y} - i\frac{\partial \Psi_1}{\partial z} - m\,\Psi_3$$

$$i\frac{\partial \Psi_4}{\partial t} = -i\frac{\partial \Psi_1}{\partial x} + \frac{\partial \Psi_1}{\partial y} + i\frac{\partial \Psi_2}{\partial z} - m\,\Psi_4$$

As an analogy, think back to your schoolroom maths and simultaneous equations. Questions such as: *if $a + b + c = 8$, while $2a - b + 3c = 11$ and $a + 3b - 4c = -2$ then find a, b and c.* The answer has to respect all three equations simultaneously. In the same way valid solutions to the Dirac equation must obey all of the four equations above. Of course, it is very much more complicated as we are dealing with derivatives, but the concept is similar.

18.4 Spin-half Is Built in

You may wonder why anyone takes the Dirac equation seriously when it seems so bizarre. Well, for one thing, it gives the correct energy levels for hydrogen to several orders of magnitude better than Schrödinger managed. However, there is another strong reason. It has spin-half baked in. We will show that any particle which conforms to the Dirac equation *must* be intrinsic spin-half. The electron was found to be spin-half in the Stern-Gerlach experiment (see Figure 17.1) so this seems too good to be a coincidence (and it is not).

The maths that follows is quite complicated, so let me describe in layman language what Dirac discovered. He found that his equation does *not* allow any spherically symmetrical solutions. When measured (as in the Stern-Gerlach experiment), there must *always* be a specific amount of angular variation in the wave function. The Dirac equation requires intrinsic spin of $\frac{\hbar}{2}$.

To show that Dirac particles are spin-half takes a little maths but is worth the effort given the importance of the result. Consider a Dirac particle in an energy eigenstate so it has precise energy **E**. We measure its angular momentum in say the z direction and find it is L_z. Measuring L_z does not affect the sum square of the angular momentum L_{sum}^2 because there is no uncertainty relationship between $\hat{L}_z$ and $\hat{L}_{sum}^2$ (refer back to Box 14.2 if you need to). As L_{sum}^2 drives the angular kinetic energy, this means that there should be no uncertainty relationship between L_z and the energy of the object **E**, so we expect the commutator $[\hat{E}, \hat{L}_z] = 0$ unless there is some other angular momentum that we are missing. Let's see...

$$\hat{E}\Psi = \left(\alpha_x \hat{p}_x c + \alpha_y \hat{p}_y c + \alpha_z \hat{p}_z c + \beta mc^2\right)\Psi \qquad \hat{L}_z \Psi = \left(\hat{x}\hat{p}_y - \hat{y}\hat{p}_x\right)\Psi \tag{18.31}$$

$$[\hat{E}, \hat{L}_z] = \left([\alpha_x \hat{p}_x, \hat{L}_z] + [\alpha_y \hat{p}_y, \hat{L}_z] + [\alpha_z \hat{p}_z, \hat{L}_z]\right)c + [\beta, \hat{L}_z]mc^2 \tag{18.32}$$

Equation 18.31 comes from the Dirac Equation 18.25 (I am switching back from natural units to include c and $\hbar$ as we will need them) and the definition of $\hat{L}_z$ from Equation 14.13. We can quickly dispense with the last two terms in Equation 18.32 because both $[\alpha_z \hat{p}_z, \hat{L}_z]$ and $[\beta, \hat{L}_z]$ are zero. Operator $\hat{p}_z$ commutes with all the operators in $\hat{L}_z$ (check back to Section 11.3 if necessary). And while the α and β factors do not commute with each other, they happily commute with operators. Just as an example below, you can see that $\alpha_x \hat{p}_y \Psi = \hat{p}_y \alpha_x \Psi$. Whether any factor comes before or after an operator does not affect the way it mixes the components of Ψ:

$$\alpha_x \hat{p}_y \Psi = -i\hbar\,\alpha_x \begin{bmatrix} \partial \Psi_1/\partial y \\ \partial \Psi_2/\partial y \\ \partial \Psi_3/\partial y \\ \partial \Psi_4/\partial y \end{bmatrix} = -i\hbar \begin{bmatrix} -\partial \Psi_4/\partial y \\ \partial \Psi_3/\partial y \\ -\partial \Psi_2/\partial y \\ \partial \Psi_1/\partial y \end{bmatrix} = -i\hbar\frac{\partial}{\partial y} \begin{bmatrix} -\Psi_4 \\ \Psi_3 \\ -\Psi_2 \\ \Psi_1 \end{bmatrix} = \hat{p}_y \alpha_x \Psi \tag{18.33}$$

Returning to Equation 18.32, when we expand the remaining terms of $[\hat{E}, \hat{L}_z]$ we find that the value of each and of the total commutator is *not* zero. Remember that we can switch the order of everything *except* we cannot switch $\hat{p}_x$ with $\hat{x}$ or switch $\hat{p}_y$ with $\hat{y}$. This means that the two terms within the bracket on the far right of Equation 18.34 are the same and cancel out. The same is true of the two terms in the bracket on the left of Equation 18.36. And for Equation 18.35 and Equation 18.37, remember that $[\hat{x}, \hat{p}_x] = i\hbar$ and $[\hat{y}, \hat{p}_y] = i\hbar$. The result in Equation 18.38 shows that $[\hat{E}, \hat{L}_z] \neq 0$.

$$[\alpha_x\hat{\mathbf{p}}_x, \hat{L}_z] = (\alpha_x\hat{\mathbf{p}}_x\hat{x}\hat{\mathbf{p}}_y - \hat{x}\hat{\mathbf{p}}_y\alpha_x\hat{\mathbf{p}}_x) - (\alpha_x\hat{\mathbf{p}}_x\hat{y}\hat{\mathbf{p}}_x - \hat{y}\hat{\mathbf{p}}_x\alpha_x\hat{\mathbf{p}}_x) \tag{18.34}$$

$$= -\alpha_x\hat{\mathbf{p}}_y(\hat{x}\hat{\mathbf{p}}_x - \hat{\mathbf{p}}_x\hat{x}) = -i\hbar\,\alpha_x\hat{\mathbf{p}}_y \tag{18.35}$$

$$[\alpha_y\hat{\mathbf{p}}_y, \hat{L}_z] = (\alpha_y\hat{\mathbf{p}}_y\hat{x}\hat{\mathbf{p}}_y - \hat{x}\hat{\mathbf{p}}_y\alpha_y\hat{\mathbf{p}}_y) - (\alpha_y\hat{\mathbf{p}}_y\hat{y}\hat{\mathbf{p}}_x - \hat{y}\hat{\mathbf{p}}_x\alpha_y\hat{\mathbf{p}}_y) \tag{18.36}$$

$$= -\alpha_y\hat{\mathbf{p}}_x(\hat{y}\hat{\mathbf{p}}_y - \hat{\mathbf{p}}_y\hat{y}) = i\hbar\,\alpha_y\hat{\mathbf{p}}_x \tag{18.37}$$

$$[\hat{\mathbf{E}}, \hat{L}_z] = ([\alpha_x\hat{\mathbf{p}}_x, \hat{L}_z] + [\alpha_y\hat{\mathbf{p}}_y, \hat{L}_z])\, c = i\hbar c\,(\alpha_y\hat{\mathbf{p}}_x - \alpha_x\hat{\mathbf{p}}_y) \tag{18.38}$$

This is a huge problem that must be fixed. Why? Imagine a free particle in an energy-momentum eigenstate. Suppose it has *no* orbital angular momentum at all: $L_x = L_y = L_z = 0$ (see Equation 14.18 to see that we can know this). Then, no – not allowed. The particle cannot have defined energy. That is what the $[\hat{\mathbf{E}}, \hat{L}_z]$ commutator means. Dirac realised at once that the particle must have some intrinsic angular momentum around the z axis in addition to L_z. We will label this S_z because it is of course spin. When you add S_z to L_z it *must* cancel out the $[\hat{\mathbf{E}}, \hat{L}_z]$ commutator, so for any of this to make sense: $[\hat{\mathbf{E}}, \hat{S}_z] = -[\hat{\mathbf{E}}, \hat{L}_z]$. Dirac quickly discovered that this requires $\hat{S}_z = \frac{\hbar}{2i}\alpha_x\alpha_y$.

For the calculation we will need to apply the rule: $\alpha_x\alpha_y\alpha_z = \alpha_y\alpha_z\alpha_x = \alpha_z\alpha_x\alpha_y$. This is true because switching two factors changes the value to its negative since $\alpha_x\alpha_y = -\alpha_y\alpha_x$. Switching another two factors returns to the original value: $\alpha_x\alpha_y\alpha_z = -\alpha_y\alpha_x\alpha_z = \alpha_y\alpha_z\alpha_x$. And do not forget that any factor squared such as α_x^2 equals one. Noting this, let's calculate $[\hat{\mathbf{E}}, \hat{S}_z]$ for $\hat{S}_z = \frac{\hbar}{2i}\alpha_x\alpha_y$:

$$[\hat{\mathbf{E}}, \hat{S}_z] = ([\alpha_x\hat{\mathbf{p}}_x, \hat{S}_z] + [\alpha_y\hat{\mathbf{p}}_y, \hat{S}_z] + [\alpha_z\hat{\mathbf{p}}_z, \hat{S}_z])\, c + [\beta, \hat{S}_z]mc^2 \tag{18.39}$$

$$= ([\alpha_x\hat{\mathbf{p}}_x, \alpha_x\alpha_y] + [\alpha_y\hat{\mathbf{p}}_y, \alpha_x\alpha_y] + [\alpha_z\hat{\mathbf{p}}_z, \alpha_x\alpha_y])\,\frac{\hbar}{2i}c + [\beta, \alpha_x\alpha_y]\,\frac{\hbar}{2i}mc^2$$

$$[\beta, \alpha_x\alpha_y] = \beta\alpha_x\alpha_y - \alpha_x\alpha_y\beta = \beta\alpha_x\alpha_y - \beta\alpha_x\alpha_y = 0 \tag{18.40}$$

$$[\alpha_z\hat{\mathbf{p}}_z, \alpha_x\alpha_y] = \alpha_z\hat{\mathbf{p}}_z\alpha_x\alpha_y - \alpha_x\alpha_y\alpha_z\hat{\mathbf{p}}_z = (\alpha_z\alpha_x\alpha_y - \alpha_x\alpha_y\alpha_z)\hat{\mathbf{p}}_z = 0 \tag{18.41}$$

$$[\alpha_x\hat{\mathbf{p}}_x, \alpha_x\alpha_y] = \alpha_x\hat{\mathbf{p}}_x\alpha_x\alpha_y - \alpha_x\alpha_y\alpha_x\hat{\mathbf{p}}_x = (\alpha_x^2\alpha_y + \alpha_y\alpha_x^2)\hat{\mathbf{p}}_x = 2\alpha_y\hat{\mathbf{p}}_x \tag{18.42}$$

$$[\alpha_y\hat{\mathbf{p}}_y, \alpha_x\alpha_y] = \alpha_y\hat{\mathbf{p}}_y\alpha_x\alpha_y - \alpha_x\alpha_y\alpha_y\hat{\mathbf{p}}_y = (-\alpha_y^2\alpha_x - \alpha_x\alpha_y^2)\hat{\mathbf{p}}_y = -2\alpha_x\hat{\mathbf{p}}_y \tag{18.43}$$

$$so\quad [\hat{\mathbf{E}}, \hat{S}_z] = ([\alpha_x\hat{\mathbf{p}}_x, \alpha_x\alpha_y] + [\alpha_y\hat{\mathbf{p}}_y, \alpha_x\alpha_y])\frac{\hbar}{2i}c = (2\alpha_y\hat{\mathbf{p}}_x - 2\alpha_x\hat{\mathbf{p}}_y)\frac{\hbar}{2i}c \tag{18.44}$$

$$= -i\hbar c\,(\alpha_y\hat{\mathbf{p}}_x - \alpha_x\hat{\mathbf{p}}_y) \tag{18.45}$$

If you compare the results of Equations 18.38 and 18.45, you can see that they do cancel out. This means that $[\hat{\mathbf{E}}, \hat{L}_z + \hat{S}_z]$ is zero. If S_z is taken into account, there is no uncertainty relationship between angular momentum and energy. Problem solved. But what is this mysterious but essential S_z, and what does it tell us about the angular momentum of the Dirac equation? We know that $\hat{S}_z\Psi = S_z\Psi$, so applying the $\hat{S}_z$ operator to Ψ will reveal the intrinsic angular momentum (spin) around the z axis for each component of the wave function:

$$\hat{S}_z\Psi = \frac{\hbar}{2i}\alpha_x\alpha_y\begin{bmatrix}\Psi_1\\\Psi_2\\\Psi_3\\\Psi_4\end{bmatrix} = \frac{\hbar}{2}\frac{1}{i}\begin{bmatrix}0&0&0&1\\0&0&1&0\\0&1&0&0\\1&0&0&0\end{bmatrix}\begin{bmatrix}0&0&0&-i\\0&0&i&0\\0&-i&0&0\\i&0&0&0\end{bmatrix}\begin{bmatrix}\Psi_1\\\Psi_2\\\Psi_3\\\Psi_4\end{bmatrix} \tag{18.46}$$

$$= \frac{\hbar}{2}\begin{bmatrix}1&0&0&0\\0&-1&0&0\\0&0&1&0\\0&0&0&-1\end{bmatrix}\begin{bmatrix}\Psi_1\\\Psi_2\\\Psi_3\\\Psi_4\end{bmatrix} = \begin{bmatrix}\frac{\hbar}{2}\Psi_1\\-\frac{\hbar}{2}\Psi_2\\\frac{\hbar}{2}\Psi_3\\-\frac{\hbar}{2}\Psi_4\end{bmatrix} \tag{18.47}$$

The intrinsic spin of the components is $\pm\frac{\hbar}{2}$. The Ψ_1 and Ψ_3 components are $+\frac{\hbar}{2}$, while Ψ_2 and Ψ_4 are $-\frac{\hbar}{2}$. Zero intrinsic spin is *not* an option. So any particle obeying the Dirac equation is spin-half. Dirac must have known that he had stumbled upon something very important.

18.5   Interpreting the Dirac Equation

The simplest way to see what is going on is to take the four parts of the Dirac equation from Equation 18.30 and to write them in operator form (in this section I will include c and $\hbar$):

$$\hat{E}\Psi_1 = \hat{p}_x c \Psi_4 - i\hat{p}_y c \Psi_4 + \hat{p}_z c \Psi_3 + mc^2 \Psi_1 \qquad (18.48)$$

$$\hat{E}\Psi_2 = \hat{p}_x c \Psi_3 + i\hat{p}_y c \Psi_3 - \hat{p}_z c \Psi_4 + mc^2 \Psi_2$$

$$\hat{E}\Psi_3 = \hat{p}_x c \Psi_2 - i\hat{p}_y c \Psi_2 + \hat{p}_z c \Psi_1 - mc^2 \Psi_3$$

$$\hat{E}\Psi_4 = \hat{p}_x c \Psi_1 + i\hat{p}_y c \Psi_1 - \hat{p}_z c \Psi_2 - mc^2 \Psi_4$$

18.5.1   Zero Momentum: Distinct Spin and Antiparticles

Consider a Dirac spin-half particle (say, an electron) in an energy-momentum eigenstate with energy **E** and *zero* linear momentum. All of the linear momentum terms are zero so the equations become simpler because $\hat{p}_x \Psi = \hat{p}_y \Psi = \hat{p}_z \Psi = 0$.

For a particle with zero linear momentum:

$$\hat{E}\Psi_1 = mc^2 \Psi_1 \qquad\qquad Spin\ S_z = +\frac{\hbar}{2} \qquad (18.49)$$

$$\hat{E}\Psi_2 = mc^2 \Psi_2 \qquad\qquad Spin\ S_z = -\frac{\hbar}{2} \qquad (18.50)$$

$$\hat{E}\Psi_3 = -mc^2 \Psi_3 \qquad\qquad Spin\ S_z = +\frac{\hbar}{2} \qquad (18.51)$$

$$\hat{E}\Psi_4 = -mc^2 \Psi_4 \qquad\qquad Spin\ S_z = -\frac{\hbar}{2} \qquad (18.52)$$

You can see that the components Ψ_1 and Ψ_2 separately represent the electron in a spin-up and spin-down state. The way the α factors are set up separates out the S_z states, so this shows intrinsic spin around the z axis. You can rearrange the α factors to make the split between Ψ_1 and Ψ_2 reflect intrinsic spin around any axis. For example switching α_z and α_x splits Ψ_1 and Ψ_2 by intrinsic spin along the x axis. By symmetry we know that the intrinsic spin in all directions will always be $\pm\frac{\hbar}{2}$.

What was more of a surprise for Dirac (and everyone else) is Ψ_3 and Ψ_4. These turn out to be *antiparticle* states with negative mass/energy. I mentioned at the end of Section 18.3.3 that we do not have to worry about the negative square root option in the energy equation. This is because these negative states appear unavoidably in the Dirac equation anyway. Dirac puzzled long and hard over what Ψ_3 and Ψ_4 might be (we will discuss his thinking process later). Let's dig a bit deeper.

18.5.2   The Dirac Equation and Minkowski Spacetime

I have noted several times that Dirac's equation works because it conforms to Minkowski spacetime. We know this because it respects the energy balance equation: $\mathbf{E}^2 = \mathbf{p}^2 c^2 + m^2 c^4$. Let's take a moment to model what this means for Dirac's wave function in terms of Minkowski spacetime.

We derived the energy balance equation way back in Section 3.2.2 so you can be forgiven if it is no longer top of your mind. If any of the following perplexes you, then do look back. The relationships between energy, momentum, time and space in Minkowski spacetime are:

$$\mathbf{E} = mc^2 \frac{dt}{d\tau} \qquad\qquad \mathbf{p} = m\frac{dx}{d\tau} \qquad\qquad As\ shown\ in\ Equation\ 3.12 \qquad (18.53)$$

We can substitute these into the Dirac energy balance (Equation 18.24), divide by mc and multiply by $d\tau$ to model it back to the invariant interval equation of Minkowski spacetime.

$$\mathbf{E} = c\alpha_x \mathbf{p}_x + c\alpha_y \mathbf{p}_y + c\alpha_z \mathbf{p}_z + \beta mc^2 \qquad (18.54)$$

$$mc^2 \frac{dt}{d\tau} = mc\,\alpha_x \frac{dx}{d\tau} + mc\,\alpha_y \frac{dy}{d\tau} + mc\,\alpha_z \frac{dz}{d\tau} + \beta mc^2 \qquad (18.55)$$

$$c\,dt = \alpha_x\,dx + \alpha_y\,dy + \alpha_z\,dz + c\beta d\tau \quad Dirac\ equation\ in\ spacetime\ intervals \qquad (18.56)$$

What happens if we square both sides of Equation 18.56? The α and β matrices work in the usual way and out pops the invariant interval equation for Minkowski spacetime:

$$c^2\, dt^2 = (c\beta\, d\tau + \alpha_x\, dx + \alpha_y\, dy + \alpha_z\, dz)^2 = c^2\, d\tau^2 + dx^2 + dy^2 + dz^2$$
$$\implies \quad c^2\, d\tau^2 = c^2\, dt^2 - dx^2 - dy^2 - dz^2 \tag{18.57}$$

Equation 18.56 is the Dirac equation modelled from a spacetime perspective. Effectively his equation is a wave function with four components that entwine and combine together in spacetime in a way that conforms with the underlying invariant interval equation. If each component is like a note, the Dirac solution is like a chord – a group of closely related interacting components working in harmony. We can apply the α and β matrices to list in Equation 18.58 the four related equations just as we did earlier in Equation 18.48:

$$\implies \quad (\textit{if stationary})$$

$$c\,dt\,\Psi_1 = c\,d\tau\,\Psi_1 + dx\Psi_4 - idy\,\Psi_4 + dz\Psi_3 \implies dt\,\Psi_1 = d\tau\,\Psi_1 \tag{18.58}$$
$$c\,dt\,\Psi_2 = c\,d\tau\,\Psi_2 + dx\Psi_3 + idy\,\Psi_3 - dz\Psi_4 \implies dt\,\Psi_2 = d\tau\,\Psi_2$$
$$c\,dt\,\Psi_3 = -c\,d\tau\,\Psi_3 - dx\Psi_2 + idy\,\Psi_2 - dz\Psi_1 \implies dt\,\Psi_3 = -d\tau\,\Psi_3$$
$$c\,dt\,\Psi_4 = -c\,d\tau\,\Psi_4 - dx\Psi_1 - idy\,\Psi_1 + dz\Psi_2 \implies dt\,\Psi_4 = -d\tau\,\Psi_4$$

Why have I gone to all this trouble? Because it shows what Dirac's equation means in terms of the spacetime metric. Take a close look at the equations. For the Dirac equation to work, two of the equations *must* have $d\tau$ related to $-dt$. This is highlighted to the right of Equation 18.58 showing the equations when $dx = dy = dz = 0$. Without these negative time components there is no solution. What does this mean? It means that for the Dirac equation to work, there exist potential particle states that are the mathematical equivalent of the wave function moving backwards through time.

Distinct half-spin states fall naturally out of Dirac's work. The simplest example is the rest frame because there is no intermingling of the components. Using Dirac's multi-component $[\Psi_1, \Psi_2, \Psi_3, \Psi_4]$ wave function, $[1, 0, 0, 0]$ is the state of a stationary $+\frac{\hbar}{2}$ spin particle and $[0, 1, 0, 0]$ that of a stationary $-\frac{\hbar}{2}$ spin particle. Brilliant! But Dirac could not ignore $[0, 0, 1, 0]$ and $[0, 0, 0, 1]$ which are their mathematical equivalent with time reversed. How would the particle in those states appear to an observer? It would have the same mass but opposite charge – an antiparticle. It took Dirac some time to fully accept this (especially after peers ridiculed the idea). He published his equation in 1928 and finally concluded in 1930 that the electron must have an antiparticle. The *positron* was discovered in 1932 (see Box 18.6).

18.5.3 Particle and Antiparticle States

So, mathematically the positron can be described as an electron moving backwards in time. Changing the direction of time gives the negative result for energy from the energy operator that appears in antiparticle solutions of the Dirac equations such as in Equation 18.52:

$$\hat{E}\Psi = i\hbar\frac{\partial\Psi}{\partial t} = mc^2\,\Psi \implies \hat{E}\Psi = i\hbar\frac{\partial\Psi}{\partial(-t)} = -i\hbar\frac{\partial\Psi}{\partial t} = -mc^2\,\Psi \tag{18.59}$$

The existence of the antiparticle states is unavoidable for a massive Dirac particle. Viewed in the particle's rest frame, each of the four components of the wave function (Ψ_1 to Ψ_4) represents a different state depending on whether it is the particle or antiparticle and its spin. In the case of the electron, the antiparticle is the positron as shown below:

$$\begin{bmatrix}\Psi_1\\\Psi_2\\\Psi_3\\\Psi_4\end{bmatrix} = \begin{bmatrix}1\\0\\0\\0\end{bmatrix} \qquad \begin{bmatrix}\Psi_1\\\Psi_2\\\Psi_3\\\Psi_4\end{bmatrix} = \begin{bmatrix}0\\1\\0\\0\end{bmatrix} \qquad \begin{bmatrix}\Psi_1\\\Psi_2\\\Psi_3\\\Psi_4\end{bmatrix} = \begin{bmatrix}0\\0\\1\\0\end{bmatrix} \qquad \begin{bmatrix}\Psi_1\\\Psi_2\\\Psi_3\\\Psi_4\end{bmatrix} = \begin{bmatrix}0\\0\\0\\1\end{bmatrix}$$

$$\textit{Electron } S_z^+ \qquad\qquad \textit{Electron } S_z^- \qquad\qquad \textit{Positron } S_z^+ \qquad\qquad \textit{Positron } S_z^-$$

What happens when an electron meets a positron? Bang! They annihilate. This means particles can be destroyed and, by symmetry, can be created as particle-antiparticle pairs. There is nothing in Schrödinger's equations or traditional quantum mechanics that can handle particle creation and destruction, so this led Dirac and others in a completely new direction, quantum field theory, which we will discuss in the next chapter.

Box 18.6   Who discovered the positron?

Carl Anderson discovered positron tracks in 1932. He later acknowledged that he had built on the work of a Caltech classmate Chung-Yao Chao. Patrick Blackett and Giuseppe Occhialini, working at the Cavendish Laboratory, spotted the positron earlier in 1932, but decided to check their results before publishing. Dimitrij Skobelzyn, a Soviet scientist, reported particles like electrons but oppositely charged as early as 1927. Anderson controversially was the only one of them recognised in the Nobel Prize for the discovery in 1936. He who publishes first, wins!

Antiparticles have mass and energy at any given moment in time, so gravity is expected to remain an attractive force. There is a strong consensus that positrons will display gravitational attraction like the electron, but this has still not been confirmed experimentally. CERN is working actively on this. If it turns out not to be the case, it will surely win a Nobel Prize.

18.5.4   Moving Frame

Let's now consider how a spin-half particle looks if we observe it from a moving frame. We will point the x axis in the direction of motion so that $\mathbf{p}_x = \mathbf{p}$. The Dirac equations become:

For a particle with momentum in the x direction:

$$\hat{E}\Psi_1 = \hat{p}_x c\,\Psi_4 + mc^2\,\Psi_1 \implies E\Psi_1 = \mathbf{p}c\,\Psi_4 + mc^2\,\Psi_1 \tag{18.60}$$

$$\hat{E}\Psi_2 = \hat{p}_x c\,\Psi_3 + mc^2\,\Psi_2 \implies E\Psi_2 = \mathbf{p}c\,\Psi_3 + mc^2\,\Psi_2$$

$$\hat{E}\Psi_3 = \hat{p}_x c\,\Psi_2 - mc^2\,\Psi_3 \implies E\Psi_3 = \mathbf{p}c\,\Psi_2 - mc^2\,\Psi_3$$

$$\hat{E}\Psi_4 = \hat{p}_x c\,\Psi_1 - mc^2\,\Psi_4 \implies E\Psi_4 = \mathbf{p}c\,\Psi_1 - mc^2\,\Psi_4 \tag{18.61}$$

If the particle when viewed stationary is a pure spin-up electron (Electron S_z^+) then when stationary, $\Psi_1 = 1$ while the other components are $\Psi_2 = \Psi_3 = \Psi_4 = 0$. If we observe the same particle from a moving frame things get more complicated. You can see from Equation 18.60 that the momentum of the electron depends on Ψ_4 so the Ψ_4 component cannot be zero. The momentum linkage between Ψ_1 and Ψ_4 is also clear in Equation 18.61.

Suppose we set the value of component $\Psi_1 = 1$ and of $\Psi_4 = a$ then we can feed the 1 of Ψ_1 and the a of Ψ_4 into Equation 18.61 to calculate a:

$$E a = \mathbf{p}c\,(1) - mc^2\,a \implies a = \frac{\mathbf{p}c}{E + mc^2} \implies \begin{bmatrix} \Psi_1 \\ \Psi_2 \\ \Psi_3 \\ \Psi_4 \end{bmatrix} = \begin{bmatrix} 1 \\ 0 \\ 0 \\ \frac{\mathbf{p}c}{E + mc^2} \end{bmatrix} \tag{18.62}$$

To check, you can feed the answer back into the equations. For example, we can feed it back into Equation 18.60 and out pops the relativistic energy equation as expected.

$$E\Psi_1 = \mathbf{p}c\,\Psi_4 + mc^2\,\Psi_1$$

$$E(1) = \mathbf{p}c\left(\frac{\mathbf{p}c}{E + mc^2}\right) + mc^2\,(1)$$

$$E^2 + Emc^2 = \mathbf{p}^2 c^2 + m^2 c^4 + Emc^2 \implies E^2 = \mathbf{p}^2 c^2 + m^2 c^4 \tag{18.63}$$

So, when the electron is observed moving, we need to take into account this small relativistic correction in the mix of components. It is approximately $\frac{v}{2c}$. There is a similar mixing of the spin-up and spin-down components that depends on the speed of motion and its direction relative to the axis of spin measurement. So an electron that is spin-up will look to some moving observers to have a small component of spin-down. Note that if you ignore these relativistic corrections, you can derive the Schrödinger equation from the Dirac equation (see Box 18.7).

This is a good moment to point out the relationship between the number of dimensions of spacetime (4) and the number of components of the Dirac wave function (4). This is not a *direct* 4:4 correspondence, but the first does drive the second. Four spacetime dimensions and mass require four Dirac matrices and that is only possible using 4x4 matrices, so the wave function must have four components. If the particle were massless, the result would be very different because the three 2x2 Pauli matrices would suffice and the wave function would need only two components, not four. This will become relevant later when we discuss chirality in Chapter 24.

Box 18.7   Deriving the Schrödinger equation from the Dirac equation

As an example, consider the Dirac equation for a free (potential $V = 0$) spin-up electron moving in the x direction (using Equations 18.60 and 18.61):

$$\hat{\mathbf{E}}\Psi_4 = \hat{\mathbf{p}}_x c\,\Psi_1 - mc^2\,\Psi_4 \quad\Longrightarrow\quad \mathbf{E}\,\Psi_4 + mc^2\,\Psi_4 = -i\hbar c\,\frac{\partial\Psi_1}{\partial x}$$

$$\Longrightarrow\quad \Psi_4 = \frac{-i\hbar c}{\mathbf{E}+mc^2}\frac{\partial\Psi_1}{\partial x} \tag{18.64}$$

$$\hat{\mathbf{E}}\Psi_1 = \hat{\mathbf{p}}_x c\,\Psi_4 + mc^2\,\Psi_1$$

$$= -i\hbar c\,\frac{\partial\Psi_4}{\partial x} + mc^2\,\Psi_1 \qquad\qquad \textit{Substitute from Equation 18.64}$$

$$= -i\hbar c\,\frac{\partial}{\partial x}\left(\frac{-i\hbar c}{\mathbf{E}+mc^2}\frac{\partial\Psi_1}{\partial x}\right) + mc^2\Psi_1$$

$$= -\frac{\hbar^2 c^2}{\mathbf{E}+mc^2}\frac{\partial^2\Psi_1}{\partial x^2} + mc^2\Psi_1$$

$$\approx -\frac{\hbar^2 c^2}{2mc^2}\frac{\partial^2\Psi_1}{\partial x^2} + mc^2\Psi_1 \qquad\qquad \textit{For non-relativistic: } \mathbf{E} \approx mc^2$$

$$\approx -\frac{\hbar^2}{2m}\frac{\partial^2\Psi_1}{\partial x^2} + mc^2\Psi_1 \qquad\qquad \textit{For non-relativistic: } \Psi_1 \approx \Psi \tag{18.65}$$

$$\hat{\mathbf{E}}\Psi \approx -\frac{\hbar^2}{2m}\frac{\partial^2\Psi}{\partial x^2} \qquad\qquad \textit{Schrödinger's equation ignores } mc^2\Psi \tag{18.66}$$

At non-relativistic speeds $E \approx mc^2$ and the Ψ_4 component is tiny so $\Psi_1 \approx \Psi$. The result of Equation 18.66 is the Schrödinger equation with no potential ($V = 0$). It measures energy over and above the rest mass energy (as explained in Subsection 11.1.1). You can do the same calculation for a spin-down particle by swapping Ψ_3 for Ψ_4 and Ψ_2 for Ψ_1.

18.6   The Dirac Equation and Hydrogen

We have gone as far as we can with the maths of the Dirac equation so I will fast forward to the result. Dirac took his equation with its relativistic corrections and applied it to the Coulomb potential to generate the energy eigenstates of an electron in the hydrogen atom. The result was stunning and a complete vindication of his equation.

$$\mathbf{E} = \frac{mc^2}{\sqrt{1 + \frac{\alpha^2}{\left(n-j-\frac{1}{2}+\sqrt{\left(j+\frac{1}{2}\right)^2-\alpha^2}\right)^2}}} \approx mc^2 - \frac{m\alpha^2 c^2}{2n^2} - \frac{m\alpha^4 c^2}{2n^4}\left(\frac{n}{j+\frac{1}{2}} - \frac{3}{4}\right) \tag{18.67}$$

The energy levels in Dirac's result depend on two quantum numbers (n and j), just as Schrödinger's result depended on two quantum numbers (n and l). The difference is that quantum number j is the sum of the angular momentum and spin. When you expand out Dirac's result, as shown on the right of Equation 18.67, the first term in the result is the rest mass (as expected). The second term is exactly the result that Rydberg found and Schrödinger confirmed. You can compare the result with that of Schrödinger's (see Equation 15.45).

But there are further terms, the largest of which depends on α^4 and is shown on the far right of Equation 18.67. The value of α, the fine structure constant, is about $1/137$ so α^4 is very small. However, this term slightly shifts the predicted emission spectrum of hydrogen versus Schrödinger's result. This discrepancy had already been detected experimentally. Dirac's equation was a much better match to the data than Schrödinger's.

18.7 Dirac Equation: Modern Formulation

The Dirac equation is often shown in perhaps a more modern form (below) that puts all four dimensions x, y, z and t on an equal footing, generally using natural units ($c = \hbar = 1$).

$$i\gamma^\mu \frac{\partial}{\partial x_\mu} \Psi = m \Psi \qquad \textit{Another form of the Dirac equation} \tag{18.68}$$

Personally, I find Dirac's original version more approachable, but there are advantages to this form. It is more succinct and easier to manipulate in advanced calculations. The logic is exactly the same, but the starting point treats all four dimensions equally by equating energy and momentum to mass as follows:

$$\mathbf{E} = \sqrt{\mathbf{p}_x^2 + \mathbf{p}_y^2 + \mathbf{p}_z^2 + m^2} \qquad \textit{Our starting point (Equation 18.7)} \tag{18.69}$$

$$m = \sqrt{\mathbf{E}^2 - \mathbf{p}_x^2 - \mathbf{p}_y^2 - \mathbf{p}_z^2} \qquad \textit{Starting point for modern formulation} \tag{18.70}$$

As the starting point is altered, the 4x4 matrices are slightly different. For example the matrices in front of $\mathbf{p}_x$, $\mathbf{p}_y$ and $\mathbf{p}_z$ must square to -1 rather than 1 to reflect the difference in signs between Equation 18.70 and 18.69. The matrices for this are usually labelled γ^0 (for t), and γ^1, γ^2 and γ^3 (for x, y and z). They are shown in Box 18.8 for reference.

$$m = \sqrt{\mathbf{E}^2 - \mathbf{p}_x^2 - \mathbf{p}_y^2 - \mathbf{p}_z^2} = \gamma^0 \mathbf{E} + \gamma^1 \mathbf{p}_x + \gamma^2 \mathbf{p}_y + \gamma^3 \mathbf{p}_z \tag{18.71}$$

$$m\Psi = i\gamma^0 \frac{\partial \Psi}{\partial t} + i\gamma^1 \frac{\partial \Psi}{\partial x} + i\gamma^2 \frac{\partial \Psi}{\partial y} + i\gamma^3 \frac{\partial \Psi}{\partial z} = i\gamma^\mu \frac{\partial}{\partial x^\mu} \Psi \tag{18.72}$$

The right side of Equation 18.72 uses abbreviated notation based on what is called the Einstein summation convention, which we will introduce later. It is often abbreviated further to $i\gamma^\mu \partial_\mu \Psi$ and it indicates that the equation is the sum of the terms for each dimension t, x, y and z.

Box 18.8   Dirac gamma matrices

$$\gamma^0 = \begin{bmatrix} 1 & 0 & 0 & 0 \\ 0 & 1 & 0 & 0 \\ 0 & 0 & -1 & 0 \\ 0 & 0 & 0 & -1 \end{bmatrix} \qquad \gamma^1 = \begin{bmatrix} 0 & 0 & 0 & 1 \\ 0 & 0 & 1 & 0 \\ 0 & -1 & 0 & 0 \\ -1 & 0 & 0 & 0 \end{bmatrix}$$

$$\gamma^2 = \begin{bmatrix} 0 & 0 & 0 & -i \\ 0 & 0 & i & 0 \\ 0 & i & 0 & 0 \\ -i & 0 & 0 & 0 \end{bmatrix} \qquad \gamma^3 = \begin{bmatrix} 0 & 0 & 1 & 0 \\ 0 & 0 & 0 & -1 \\ -1 & 0 & 0 & 0 \\ 0 & 1 & 0 & 0 \end{bmatrix}$$

18.8 The Aftermath: Physics Falls Apart Again

The Dirac equation caused an enormous stir. It was so very unexpected and highlighted so many questions and problems. Before we address the problems, let's summarise the chapter so far and bathe in the glory of Dirac's amazing success.

Dirac's seemingly crazy starting point was to eliminate the squares in the relativistic energy-momentum relationship by developing a formula for **E** that is based on $\mathbf{p}_x$, $\mathbf{p}_y$, $\mathbf{p}_z$ and m:

$$\mathbf{E} = \sqrt{\mathbf{p}_x^2 + \mathbf{p}_y^2 + \mathbf{p}_z^2 + m^2} = \alpha_x \mathbf{p}_x + \alpha_y \mathbf{p}_y + \alpha_z \mathbf{p}_z + \beta m \tag{18.73}$$

For this to be true, the factors α_x, α_y, α_z, and β must have the relationship that each squares to 1 (e.g. $\alpha_x^2 = 1$) and when multiplied with each other the order of multiplication turns the value from positive to negative (e.g. $\alpha_x \alpha_y = -\alpha_y \alpha_x$). This requires matrices. The smallest type of matrix that can give us four such factors is a 4x4 matrix. This means in turn that any wave function Ψ that conforms to the Dirac equation must have four components.

When you analyse the four components you find that two of them relate to a particle (one spin-up, one spin-down) and the other two relate to equivalent spin-up/spin-down states, but of an *antiparticle*. In the case of the electron, this is the positron, but it was unknown at the time.

Faced with such a tortuous piece of maths plus such a bizarre result, one's natural reaction is to burst out laughing and tell Dirac to stop wasting time on this nonsense, if it were not for two things. *First*, the maths behind the Dirac equation requires any such particle to be spin-half exactly as experiments had shown to be the case for the electron. *Second*, and this was huge, the Dirac equation correctly predicts the hydrogen spectrum to an accuracy several orders of magnitude better than the Schrödinger equation.

You cannot explain away success like this. The physics community just *had* to take Dirac's equation to heart and that brings with it a lot of extra baggage, particularly the weird sounding antiparticle components of the wave function – problems, problems, problems... (see Box 18.9 and Box 18.10 for some common puzzlers).

Box 18.9   FAQ: Frequently Asked Questions

Will an electron, observed at speed, sometimes be seen as a positron? – In Equation 18.62 an electron with momentum has a small negative energy component. Is there a chance it is a positron? *No! Every smart student should ask this. Good luck finding a clear answer on the web, but no, that would violate charge conservation. The energy footprint or vibration that we call an electron has some of the maths of a positron built in. But an electron is an electron and a positron is a positron until they meet, then bang!*

Is there only one electron in the universe bouncing forwards and backwards through time? Another great question but – *No! If it were so, electrons and positrons should be equal in number.*

Dirac had doubts himself and first sought to explain away the antiparticle states as being the proton, but this clearly will not work. The mass of the proton is almost 2000x that of the electron. If you run through the sequence of equations in Box 18.7 with say mass m for the Ψ_1 component and a much larger mass M for component Ψ_4, you will find the Schrödinger approximation results in a mismatch between the two masses that does not work at all. This is shown below based on the result in Equation 18.65 of Box 18.7.

$$\hat{\mathbf{E}}\,\Psi \approx -\frac{\hbar^2}{2M}\frac{\partial^2 \Psi}{\partial x^2} + mc^2\Psi \qquad \textit{Mismatch if mass of } \Psi_1 \textit{ and } \Psi_4 \textit{ are different} \tag{18.74}$$

It took a couple of years for Dirac to conclude that the antiparticle states really *are* antiparticle states. To reconcile this with the existing framework of quantum mechanics, he proposed an infinite number of negative energy states filled with an infinite number of electrons. An electron can be excited from the first negative energy state ($-mc^2$) up to the first positive energy state ($+mc^2$) with an infusion of energy from, say, a photon. This leaves a hole in the negative energy sea of states that acts exactly like an antiparticle to the electron (same mass, opposite charge). At some stage an electron will fall back into that hole releasing $2mc^2$ energy. His theory is called the *Dirac sea*.

While this provides a workable mathematical structure, it is far from elegant. An infinite number of electrons hidden in negative energy states? Yuk! That sounds messy. Heisenberg's response was: *I regard the Dirac theory... as learned trash which no one can take seriously.*

The underlying problem was that the antiparticle states were there for all to see and the positron subsequently was found by Anderson and others. Yet nothing in any of the existing theories allowed for particle creation or destruction. Dirac was trying to shore things up with his electron sea, but the structure of quantum mechanics was like a boat that had been holed beneath the waterline. For a while everyone tried to bail out water and keep the structure afloat, but with time, it became clear that what was needed was a new boat. The ensuing chaos and frustration among physicists is best conveyed through a couple of quotes from Heisenberg reflecting on the new challenge:

The saddest chapter of modern physics is and remains the Dirac theory.

Until that time I had the impression that in quantum theory we had come back into the harbor, into the port. Dirac's paper threw us out into the open sea again. Everything got loose again and we got into new difficulties. Of course at the same time, I saw that we had to go that way. There was no escape from it because relativity is true.

Box 18.10   Could antimatter galaxies and stars exist?

Dirac in his 1933 Nobel Prize acceptance speech mused on the possibility of antimatter stars:

If we accept the view of complete symmetry between positive and negative electric charge so far as concerns the fundamental laws of Nature, we must regard it rather as an accident that the Earth (and presumably the whole solar system), contains a preponderance of negative electrons and positive protons. It is quite possible that for some of the stars it is the other way about, these stars being built up mainly of positrons and negative protons. In fact, there may be half the stars of each kind. The two kinds of stars would both show exactly the same spectra, and there would be no way of distinguishing them by present astronomical methods.

Interactions between matter and antimatter have been spotted close to the core of the Milky Way. The antimatter is believed to be produced by collisions involving black holes or neutron stars. It crashes into dust and stars generating detectable bursts of gamma radiation. But overall, our galaxy is *matter* with only tiny traces of antimatter.

On a broader scale, the universe appears to be too quiet to allow larger scale matter-antimatter interaction. Furthermore, if isolated antimatter galaxies exist, the antimatter supernovas would produce a different signature of particle emissions from that produced by their matter counterparts. So, the answer appears to be *no* to widespread antimatter galaxies and stars.

But there is an underlying symmetry that suggests the Big Bang should have produced equal amounts of matter and antimatter. What broke the symmetry? Where did all the antimatter go? And yes, the answer to that question does – matter.

Modern physics has found hints of possible answers, but nothing certain. If you know for sure, ring Stockholm and get ready to polish your gold Nobel medal – sadly only 175 grams of 18 carat gold since 1980, down from 200 grams of 23 carat gold earlier – cheapie cheapie!

The first person to drive forwards towards a new solution was, needless to say, that genius Dirac. This brings us to the topic of the next chapter, quantum field theory, and as you can see in the final line of the second of Heisenberg's quotes, yet again it is the requirements of *relativity* that will guide us on the path ahead.

Chapter 19

Quantum Field Theory

CHAPTER MENU
19.1 Changing the Question
19.2 Quantum Fields Win the Day
19.3 Non-relativistic Path Integrals and Action
19.4 QFT Path Integrals: A Relativistic Twist
19.5 Energy and Time
19.6 QFT Field Development Pathways
19.7 The Klein-Gordon Lagrangian as a Model
19.8 Global Gauge Invariance to Phase
19.9 Summary

Box 19.1   Dirac's gown
Dirac left Cambridge in 1972 to take up a post in Florida. He was known to be absent-minded and left his Cambridge gown with a note: *Prof Dirac's gown. Please take it to the Master and ask him to keep it until the next time I come to Cambridge.* Dirac never returned. He died in Florida in 1984. The cause of death was not reported. In his honour, his gown still hangs in the Master's Lodge at St John's College, Cambridge.

QFT is often a graduate course so way more advanced than anything we have handled so far. Why then, you might ask, have I devoted a significant chunk of this book to the subject? The answer is simple. If I don't cover QFT then I am not giving you any clue of how modern physics stands. It would be like telling a long joke and then omitting the punch line. So we will follow Dirac's brilliant mind further (check out Box 19.1 for the tale of his gown).

Before we jump in, it is important to put into context everything you have read so far in this book. You are about to learn that physicists have concluded things are fundamentally different. Up to now we have discussed particles as having wavelike features. The reality, it transpires, is that we are not really studying particles at all. What we call particles are actually wave patterns that come in quantum lumps, quanta of excitation that we interpret as the presence of particles.

But don't be disheartened. The QFT view of physics does not undermine what we have studied so far. The difference becomes apparent only when relativistic effects dominate, so sit back, relax and let the QFT-world described in these chapters flow over you. The status of non-relativistic versus relativistic quantum mechanics is similar to that of Newtonian mechanics versus special relativity. Newtonian mechanics is simpler and works for any normal-world problem. In the same way, the non-QFT stuff that we have studied will work just fine for any quantum problem, unless you are planning to work at the Large Hadron Collider.

This is not easy stuff. For those students who go on to study QFT seriously, I apologise for all the shortcuts I have been forced to use. I can only hope that this is a useful road map when you start reading more detailed textbooks on the subject. For the rest of you, I will try to give you a warm cosy feeling that you understand some little pieces of QFT. I can hope for no more than that. Box 19.2 gives you an early indication of the excitement to come.

Quantum Untangling: An Intuitive Approach to Quantum Mechanics from Einstein to Higgs, First Edition. Simon Sherwood.
© 2023 John Wiley & Sons Ltd. Published 2023 by John Wiley & Sons Ltd.

Box 19.2   Spoiler alert!

The remaining chapters contain some heavy stuff so here is a taste of what is to come. You are going to learn that the laws of physics afford the quantum field of the electron (spin-half) a type of phase flexibility that *requires* the existence of the photon (spin-1). How cool is that?

19.1   Changing the Question

Throughout this book we have tried to satisfy the needs of special relativity, but we have fallen short. Dirac realised that we need to change the question we are asking. Our approach so far has been to ask: *when, where, what*? We have looked at how the wave function of a particle evolves over time (*when*), the spatial shape of that wave function as it develops (*where*) and then drawn conclusions about the particle's behaviour (*what*). This gives time (*when*) and space (*where*) central roles, and that creates a conflict. Clearly, time and space are different, but relativity tells us they must be, to some extent, interchangeable.

QFT avoids this problem. The question is simply: *what*? What is the probability of going from a given initial state (such as two particles heading into a possible reaction with certain momentum) to a given final state (particles coming out with certain momentum). We calculate the total *probability* of going from the initial state to the final state (typically labelled $< f | i >$ with $| i >$ and $| f >$ as initial and final states). No assumption is made about *when* or *where* any interaction occurs (see Box 19.3).

In QFT, neither time nor space are operators. They are just coordinates that you integrate over in a calculation. In fact Dirac realised that time and space, even as a coordinate system, are fickle bedfellows. They *change*. There is no obvious relationship between the time and space coordinates of state $| i >$ and state $| f >$. So, instead of focussing on spacetime, we frequently use energy-momentum coordinates (energy-momentum space). These contain the full quantum picture because they are the Fourier transforms of the spacetime quantities and, importantly, energy and momentum are conserved (for a reminder on the link between spacetime and energy-momentum coordinates, flick back to Figure 3.1). This makes them much easier to work with. They are reliable constant friends – whenever, wherever and however an interaction may occur.

Box 19.3   Calculations in QFT

(1) Calculate the probability amplitude of a certain interaction path that I will call S. Make no assumption about exactly when or where it happens. This is: $< f | S | i >$

(2) The square of the absolute value of this amplitude is the probability of it occurring:

$$Probability_{<f|S|i>} \ = \ | < f | S | i > |^2$$

(3) Use Feynman diagrams to consider alternative interaction paths that give the same outcome and then combine the probabilities (we will cover this later) to arrive at $Probability_{<f|i>}$

Some of the things in this chapter are likely to strike you as overly complicated and obscure, but time and space *must* be put on an equal footing. To date, this is the best approach found.

19.2   Quantum Fields Win the Day

Dirac's four-component equation and the subsequent discovery of the positron show that quantum mechanics must allow for the creation and destruction of particles. And Feynman's path integral shows particles seemingly directed by the combined effect of a vibrating field of wave functions that samples every possible path. How then can we treat particles as individual discrete lumps as Schrödinger's equation does? What if we alter our perspective? Instead of putting particles at the core of our theory directed by some vibrating field, let's put the vibrating field at the core and assume that the vibrations are what we call particles.

Let me be clear upfront that this is *not* a physical field vibrating in our 3+1 dimensional universe, just as the complex wave function we studied earlier is not a physical oscillating object. The quantum fields we will be discussing in this chapter are a mathematical representation that can be used to describe (and predict) the behaviour of particles and their interactions.

How does a vibrating quantum field behave? Step forward the Quantum Harmonic Oscillator (QHO). The QHO describes the behaviour of a generalised quantum field that is disturbed from equilibrium. If you cannot remember why, check back to Subsection 13.1.2. We will be referring to quite a lot of things in Chapter 13 so you may want to keep it handy.

The QHO vibrates at distinct energy levels that are separated by $E = \hbar\omega$. Only certain waves are stable in the same way as a guitar string will only play certain notes. Give the QHO a small kick and it will move up to the next level of excitation and absorb $\hbar\omega$ energy. Similarly it may drop to a lower level of excitation and release energy, but always in multiples of $\hbar\omega$. Figure 13.3 shows the pattern of QHO excitation.

The parallel with light photons is clear. Consider a beam of light of angular frequency ω (the link to frequency is $\omega = 2\pi f$). Light is the excitation of an electromagnetic field. If that is a quantum field then the excitation will be in lumps of $\hbar\omega$ which is exactly the energy of photons: $E = \hbar\omega$. A change in the excitation level of the field (i.e. a change in the vibration) releases a lump of energy that we call a photon.

QFT generalises this. Vibrating quantum fields sit behind everything. A change in vibration releases a lump of energy that we call a particle. A photon is a step in the excitation of the electromagnetic field. An electron is a step in the excitation of the electron field. Why are all electrons the same? Because they are rungs of energy on the ladder of excitation of the same electron field. This is called *indistinguishability* and is true for all elementary particles. Vibrating fields are the reality, particles are the illusion; particles are our interpretation of steps up and down in the excitation of quantum fields.

19.2.1 The Quantum Field Structure

Each particle is associated with a distinct field. You can think of each quantum field as having a structure similar to a multitude of QHO oscillators. Let's follow this analogy and focus on one of these QHO-like oscillators that I will call the $\mathbf{p}_1$ oscillator (note Box 19.4 on variable labels). Its vibration is related to particles of the field with momentum $\mathbf{p}_1$. If there is one such particle then the oscillator is in excitation level $n = 1$. If it is suitably excited it will jump up to $n = 2$ and we interpret that as two particles with momentum $\mathbf{p}_1$. We can excite it further up to perhaps $n = 5$. From the perspective of a quantum field we say this particular oscillator has jumped to the fifth energy level. From the perspective of classical physics, we say there are five particles with momentum $\mathbf{p}_1$. Note that this must be a boson field (integer-spin particles) as the Pauli exclusion principle does not allow fermions (spin-half particles) to be in the same state, but more on that later.

Box 19.4 p and E versus k and ω

Many QFT texts use wave number k in preference to momentum $\mathbf{p}$ and angular frequency ω in preference to energy E. For most of this chapter I use natural units ($\hbar = c = 1$) so they are equivalent. I prefer sticking with momentum and energy because they are more intuitive.

$$\mathbf{p} = \hbar k \quad E = \hbar\omega \quad \textit{so in natural units} \quad \mathbf{p} = k \quad E = \omega$$

In addition to the $\mathbf{p}_1$ oscillator, there is a $\mathbf{p}_2$ QHO-like oscillator. The energy level of the vibrations of that oscillator is the number of particles with momentum $\mathbf{p}_2$. In fact, the overall quantum field is composed of an infinite number of these oscillators. We interpret the excitation level of each oscillator as the number of particles with that particular momentum.

Equation 19.1 shows how to represent the state of the quantum field if it is composed of 15 particles: 3 particles with momentum $\mathbf{p}_1$, none with $\mathbf{p}_2$, 7 with $\mathbf{p}_3$, 3 with $\mathbf{p}_4$ and 2 with $\mathbf{p}_5$... (I am showing only 5 momentum levels to keep it manageable):

$$\Psi = |3, 0, 7, 3, 2...> \tag{19.1}$$

More generally, we can say that the quantum field is the combined sum of the vibration of all the oscillators from the one for momentum $\mathbf{p} = -\infty$ up to the one for $\mathbf{p} = +\infty$. What is the combined wave function of the quantum field? We know the wave function for the free particle wave function for each momentum. In Equation 19.2 this is shown on the left for the quantum field of Equation 19.1 and more generally on the right using the symbol α as the coefficient at each momentum.[1]

$$\Psi = 3e^{-ip_1x} + 7e^{-ip_3x} + 3e^{-ip_4x} + 2e^{-ip_5x}... \implies \Psi = \sum_{p=-\infty}^{+\infty} \alpha_{\mathbf{p}}\, e^{-ipx} \tag{19.2}$$

You may be puzzled by this madness. It all looks horribly complicated, but here is the clever bit. The quantum field Ψ has a neat mathematical structure. Each oscillator has a different wavelength. Think back to the discussion of Fourier analysis in Section 10.1. You can decompose any waveform or function into the sinusoids that make it up. In the same way, the quantum field Ψ contains all the information on the excitation state of each oscillator for each state of momentum and the composition is uniquely defined by Ψ. If you work with Schrödinger's equation, you must work with only one particle at a time. The quantum fields of QFT can each handle as many as you want.

The model above is useful and intuitive, but things are actually a bit more complicated (sorry!). Each oscillator in the quantum field is unlikely to be in an exact energy eigenstate so it will not indicate a precise number of particles. It will be in a superposition of energy eigenstates. This is an important point so check back to Section 13.3.4 if you need to. This means that the quantum oscillator for a particular momentum describes the relative probability of finding different numbers of particle at that momentum. The quantum field is the sum of all these oscillators so it contains all the information on the probability of finding various numbers of particles in various states of momentum. In quantum mechanics, it seems nothing is ever certain – not even the number of particles – this is a far cry from Schrödinger's equation with its eternal indestructible particles (and not without problems if you try to include gravity in the model, see Box 19.5).

Box 19.5 Infinite energy in the vacuum – Oops!

Quantum fields are composed of an infinite number of oscillators, but each oscillator's ground state energy is $\frac{\hbar\omega}{2}$ implying that the vacuum has infinite energy. In QFT we sweep this under the carpet by considering only *changes* in energy; but how do you reconcile this when infinite energy means infinite gravity? Oops!

19.2.2 Quantum Fields and Spin

The quantum field described above is a *scalar* field. The field has a *value* at every point in spacetime. We can use the height of the water in a swimming pool as an intuitive model. At any point on the surface of the pool, the height of the water is a numerical value. The fluctuations in the height reflect the underlying undulations in the pool just as the values of a quantum field reflect the underlying undulations of its oscillators. If you take all the values for the height of the water in the pool and perform a Fourier transform you can describe it as the sum of a combination of sinusoidal waves. In the same way, the value of a quantum field contains the excitation level of the underlying oscillators which, from a particle perspective, depends on the likely number of particles in each quantum state.

Something is missing: spin. A quantum field has to contain all the information about the particles it describes. This is obvious because the quantum field *is* the particles it describes. So for a spin-1 particle such as the photon, the quantum field must be a *vector* field, not a scalar field. At each point of spacetime it must have a value and a direction (you can imagine this as a little arrow).

In this chapter we will start with scalar quantum fields and extend from there. Just remember that the underlying quantum field must reflect the spin: scalar fields for spin-0 particles, vector fields for spin-1 particles and so-called spinor-fields for spin-half particles. If you have the courage to study general relativity you will find that curved spacetime is an *order-2 tensor* field. The difference between this and a vector is complicated and beyond our scope, but this is why if we find a particle associated with gravity (the graviton), we expect it to be spin-2.

19.2.3 Creation and Annihilation

It is time to get to grips with the maths. For the QHO, we developed some mathematical operations to move the QHO from one level of excitation to the next. These creation and annihilation operators ($\hat{a}^+$ and $\hat{a}^-$) move up and down energy levels of the QHO (Subsection 13.3.2). Similar operators are used in QFT but each step up and down is the creation or annihilation of a *particle*. Every particle interaction is treated as annihilation of the original particle and creation of a new one. If say an electron changes momentum then the quantum field changes. The excitation level of the quantum oscillator relating to its original momentum steps down (annihilation) and the excitation of the oscillator relating to the new momentum steps up (creation).

This can seem tortuous, but if a photon is absorbed and subsequently emitted by an atom, would you say it is the *same* photon? What if it is emitted at a different wavelength? Why is an electron or any other particle different? Going into an interaction, the electron is an excitation in the electron field. Coming out, it has different momentum and so it is a different excitation in the field. The original excitation disappears and is replaced by a new excitation. The idea that it is the *same* electron no longer makes much sense.

To create a particle with momentum **p**, we apply the creation operator for that momentum. If we do that to the vacuum which is shown typically as $|0>$, then the result will be one particle in the **p** momentum state. Remember always to apply the $\sqrt{n+1}$ factor when using the creation operator although it changes nothing in this simple example (you may need a reminder from Equation 13.44). The result can be written showing a string of empty momentum states except for one particle in the **p** momentum state:

$$\hat{a}_{\mathbf{p}}^{+} e^{-i\mathbf{p}x} |0> = \sqrt{n+1} |n+1>_{\mathbf{p}} = |..,0,0,1_{\mathbf{p}},0,0..> \tag{19.3}$$

If you are dealing with bosons such as the photon, you can apply the creation operator again and again to add further particles in the same momentum state. However, fermions such as the electron are different because of the Pauli exclusion principle, so be careful. If you act with the creation operator to create a fermion in the momentum state **p**, you cannot act with it again to add another identical fermion in the same momentum state. This, of course, is why there can be only two electrons in a particular atomic angular momentum state, one spin-up and the other spin-down.

How do we create a particle at an exact position x? We want to work in energy-momentum coordinates so let's express this in terms of momentum **p**. A particle with an exact position x is in a superposition of *all* momentum states (if you are not completely familiar with this, then shame on you, and back to Section 7.2). In order to create a particle at x, we need to add a teeny weeny bit of amplitude to the quantum field at every value of momentum. This is shown in Equation 19.4. Compare it with Equation 19.2. All we have to do is to make the quantum field into an operator by injecting creation operators that add the equivalent of a little bit of a particle at every momentum. It also needs a normalisation factor which I am ignoring. The annihilation operator is the adjoint of the creation operator (basically the complex conjugate equivalent) so the adjoint quantum field operator annihilates a particle at x as shown in Equation 19.5.

$$\text{Create particle at } x: \quad \sum_{\mathbf{p}=-\infty}^{+\infty} \hat{a}_{\mathbf{p}}^{+} e^{-i\mathbf{p}x} = \hat{\Psi}^{+} \quad \text{The quantum field creation operator} \tag{19.4}$$

$$\text{Annihilate particle at } x: \quad \sum_{\mathbf{p}=-\infty}^{+\infty} \hat{a}_{\mathbf{p}}^{-} e^{+i\mathbf{p}x} = \hat{\Psi}^{-} \quad \text{Adjoint operator to } \hat{\Psi}^{+} \tag{19.5}$$

19.2.4 Bosons Like to Party

What happens if a boson is created where there are already bosons present? This is shown in Equation 19.7. In this example, the quantum field creation operator acts on a quantum state that already has 99 particles in one momentum state (shown as b) and none in any other. The result is a small chance of finding the boson in any particular momentum state (which I show as 1, but remember it is tiny) and a much higher chance of its having momentum b.

$$\hat{\Psi}^{+} |0_a, 99_b, 0_c, 0_d, 0_e...> = \sum_{\mathbf{p}=-\infty}^{+\infty} \hat{a}_{\mathbf{p}}^{+} \sqrt{n+1} |n+1>_{\mathbf{p}} \tag{19.6}$$

$$\text{Amplitudes for new boson's state} = 1|a> + \sqrt{100}|b> +1|c> +1|d> ... \tag{19.7}$$

The probability is the square of the amplitude so it is 100 times more likely to be b than any other value. Why? You should not be surprised that a small increase in amplitude can lead to a much larger increase in probability. If you increase the amplitude ten-fold, the probability increases a hundred-fold (its square). You don't need the detailed maths behind this. The creation operator coefficient of $\sqrt{n+1}$ instantly gives the answer. If there are n bosons already in the state, it is $(n+1)$ times more likely that the new boson will be in that state. Bosons like to hang out together like partying students. Imagine that you get a bunch of bosons in a single momentum state, then create more and more, until there are billions. With each additional boson the chance increases that the next will be in the same state.

The ability of bosons to occupy the same state can be useful. For example, this is true of the photons in a laser beam (with more jokes in Box 19.6). So, what about the idea of creating an electron laser? *Forget it!* One electron cannot be in an identical state to another. The electron is a fermion and fermions don't party!

Box 19.6   Just for laughs: Some laser jokes…

What did the laser say when it was turned off? *I don't know. It was incoherent.*

The surgeon asks a patient recovering from laser eye surgery whether he wants the good news or the bad news first. The patient says: *Tell me the good news.* The surgeon replies: *Well, you're about to get a new dog.*

19.2.5   Conservation of Energy and Momentum

Some of you may be scratching your heads and wondering a bit about all this creating and annihilating. If we ignore the ongoing existence of a particle and just create a new one, then how are energy and momentum conserved? Well, it turns out that energy is automatically conserved when we integrate an event over time and likewise momentum is conserved when we integrate over space.

For example, consider an event where a particle decays into two particles. Let's call the in-going particle a and the outgoing ones b and c. In QFT terms the process is that a is annihilated and b and c are created. I have added a constant g related to the chance of the decay happening (this is called the *coupling constant*). Let's first show that momentum is conserved when we integrate over space. The event in QFT is described as:

$$g \; \hat{a}_c^+ e^{-i\,\mathbf{p_c}x} \; \hat{a}_b^+ e^{-i\,\mathbf{p_b}x} \; \hat{a}_a^- e^{i\,\mathbf{p_a}x} \quad \textit{annihilation of a; creation of b and c} \tag{19.8}$$

The annihilation and creation operators work their magic changing the field state $|a> \;\rightarrow\; |bc>$. To get the amplitude of the pathway, we integrate over all possible x. This may seem odd. How can the change in the state of the field happen absolutely anywhere? But think back to the path integral. That is exactly what is going on. The quantum field samples every route. We must integrate over every possibility:

$$<bc|a> = \int_{-\infty}^{+\infty} g \; e^{-i\,\mathbf{p_c}x} e^{-i\,\mathbf{p_b}x} e^{i\,\mathbf{p_a}x} \; dx$$

$$= g \int_{-\infty}^{+\infty} e^{i\,(\mathbf{p_a}-\mathbf{p_b}-\mathbf{p_c})x} \; dx \tag{19.9}$$

$$= 2\pi g \, \delta \left(\mathbf{p_a} - \mathbf{p_b} - \mathbf{p_c}\right) \tag{19.10}$$

First, don't worry about the 2π factor. That is a feature of this sort of integration and it is normalised out in calculations. Equation 19.10 has a delta function. The value is 0 unless $\mathbf{p_a} - \mathbf{p_b} - \mathbf{p_c} = 0$ in which case it is 1. This means that there is only a probability amplitude for the event happening, *providing* momentum is conserved.

The step from Equation 19.9 to 19.10 deserves some explanation for mere mortals (congratulations to all you geniuses who think it is obvious). If $\mathbf{p_a} - \mathbf{p_b} - \mathbf{p_c}$ is not zero, then the function in Equation 19.9 is a sinusoid varying up and down with x. Think of a sine wave. Its amplitude is as often and as much below zero as above zero. Therefore, if you integrate it over x the answer must be zero unless $\mathbf{p_a} - \mathbf{p_b} - \mathbf{p_c} = 0$, when the sinusoidal relationship with x disappears.

You can do exactly the same exercise in the y and z direction to show that $\mathbf{p}_y$ and $\mathbf{p}_z$ are conserved. And you can do the same integrating over time (using the $\mathbf{E}t$ rather than the $\mathbf{p}x$ component of the equation). Integrating over time ensures conservation of energy.

The amazing mathematician Emmy Noether (Box 19.7) showed that there is a relationship between symmetry and conservation laws. She proved that where there is a symmetry there is a measurable conserved quantity. In the case above, the symmetry that the event can happen anywhere in *space* creates conservation of *momentum* and that it can happen at any *time* creates conservation of *energy*.

Box 19.7   Symmetry and conservation: Emmy Noether (1882–1935)

Einstein: *Noether was the most significant creative mathematical genius thus far produced since the higher education of women began.* Wiener: *Miss Noether is … the greatest woman mathematician who has ever lived.*

Strong praise indeed. Noether's theorem proved that for every symmetry there is an associated conservation law, a conserved observable quantity. This was a key step in helping Einstein to sort out the maths of general relativity. It is also a central tenet of quantum mechanics. But consider the obstacles this genius faced as a woman:

The challenge to study mathematics as one of only two female students in the whole of Erlangen University. The numbers are not surprising as the University's stated position was that mixed-sex education would *overthrow all academic order*. The challenge to get a position at Göttingen University in 1915 in spite of being a leading mathematician of her time. One faculty member's reaction was: *What will our soldiers think when they return to the university and find that they are required to learn at the feet of a woman?* The challenge to get paid. As a woman, her teaching went unpaid for the first 8 years.

On top of this she suffered Nazi persecution as a Jew and was summarily dismissed in 1933. She relocated to the USA where sadly she died of an ovarian cyst at only 53 years old. David Hilbert summed up the frustration of many physicists at her lack of advancement:

I do not see that the sex of the candidate is an argument against her admission – after all, the senate (university) is not a bath house.

19.3   Non-relativistic Path Integrals and Action

Seeing the quantum field as something that operates by creating and annihilating particles opens up a totally new perspective on the *Action* of the field. It is time to revisit that topic. I need you to cast your mind way back to Chapter 6. In that chapter we followed Feynman's argument that particles strictly do not travel in a straight line from *A* to *B*. Their behaviour is actually the sum of the wave function along all possible routes. Each different route results in the wave function arriving with a different phase.

We illustrated the idea with little phase-gauges that spin around keeping track of the phase of the wave function along a pathway. The amplitudes of the pathways interfere. As a result, the amplitudes of nearly all the routes cancel out except for those closest to the path of *stationary Action*. These closest pathways constructively interfere (adding up) because the Action of each is very similar. This is such an important concept that I have reproduced Figure 6.3 from Chapter 6 as Figure 19.1. It represents a particle fired from a source to a screen. Various possible paths of travel are shown with the closest ones to the direct route coloured blue. For each path, the phase of the arriving wave function is shown on the line of phase-gauges

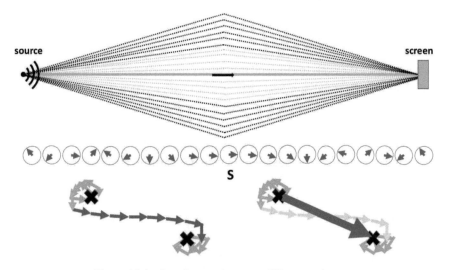

Figure 19.1   Interference between different pathways.

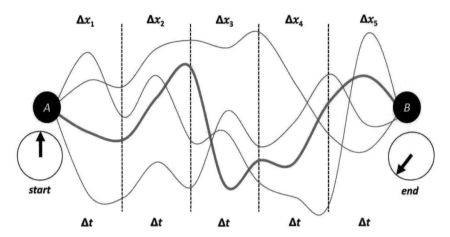

Δx_1 Δx_2 Δx_3 Δx_4 Δx_5

start   *end*

Δt Δt Δt Δt Δt

Figure 19.2   Path integral calculation (non-relativistic Action).

using an illustrative model with the Action proportional to the length of each path. At the bottom of the figure the results from each path are added together. The total result is not much different from the sum contribution of the blue pathways close to the direct route so the particle appears to travel directly from source to screen. In Section 6.5, I showed why the classical approximation of *Action* (the amount of change in the phase of the wave function) is the difference between kinetic energy $\frac{p^2}{2m}$ and potential energy V. This is called the *classical Lagrangian*.

One major difference is that QFT does *not* include any potential V. Potential energy is treated as a series of interactions. For example, in the case of the electromagnetic field, one quantum field (the electron field) interacts with another quantum field (the photon field). Each field has its own Lagrangian and each field is quantised. As a result, QFT reflects the quantisation of photons. This is sometimes referred to as *second quantisation* because the *first* is the quantisation of the electron and the *second* is the quantisation of the electromagnetic field (the photon field).

We need to develop relativistic Lagrangians for QFT fields. Our starting point is the path integral using *non-relativistic* quantum mechanics (i.e. not QFT). For the rest of this chapter, I am going to include $\hbar$ and c factors for clarity. Much of the following repeats work from Chapter 6, but bear with me, it is for a good reason. To work out the combined amplitude for the wave function of a particle along all possible routes between two points A and B, we first calculate the phase of the wave function along one path by dividing that path into tiny steps, each separated by Δt of time as shown in Figure 19.2. The Action for a single step along a path is shown in Equation 19.11.

$$\phi = e^{\frac{i}{\hbar}(\mathbf{p}x - \mathbf{E}t)} \implies \textit{Action over } \Delta t = \frac{i}{\hbar}(\mathbf{p}\Delta x - \mathbf{E}\Delta t) = \frac{i}{\hbar}(\mathbf{p}\frac{\Delta x}{\Delta t} - \mathbf{E})\Delta t$$

$$= \frac{i}{\hbar}(\mathbf{p}v - \mathbf{E})\Delta t \quad v \textit{ is velocity} \tag{19.11}$$

The Action over the whole of a pathway is the integral over time t between A and B. By integrating, we are making each Δt step infinitely small. Applying the usual non-relativistic approximations ($\mathbf{p} \approx p = mv$ and $\mathbf{E} \approx mc^2 + \frac{p^2}{2m}$) gives the result in Equation 19.12. Back in Chapter 6, I argued that, while p is a variable, mc^2 is not. This means that, whatever the pathway, $\frac{mc^2 t}{\hbar}$ is a constant (I have labelled this constant with a capital C in Equation 19.13 for emphasis). It affects the Action of all pathways equally so C does not alter the interference pattern between pathways. Therefore, it can be ignored. The result is $\mathscr{L} = \frac{p^2}{2m}$. This is exactly the same as Equation 6.8 for a free particle (zero potential). Adding in a potential gives $\mathscr{L} = \frac{p^2}{2m} - V$.

$$\textit{Pathway Action} = \frac{1}{\hbar}\int_0^t \left(\frac{p^2}{m} - mc^2 - \frac{p^2}{2m}\right) dt = \frac{1}{\hbar}\int_0^t \left(\frac{p^2}{2m} - mc^2\right) dt \tag{19.12}$$

$$= \frac{1}{\hbar}\int_0^t \left(\frac{p^2}{2m}\right) dt - \frac{mc^2 t}{\hbar} = \frac{1}{\hbar}\int_0^t \left(\frac{p^2}{2m}\right) dt - C \tag{19.13}$$

The last step (phew!) is to add up the amplitudes for each and every possible path between A and B. To do this, we use the Lagrangian and integrate over every possible spatial pathway: $(x_1, y_1, z_1), (x_2, y_2, z_2), (x_3, y_3, z_3)$... the details do not matter to us here. Typically this path sum is abbreviated as $D(x)$:

$$\text{Sum of paths integral:} \quad \int D(x)\, e^{\frac{i}{\hbar} \int \mathcal{L} dt} \tag{19.14}$$

19.4 QFT Path Integrals: A Relativistic Twist

I have stepped through all of this again to highlight some major differences between QFT Lagrangians and non-relativistic Lagrangians. In QFT, the approach we have taken to arrive at Equation 19.13 *cannot* and *does not* work. We must take a *relativistic* approach.

By taking the sum of all possible pathways in Figure 19.2 with the integral in Equation 19.13, we consider *every* possible value for momentum. That is every possible value of $\frac{\partial \Psi}{\partial x}$, $\frac{\partial \Psi}{\partial y}$ and $\frac{\partial \Psi}{\partial z}$ from $-\infty$ to $+\infty$. What about time t? If we are to put time and space on an equal footing, then the path integral must take into account every possible value of $\frac{\partial \Psi}{\partial t}$ from $-\infty$ to $+\infty$.

This may be another of those moments to have a cold shower and cool your brain before reading further. For those values of $\frac{\partial \Psi}{\partial t}$ that are less than $m_e c^2$ for an electron, the rest mass must be *less* than m_e. The *non-relativistic* calculation that we did earlier ignores the rest mass term because it is based on an *exact* relationship $E_{rest} = mc^2$ which is always the same for all pathways. In the QFT path integral, we *must* consider all possible values of rest mass energy – how can that work?

Let's first consider momentum in our non-relativistic model. The pathway of stationary Action is the one where the value of momentum is constant; see Equation 6.11 (remember QFT does not include any potential V). Simply put, if the spacetime distance between points A and B is x and time t, then that pathway has constant momentum $p = mv = \frac{mx}{t}$. The path integral covers all possible values and changes to p but the further a pathway diverges from constant p, the higher the Action. The sum of the integral closely resembles the pathway of stationary Action so the behaviour of the particle appears to be $p = \frac{mx}{t}$ even though it is really the result of the combination of the resulting wave function from every possible pathway. Using layman language, we might say that the Lagrangian *punishes* divergence from $p = \frac{mx}{t}$ with extra Action.

The story is very similar for energy in the relativistic version. We know that the energy of an electron follows the relationship $\mathbf{E}^2 = m_e^2 c^4 + \mathbf{p}^2 c^2$. This then is the pathway of stationary Action. The path integral covers all possible values of $\mathbf{E}$ but the further a pathway diverges from $\mathbf{E}^2 = m^2 c^4 + \mathbf{p}^2 c^2$, the higher the Action. In crude layman language, the QFT relativistic Lagrangian *punishes* divergence from $\mathbf{E}^2 = m^2 c^4 + \mathbf{p}^2 c^2$ with extra Action.

A subtle but very important difference emerges. In the non-relativistic path integral, we considered only pathways that conform *always* and *exactly* to $\mathbf{E}^2 - \mathbf{p}^2 c^2 = m^2 c^4$. These pathways are described as *on-shell* because $\mathbf{E}_{rest} = mc^2$. In a QFT path integral, all possible development pathways must be factored in, including those which are *off-shell* where $\mathbf{E}^2 - \mathbf{p}^2 c^2 \neq m^2 c^4$. These pathways have higher Action and so tend to cancel out with surrounding off-shell pathways. However, off-shell pathways with very small Action (for example, they may be off-shell for only a very short time) impact the total probability amplitude of the result. Off-shell field excitations matter!

This is why you get *quantum tunnelling* in quantum mechanics. Short-lived off-shell pathways contribute to the path integral so particles have a small but tangible probability of tunnelling across potential barriers which classically would be impenetrable (see Section 12.4). These off-shell pathways play a significant role in *interactions*. They lead to the quantum tunnelling that in turn leads to covalent bonds in molecules and the nucleon-nucleon bonds in the nucleus. As described in the material following Subsection 16.4.2, they are called (controversially) *virtual particles*, but are better thought of as possible *development pathways* whose contribution to the path integral must be included in calculations.

19.5 Energy and Time

How far can a contributing virtual pathway drift off-shell and for how long? We can approximate the path integral by adding up the contribution of the pathways whose Action is similar to the pathway of stationary Action. Figure 19.1 shows that adding only the blue pathways gives close to the correct result. The same is true for quantum fields. The less the drift and the shorter the duration off-shell, the larger the contribution of the pathway to the total probability amplitude. This relationship is embodied in the *propagator* that we will discuss later. However, we can take a shortcut and use the *non-relativistic* path integral to develop a rough but useful rule of thumb.

We use exactly the same argument as used back in Section 6.7. Using Figure 19.1, we approximate the path integral using only the (blue) pathways. Their phase-gauges are less than about 1 radian ($\approx$ 60 degrees) different from the pathway of stationary Action (*S*). Once the difference is a half-turn to a full turn of the phase-gauge (π to 2π radians), the contribution of the pathway, after cancelling with surrounding pathways, becomes vanishingly small.

If a pathway drifts $\Delta\mathbf{E}$ off-shell for time Δt, then the difference in phase versus an on-shell pathway is as shown in Equation 19.15. This difference in phase must be less than (very approximately) 1 to 2π radians, giving the rough relationship that $\Delta\mathbf{E}\,\Delta t\ <\ \hbar$ or h shown in Equation 19.16. As we are dealing with probability, your answer depends on what level of sensitivity you set.

$$\phi\ =\ e^{\frac{i}{\hbar}(\mathbf{p}x - \mathbf{E}t)} \quad\Longrightarrow\quad \textit{Phase difference}\ =\ \frac{i}{\hbar}\,\Delta\mathbf{E}\,\Delta t \tag{19.15}$$

$$\Delta\,\textit{phase}\ =\ \frac{1}{\hbar}\,\Delta\mathbf{E}\,\Delta t\ <\ 1\ \textit{to}\ 2\pi \quad\Longrightarrow\quad \Delta\mathbf{E}\,\Delta t\ <\ \hbar\ \textit{or}\ h\ \textit{approx... order of...} \tag{19.16}$$

Let me emphasise that this is a very rough and ready rule of thumb. It tells us that there is a limit to the off-shell drift we need to factor in to our calculations. The higher the energy drift $\Delta\mathbf{E}$, the less time Δt it can endure before the Action of the pathway makes its contribution insignificant to the total path integral. $\Delta\mathbf{E}$ is the energy gap. Δt is the length of time that the energy gap endures (do *not* confuse this with uncertainty in time). These are the virtual pathways that allow quantum tunnelling and are important in the analysis of electromagnetic and other interactions (see example in Box 19.8).

Box 19.8   Off-shell drift and the weak force

The *weak force* or *weak interaction* gives an interesting insight into off-shell drift. It is responsible for radioactive decay and plays a key role in nuclear fission. The force is mediated by massive (and therefore, high energy) W_+, W_-, and Z bosons. Because of the energy required, the mediating boson pathways are typically off-shell, so only very short-lived pathways contribute to the field's development. As a result, the weak force operates over an extremely short range, less than the diameter of a proton. We discuss the weak force further in Chapters 24 and 25.

19.6   QFT Field Development Pathways

The non-relativistic calculation that we made earlier (resulting in Equation 19.14) is a good example of the approach I described earlier as *when, where, what* analysis. First we integrated the phase change over time (*when*), then we considered every possible path through space (*where*). From that we calculated the sum of amplitudes to determine the behaviour of a particle or object (*what*). This gives the non-relativistic sum of paths integral in Equation 19.14.

In relativistic analysis we simply ask *what*. We take an initial state of the quantum field Ψ_i (this might be two particles heading towards each other) and compare it with a final state of the quantum field Ψ_f (this might be two particles heading away from each other). How might Ψ_i migrate into Ψ_f? Perhaps it involves one interaction, or perhaps two interaction steps, or perhaps three, or more. Whichever, we have no care for when or where those steps might occur – we simply integrate over all of time and space to cover every eventuality. The sum of paths integral from Ψ_i to Ψ_f becomes:

$$\textit{Sum of paths integral in QFT:} \quad \int_{\Psi_i}^{\Psi_f} D(\Psi)\, e^{\frac{i}{\hbar}\int \mathscr{L}\mathrm{d}^4 x} \tag{19.17}$$

The major change is that we compare the Action of different development pathways of the quantum field, rather than particle pathways. In Equation 19.17, the $D(\Psi)$ indicates summing over all possible development pathways between the two states of the quantum field. The symbol $\mathrm{d}^4 x$ means that the Lagrangian (more correctly now called the Lagrangian density) is integrated over space and time so they are now on an equal footing in the calculation. The integral is: $\int\int\int\int \mathscr{L}\mathrm{d}t, \mathrm{d}x, \mathrm{d}y, \mathrm{d}z$.

At this stage, I expect some of you are sobbing in despair. You may think this looks even more of a nightmare calculation, and you are right. Thankfully, the bongo-playing Feynman and others developed a rule book full of shortcuts. We will come to that in the next chapter. First we need to get to grips with the QFT Lagrangians that sit behind Feynman's rule book.

19.7 The Klein-Gordon Lagrangian as a Model

In the next two chapters, I hope to give you a feel of quantum electrodynamics (QED), the interaction of electrons and photons. For electrons we will need the Dirac equation, but that brings with it the complexity of matrices and spin so we are going to start off with something simpler, the Klein-Gordon Lagrangian. It is an excellent model that we can build on later (see Box 19.9).

Box 19.9   The Higgs boson and the Klein-Gordon Lagrangian

Take note that this analysis of the Klein-Gordon Lagrangian also will prove useful later when we come to discuss the Higgs field, which is scalar (spin-zero).

We discussed the Klein-Gordon equation back in Subsection 8.3.1. It is the relativistic energy-momentum relationship expressed using second derivatives of the wave function. This is shown below in one spatial dimension:

$$\mathbf{E}^2 - \mathbf{p}^2 c^2 = m^2 c^4 \implies \frac{\partial^2 \Psi}{\partial t^2} - c^2 \frac{\partial^2 \Psi}{\partial x^2} = -\frac{m^2 c^4}{\hbar^2}\Psi \quad \textit{Klein-Gordon} \tag{19.18}$$

We will use it as a relativistic approximation for the electron in the same way that we used the Schrödinger equation as a non-relativistic approximation. We need the Lagrangian for the Klein-Gordon quantum field. I am going to cheat by revealing the answer below (showing only one spatial dimension) and then demonstrating that it works.

$$\mathscr{L}_{KG} = \frac{1}{2}\frac{\partial \Psi^*}{\partial t}\frac{\partial \Psi}{\partial t} - \frac{1}{2}c^2 \frac{\partial \Psi^*}{\partial x}\frac{\partial \Psi}{\partial x} - \frac{1}{2}m^2 c^4\,\Psi^*\Psi \quad \textit{Klein-Gordon Lagrangian} \tag{19.19}$$

Back in Section 6.6, we used the Euler-Lagrange equation to derive equations of motion from the classical Lagrangian. We can do the same here using a slightly more complete version of Euler-Lagrange (Equation 19.20) to cover the spatial derivatives as well as the time derivative. We can apply it in terms of the complex conjugate Ψ^* as shown in Equation 19.21. Differentiating and cancelling out the $\frac{1}{2}$ factors gives Equation 19.22 which results in the Klein-Gordon equation.

$$\frac{\partial}{\partial t}\left(\frac{\partial \mathscr{L}}{\partial \frac{\partial \Psi}{\partial t}}\right) + \frac{\partial}{\partial x}\left(\frac{\partial \mathscr{L}}{\partial \frac{\partial \Psi}{\partial x}}\right) = \frac{\partial \mathscr{L}}{\partial \Psi} \quad \textit{Euler-Lagrange equation} \tag{19.20}$$

$$\frac{\partial}{\partial t}\left(\frac{\partial \mathscr{L}}{\partial \frac{\partial \Psi^*}{\partial t}}\right) + \frac{\partial}{\partial x}\left(\frac{\partial \mathscr{L}}{\partial \frac{\partial \Psi^*}{\partial x}}\right) = \frac{\partial \mathscr{L}}{\partial \Psi^*} \tag{19.21}$$

$$\frac{\partial}{\partial t}\left(\frac{\partial \Psi}{\partial t}\right) - \frac{\partial}{\partial x}\left(c^2 \frac{\partial \Psi}{\partial x}\right) = -m^2 c^4\,\Psi \tag{19.22}$$

$$\frac{\partial^2 \Psi}{\partial t^2} - c^2 \frac{\partial^2 \Psi}{\partial x^2} = -m^2 c^4\,\Psi \quad \textit{Klein-Gordon equation}$$

In case you need it, here is a reminder why the Lagrangian is so important. Imagine summing up the outcome of every possible development of the quantum field. The result depends on how the development pathways interfere and add up. That is driven by the Action, which is mathematically reflected in the Lagrangian. The maths of path integrals tells us that the Action sits behind every development in the quantum field, everything that happens, and we quantify that with the Lagrangian.

The Klein-Gordon *Lagrangian* tells us that the development pathway of stationary Action obeys the Klein-Gordon *equation* and that means that the pathway of stationary Action obeys the relationship shown in Equation 19.18: $m^2 c^4 = \mathbf{E}^2 - \mathbf{p}^2 c^2$. This, of course, is the energy equation of an object with rest energy $\mathbf{E}_{rest} = mc^2$. Pathways that are off-shell have higher Action and tend to cancel out so the Klein-Gordon Lagrangian means that excitations in the quantum field cannot stray too far off-shell or for too long.

19.8 Global Gauge Invariance to Phase

The Klein-Gordon field is complex-valued, that is, the amplitude of the quantum field at each point of spacetime is a complex number. This raises an interesting question that turns out to be important. What happens if we change the phase of the quantum field?

If you put your non-QFT hat back on for a moment, you will see why this might be significant. In the first half of this book I stressed that the free particle wave function could not be a particular phase, such as a sine or cosine wave, because that would create special points where the amplitude of the wave function is zero. Theoretically an observer could identify these special zero-points. However, relativity tells us that, free of any potential or constraint, all space is equivalent. There can be no special points. We solved this dilemma by introducing a complex version of the free particle wave function which has no special points because the value of any measurable is the same, whatever the phase: $|\phi|^2 = \phi^*\phi$.

Another way to think about this is through the Action of particles. Consider the phase of a particle based on the non-QFT model in Figure 19.2. What matters is the *change* in phase from point A to B. The starting phase of the particle at A is irrelevant. Altering *all* the phase-gauges *everywhere* by the same amount changes nothing. This is called *global gauge invariance* to a change of phase (the name has a different origin, but fits well with the model, see Box 19.10).

The Klein-Gordon quantum field is a very different mathematical object from the free particle wave function, but it also displays this global gauge invariance. This is not surprising. Our starting point in QFT was an infinite array of QHOs (Section 19.2.1). A superposition of QHOs is unaffected by a global phase change (see Box 13.7). We know that all quantum systems near equilibrium resemble the QHO (see Section 13.1.2). Therefore, you should be deeply suspicious of any proposed quantum field Lagrangian that is not globally gauge invariant to a change in phase!

We can show the effect on the Klein-Gordon Lagrangian of a global change of phase by multiplying it with $e^{i\theta}$. The Lagrangian is unaffected *providing* Ψ always acts alongside Ψ^* and $\partial\Psi$ alongside $\partial\Psi^*$. The shift of θ in phase means multiplying Ψ by $e^{i\theta}$ and multiplying its adjoint Ψ^* by $e^{-i\theta}$ (the complex conjugate equivalent). The two effects cancel out for all the terms in the Lagrangian as shown in Equation 19.23 and below. The global phase change has no effect on the overall Klein-Gordon Lagrangian.

$$\mathscr{L}_{KG} \implies \frac{1}{2}\frac{\partial e^{-i\theta}\Psi^*}{\partial t}\frac{\partial e^{i\theta}\Psi}{\partial t} - \frac{1}{2}c^2\frac{\partial e^{-i\theta}\Psi^*}{\partial x}\frac{\partial e^{i\theta}\Psi}{\partial x} - \frac{1}{2}m^2c^4\,e^{-i\theta}\Psi^*e^{i\theta}\Psi \tag{19.23}$$

$$= e^{-i\theta}e^{i\theta}\frac{1}{2}\frac{\partial\Psi^*}{\partial t}\frac{\partial\Psi}{\partial t} - e^{-i\theta}e^{i\theta}\frac{1}{2}c^2\frac{\partial\Psi^*}{\partial x}\frac{\partial\Psi}{\partial x} - e^{-i\theta}e^{i\theta}\frac{1}{2}m^2c^4\,\Psi^*\Psi$$

$$= \frac{1}{2}\frac{\partial\Psi^*}{\partial t}\frac{\partial\Psi}{\partial t} - \frac{1}{2}c^2\frac{\partial\Psi^*}{\partial x}\frac{\partial\Psi}{\partial x} - \frac{1}{2}m^2c^4\,\Psi^*\Psi \qquad \textit{Klein-Gordon Lagrangian}$$

Box 19.10 Hermann Weyl and gauge invariance

Hermann Weyl (1885–1955) has already appeared in our story as the brilliant mathematician who helped Schrödinger solve his equation for hydrogen... and helped Schrödinger's wife with other things (see Box 15.7). Weyl developed the idea of gauge invariance in 1929. In German it goes under the even catchier name *Eichinvarianz*. As a concept, it has grown to be a cornerstone of QFT. Dirac, in an interview with the Wisconsin State Journal was asked: *Do you ever run across a fellow that even you can't understand?* Dirac answered: *Yes... Weyl.*

The Dirac Lagrangian also describes a quantum field and therefore is also global gauge invariant to a change in phase. This has profound consequences. Ψ and Ψ^* are the creation and annihilation operators $\hat{\Psi}^-$ and $\hat{\Psi}^+$ that respectively annihilate and create an electron. For the Lagrangian to be invariant to a global phase change, it turns out that $\hat{\Psi}^-$ and $\hat{\Psi}^+$ must always act in unison. The annihilation of an electron must always be accompanied by the creation of another electron. What does this mean? Consider the hypothetical scenario of an electron moving through spacetime with *no* interactions with other fields. Each step of the journey involves electron annihilation, but this is always accompanied by electron creation. As conservation of energy and momentum apply, the electron will always remain on-shell and will exist for ever, except...

...the Dirac equation gives another legitimate particle state, the positron, that is the *mathematical* equivalent of the electron travelling backwards in time (see Equation 18.58). This means that the annihilation of an electron is the mathematical equivalent of the creation of a positron. It is worth pausing to think about this. Imagine an electron moving away from another electron (being repelled) and then annihilating. What would this look like if we *reverse* time? Instead of a particle annihilating, you would see a particle being created out of nowhere. Following that particle backwards in time, it would get closer to the other electron so you would conclude the new particle is attracted by the electron and has a positive charge.

This means that the operation $\hat{\Psi}^+$ can be the creation of an electron *or* the annihilation of a positron. Similarly $\hat{\Psi}^-$ can be the annihilation of an electron *or* the creation of a positron. Taken together (as they always must be) the act $\hat{\Psi}^+\hat{\Psi}^-$ can be the annihilation and creation of an electron, the annihilation and creation of a positron, the creation of an electron/positron pair or the annihilation of an electron/positron pair. Hey ho – particles can be created and destroyed! In a later chapter, you will learn that these acts of creation and annihilation involve interaction with other particles such as the photon.

The fact that a global phase change does not affect a quantum field is a *symmetry* which, as Emmy Noether proved, is always associated with a measurable conserved quantity. In this case, it is *conservation of charge*. For example, if an electron is created (-1 charge), either an electron is destroyed or a positron is created. As both events are $+1$ charge, the net charge addition is zero.

19.9   Summary

In this chapter, we have touched on some of the basic features of QFT. Events are expressed in terms of particle annihilation and creation. We used the Quantum Harmonic Oscillator for the maths of how quantum fields function near equilibrium and out popped the principle behind the laser – bosons like to party – but fermions don't. These creation and annihilation events do not happen at any specific time or place. This seems counter-intuitive, but fits with what we have learned from path integrals: the evolution of a quantum field is the sum of every possible twist and turn through spacetime. Into the bargain, integrating over space and time automatically incorporates the conservation of momentum and energy.

We adopted the Klein-Gordon Lagrangian as a starting model and discussed a subtle but important distinction between the non-relativistic and relativistic path integrals. In the non-relativistic analysis we mandated that $\mathbf{E}_{rest} = mc^2$ *always* and *exactly*. In the case of QFT it is the Action of the field that produces this on-shell balance. We can approximate a total path integral by adding up the contribution of the pathways with low Action. If pathways are not too far off-shell and/or are off-shell for only a limited duration, we need to include them in our calculations. These development pathways allow quantum tunnelling which is very important in interactions. For example, it leads to covalent bonding in molecules and the binding of nucleons together in the atomic nucleus (Subsection 16.4.2).

We considered the implication for quantum fields of global gauge invariance to phase change. As a result the creation and annihilation operators in the Lagrangian must operate hand in hand. In the case of the electron field for example, when an electron is created ($\hat{\Psi}^+$), either an electron must be destroyed or a positron created ($\hat{\Psi}^-$). Charge is conserved.

You may wonder where all of this takes us. What do these quantum fields tell us? Why is this QFT stuff important? Why do we care that the quantum fields are not sensitive to certain changes in phase? Please be patient. All will be revealed.

To that end, I want to finish this chapter with a different question. Changing the phase of an *entire* quantum field has no physical effect (a *global gauge change*). However, what if the phase of only part of the quantum field changes (a *local gauge change*)? Absolute phase is neither defined nor physically measurable, so might local inconsistencies in phase create problems? Nature appears to take this seriously. It has built-in structures that offset any local gauge change in the phase of the quantum fields. In the case of the electron, this built-in structure is the electromagnetic field. Don't stop reading – the best is yet to come.

Note

1 For simplicity, I have used time $t = 0$ in the free particle wave function. We will add back time dependency later.

Chapter 20

Local Gauge Invariance

CHAPTER MENU
20.1 Introduction to Local Gauge Invariance
20.2 The Infinity Swimming Pool – an Analogy
20.3 Refresher in Electromagnetics (EM)
20.4 The EM Quantum Field and Lagrangian
20.5 EM Gauge Invariance
20.6 U(1) Local Gauge Invariance: Putting Together the Pieces
20.7 The Dirac Lagrangian
20.8 Interaction and the Pathway of Stationary Action
20.9 The Photon Must Be Massless
20.10 Summary

In this chapter, we will start to discuss the way that quantum fields are coupled and the role of *local gauge invariance*. Don't worry if you are not sure what this is yet. Let me assure you that it is an absolutely essential topic. It is the key to understanding how quantum fields interact in the Standard Model of particle physics. Below in Box 20.1 is another little anecdote about Dirac.

Box 20.1   Dirac and dancing
Dirac once asked Heisenberg why he danced and got the unsurprising answer that it was a pleasure to dance with nice girls. After about five minutes of silence, Dirac asked:
Heisenberg, how do you know beforehand that the girls are nice?

20.1   Introduction to Local Gauge Invariance

In the last chapter, we discussed the *global* gauge invariance of fermionic fields (such as the electron field) to phase. This symmetry is called U(1) and is based on the underlying symmetry of a circle. To understand why, refer back to Figure 9.1 that shows the unit circle in the complex plane. If you multiply any complex number by $e^{i\theta}$, it rotates it by θ in the complex plane. So, a change of phase is equivalent to a rotation around a circle. U(1) symmetry means that multiplying by $e^{i\theta}$ changes nothing, in the same way that rotating a circle around its centre changes nothing.

What is the difference between *global* and *local* gauge invariance to phase? Let me illustrate the concept using a *non-QFT* example, the path integral of a particle travelling between two points as shown in Figure 20.1.

The top box shows two possible pathways for a particle travelling from A to B. On the upper pathway, I have shown a couple of phase-gauges along the journey where they point to 12 o'clock and assumed that a particle following this pathway

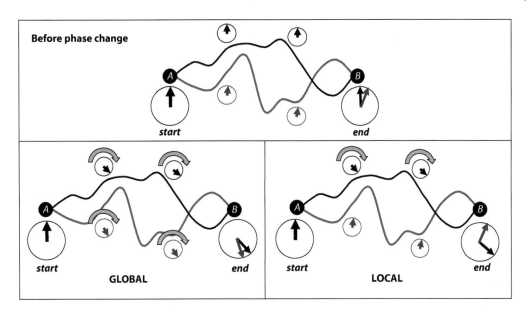

Figure 20.1 Global versus Local gauge change.

would arrive at 12 o'clock at *B*. A particle moving along the lower pathway arrives with a slightly different phase. The result is that the phases of the probability amplitudes along the two paths are similar and would constructively interfere.

The same pathways are shown on the bottom left of the figure with a *global* gauge change in phase. *All* the phase-gauges are shifted by the same amount. This makes no difference to the interference pattern because the phase of both pathways is changed by the same amount. The *global* gauge change doesn't impact anything measurable. We achieve this mathematically by multiplying the (complex) wave function with $e^{i\theta}$ where θ is a constant.

The bottom right corner of Figure 20.1 shows a *local* gauge change. The phase-gauges are not all changed by the same amount. In the example shown, along the upper path there is a gauge change but along the lower path there is not. The phase difference between the two particle pathways arriving at *B* changes. This will affect the outcome. We can say with confidence that these particle pathways show *global* U(1) gauge invariance, but do not show *local* U(1) gauge invariance.

What about the Klein-Gordon Lagrangian? We know it shows *global* gauge invariance to a phase change (see Section 19.8) so the Klein-Gordon *field* has *global* U(1) gauge invariance. What about *local* U(1) gauge invariance? How can we determine this?

$$\mathscr{L}_{KG} = \frac{1}{2}\frac{\partial \Psi^*}{\partial t}\frac{\partial \Psi}{\partial t} - \frac{1}{2}\frac{\partial \Psi^*}{\partial x}\frac{\partial \Psi}{\partial x} - \frac{1}{2}m^2\,\Psi^*\Psi \tag{20.1}$$

To assess the impact of a *local* phase change on the Klein-Gordon Lagrangian we multiply it by $e^{i\theta}$ but this time we allow θ to vary in time and space, that is, we multiply the wave function by $e^{i\theta(x,t)}$. This means that the phase change $\theta(x,t)$ is a function of x and t (using only one spatial dimension for simplicity). It can be one value at one location or time, another at another, zero at another etc. We then repeat the calculations in Equation 19.23. I will use again the notation $\Psi_{<\theta>}$ for the the wave function Ψ after a phase shift of θ.

Below, you can see that the mass term in the Lagrangian is *not* affected by a local phase change. We will work in one spatial dimension and in natural units ($c = 1$):

$$m^2\,\Psi^*_{<\theta>}\,\Psi_{<\theta>} \;=\; m^2\,e^{-i\theta(x,t)}\,\Psi^*\,e^{i\theta(x,t)}\,\Psi \;=\; m^2\,\Psi^*\Psi \tag{20.2}$$

However, the derivatives *are* affected. Again, we can model the effect of a local phase change by comparing Ψ with $e^{i\theta(x,t)}\Psi$ (which is Ψ with a local phase change). In this case a new term appears in each derivative because of the variation of θ with x and t. As an example, the result of a local phase change on the time t derivative term is shown below. The spatial derivatives are also affected.

$$\frac{\partial \Psi_{<\theta>}}{\partial t} = \frac{\partial (e^{i\theta(x,t)}\Psi)}{\partial t} = e^{i\theta} \frac{\partial \Psi}{\partial t} + \frac{\partial e^{i\theta}}{\partial t} \Psi = e^{i\theta}\left(\frac{\partial \Psi}{\partial t} + i\frac{\partial \theta}{\partial t}\Psi\right) \tag{20.3}$$

$$\frac{\partial \Psi^*_{<\theta>}}{\partial t} = e^{-i\theta}\left(\frac{\partial \Psi^*}{\partial t} - i\frac{\partial \theta}{\partial t}\Psi^*\right)$$

$$\frac{\partial \Psi^*_{<\theta>}}{\partial t}\frac{\partial \Psi_{<\theta>}}{\partial t} = \left(\frac{\partial \Psi^*}{\partial t} - i\frac{\partial \theta}{\partial t}\Psi^*\right)\left(\frac{\partial \Psi}{\partial t} + i\frac{\partial \theta}{\partial t}\Psi\right) \neq \frac{\partial \Psi^*}{\partial t}\frac{\partial \Psi}{\partial t} \tag{20.4}$$

You can see in Equation 20.4 the appearance of derivative terms of θ that do not cancel out so, although the Klein-Gordon Lagrangian shows *global* U(1) gauge invariance, it does *not* (as written so far) show *local* U(1) gauge invariance. A local phase change affects the Klein-Gordon Lagrangian, which means it changes the Action, which means it changes the interference between pathways and therefore changes measurable outcomes.

So what? There are reasons from special relativity to expect that global phase changes should not affect the nature of a quantum field, but even if this is the case (which it is), isn't global gauge invariance sufficient? You might think so, but the universe would disagree. It turns out that the laws of physics require fermion quantum fields to have *local* gauge invariance to phase changes. An underlying local U(1) symmetry is built into all quantum fields.

Why? A sensible professor would glare at at you and say this is just how things are. Most likely that is the best answer, but bear in mind that relying on global gauge invariance would require all the phase-gauges in each quantum field to remain synchronised everywhere in the universe. Perhaps not so easy when absolute phase is neither defined nor measurable.

How? Just as is the case with the Klein-Gordon Lagrangian, the Dirac Lagrangian does not have local gauge invariance to phase change. However, it does have it when you include its interaction with the electromagnetic field. But first we need to talk about swimming pools.

20.2 The Infinity Swimming Pool – an Analogy

Back in Subsection 19.2.2, I compared the amplitude of a quantum field with the height of water in a swimming pool. The underlying sinusoidal wave movements in the pool add up to the overall height of the water. Similarly, at any point in spacetime, the quantum field is a superposition of an infinite number of oscillators in various states of excitement that add together to give one total amplitude (in this case, a complex number). We are going to take this analogy a bit further.

Suppose that the swimming pool is half full of water. If you could lift the bottom of the pool it would raise the height of all the water by the same amount so the underlying pattern of sinusoidal waves in the water would not change. In terms of QFT, this is the equivalent of a *global* phase change that does not affect the Klein-Gordon or Dirac Lagrangians.

Now, imagine instead that we raise the bottom of only one corner of the pool or, if you prefer, somebody gets into the pool at one corner. The height of the water in that corner suddenly increases. This is the QFT equivalent of a *local* phase change. In the pool it creates a height gradient between the water in the corner and surrounding areas and affects the wave patterns. Similarly, it affects the quantum fields by changing the derivative (gradient) terms in the Lagrangian, altering the interference patterns and affecting the development of the field.

For swimming pools, this is a problem. Many people prefer to swim with a view, but a swimmer gets out, the water level drops and the view is lost. The answer is the *infinity* swimming pool. It has a *balancing tank* (see Figure 20.2). When a swimmer gets out, water is pumped into the pool from the balancing tank to compensate. If a swimmer jumps in at one corner of the pool, the water in that corner overflows from the pool into a pipe that goes to the balancing tank. In this way the balancing tank absorbs variations in the height of the water in the pool.

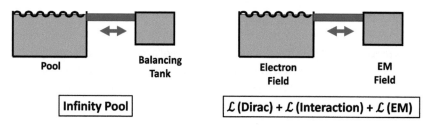

Figure 20.2 The infinity pool as a model of local gauge invariance.

Nature uses a similar mechanism as shown on the right side of Figure 20.2. Think of the electron-field as the swimming pool, the electromagnetic (EM) field as the balancing tank and the interaction between the two fields as the equivalent of the water flow between the pool and balancing tank. There are three terms in the overall Lagrangian that describe the combined electron field and EM field: that of the electron field $\mathscr{L}_D$ (D for Dirac), that of the interaction term $\mathscr{L}_I$ (I for interaction) and that of the EM field $\mathscr{L}_{EM}$ (EM for electromagnetic).

If there is a local phase change in the electron field it affects $\mathscr{L}_D$ because it changes the derivative terms. However, the local phase change has exactly the opposite effect on the interaction term $\mathscr{L}_I$, so the two cancel out. There is no effect on the EM field Lagrangian $\mathscr{L}_{EM}$. In total, the local phase change makes no difference to the overall Lagrangian $\mathscr{L}_D + \mathscr{L}_I + \mathscr{L}_{EM}$.

A local phase change is the equivalent of someone getting in or out at the corner of the infinity pool. The height of the water in the corner changes ($\mathscr{L}_D$ changes), and water flows in or out from the balancing tank to compensate ($\mathscr{L}_I$ offsets the change in $\mathscr{L}_D$). The balancing tank absorbs the change ($\mathscr{L}_{EM}$ is unaffected). Hey presto! Nature has conjured up the perfect balancing tank for the electron field – it is the EM field – and overall, we have U(1) local gauge invariance.

This is *not* a coincidence but is by design. Similar mechanisms exist for all the spin-half Dirac fields of the fermions (such as electrons and quarks). These fields display various symmetries, all of which are enabled by interacting with spin-1 vector fields of gauge bosons (the photon, W/Z bosons and gluons). These account respectively for the electromagnetic, weak and strong forces. We will discuss this further in the last few chapters of this book, but for now we will focus on the relationship between the electron and EM fields.

20.3 Refresher in Electromagnetics (EM)

The Schrödinger equation uses a smooth potential V for the electromagnetic interaction, but the EM field is quantised (photons) according to the formula $\mathbf{E} = \mathbf{p}c$ (or $\mathbf{E} = \mathbf{p}$ in natural units). We need to incorporate this in the EM field's Lagrangian. This is referred to as *second quantisation*. The Schrödinger equation quantises what classically we called particles (first quantisation). QFT takes the next step. It quantises what classically we called fields (second quantisation).

Before we address the EM quantum field and Lagrangian, I am going to offer a refresher in EM. Those of you who know your stuff can relax and have a nice cup of tea. Others may find this too much detail. If so, skim this material or skip ahead to Section 20.4, but you may want to check out the material covering the Einstein summation convention in Box 20.4 and Box 20.5.

20.3.1 EM Refresher (1): The Basics

Classically, the electric and magnetic fields were treated as distinct entities, albeit related. The electric field $\vec{E}$ has a magnitude and direction so it is expressed by a vector. It exerts a force on a particle with charge q that is $\vec{F} = q\vec{E}$. Be careful not to confuse the electric field E with energy $\mathbf{E}$.

The magnetic field $\vec{B}$ is also a vector. It exerts a force only if the charged particle is moving across the magnetic field. The force is perpendicular to both the magnetic field vector and the particle velocity: $\vec{F} = q(\vec{v} \times \vec{B})$ where $\times$ is the cross-product and $\vec{v}$ is the particle velocity. If you do not remember what the cross-product means, refer back to Box 14.1. Taken together:

$$\vec{F} = q(\vec{E} + (\vec{v} \times \vec{B})) \tag{20.5}$$

Box 20.2   Some useful vector definitions:

Divergence: $\nabla \cdot E = \dfrac{\partial E_x}{\partial x} + \dfrac{\partial E_y}{\partial y} + \dfrac{\partial E_z}{\partial z}$

Curl: $\nabla \times E = \left(\dfrac{\partial E_z}{\partial y} - \dfrac{\partial E_y}{\partial z}\right)\hat{x} + \left(\dfrac{\partial E_x}{\partial z} - \dfrac{\partial E_z}{\partial x}\right)\hat{y} + \left(\dfrac{\partial E_y}{\partial x} - \dfrac{\partial E_x}{\partial y}\right)\hat{z}$

Gradient: $\nabla E = \dfrac{\partial E}{\partial x}\hat{x} + \dfrac{\partial E}{\partial y}\hat{y} + \dfrac{\partial E}{\partial z}\hat{z}$ *where $\hat{x}$, $\hat{y}$ and $\hat{z}$ are unit vectors*

It became clear through experimentation that there is a strong connection between the two fields. That clever chappy James Maxwell summarised it with four equations. The version shown below is for a region containing no charge (I have dropped the vector arrow signs).

$$\nabla \cdot E = 0 \qquad \nabla \cdot B = 0 \qquad \nabla \times E = -\frac{\partial B}{\partial t} \qquad \nabla \times B = \mu_0 \epsilon_0 \frac{\partial E}{\partial t} \tag{20.6}$$

The first two say that the *divergence* of an electric or magnetic field is zero (see Box 20.2 for a refresher on the definition of some vector operations). If you imagine electric or magnetic field lines, it means that every line that enters a region also leaves the region, that is, none end in the region. For the electric field, this is true only if there is no charge in the region.

The last two equations relate the *curl* of the electric field and the rate of change of the magnetic field, and vice versa. The curl of a vector field is perhaps best described as the rotation or circulation in it. In layman's terms, the last two equations say that a moving electric field generates a magnetic field that swirls around it and, vice versa, a moving magnetic field produces an electric field that swirls around it. The value of the constants μ_0 and ϵ_0 came from experiments.

Maxwell reasoned the following. A moving electric field creates a magnetic field and a moving magnetic field creates an electric field, so there might be a speed at which the electric field creates a large enough magnetic field to regenerate the electric field that again generates the magnetic field and so on ad infinitum. He did the maths, and eureka! Light. A short derivation is shown in Box 20.3 for those familiar with vector notation and arithmetic. The result is that the speed of light c has to be $\frac{1}{\sqrt{\mu_0 \epsilon_0}}$ which is about 300,000 kilometres per second.

Box 20.3   Calculating the speed of light in a vacuum – in case you feel the urge

We use the vector identity:

$$\nabla \times (\nabla \times \vec{E}) = \nabla(\nabla . \vec{E}) - \nabla^2 \vec{E} \tag{20.7}$$

Evaluate the left side of Equation 20.7 using Maxwell's two equations for curl:

$$\nabla \times (\nabla \times E) = \nabla \times \left(-\frac{\partial B}{\partial t}\right) = -\frac{\partial}{\partial t}(\nabla \times B) = -\mu_0 \epsilon_0 \frac{\partial^2 E}{\partial t^2}$$

Substitute back into Equation 20.7 remembering that $\nabla \cdot \vec{E} = 0$ so $\nabla(\nabla . \vec{E}) = 0$:

$$-\mu_0 \epsilon_0 \frac{\partial^2 E}{\partial t^2} = -\nabla^2 E \qquad \Longrightarrow \qquad \mu_0 \epsilon_0 \frac{\partial^2 E}{\partial t^2} - \nabla^2 E = 0$$

The result is the wave function of a plane wave with velocity $\frac{1}{\sqrt{\mu_0 \epsilon_0}}$.

Einstein came to the realisation 40 years later that the electric and magnetic fields are different aspects of a more general EM field. In fact, the same electromagnetic phenomena can for one observer appear to be the result of an electric field, and for another the result of a magnetic field. This was one of the themes that drove him to his theory of special relativity.

20.3.2   EM Refresher (2): The Vector Potential

This section will be hard going for anyone unfamiliar with vectors so let me summarise what is covered. Rather than describing EM with two fields, we use one field called the *vector potential*, typically labelled A_μ (which forms the EM quantum field which we will get to later). The little μ subscript is to remind you this is a 4-vector. It has spatial components *and* a time component. The relationship between A_μ and both the electric and magnetic fields can be encapsulated in a neat summary notation $F_{\mu\nu}$ and the Maxwell equations tell us that $\partial_\mu F^{\mu\nu} = 0$. Now, on to the maths.

We need only four components (A_t, A_x, A_y, A_z) to define the EM vector potential A_μ versus the six components of the electric and magnetic fields $(E_x, E_y, E_z, B_x, B_y, B_z)$. The relationship is shown below in natural units ($c = 1$). Note that many texts use φ for the time component A_t.

$$E_x = -\frac{\partial A_x}{\partial t} - \frac{\partial A_t}{\partial x} \qquad E_y = -\frac{\partial A_y}{\partial t} - \frac{\partial A_t}{\partial y} \qquad E_z = -\frac{\partial A_z}{\partial t} - \frac{\partial A_t}{\partial z} \qquad (20.8)$$

$$B_x = \frac{\partial A_z}{\partial y} - \frac{\partial A_y}{\partial z} \qquad B_y = \frac{\partial A_x}{\partial z} - \frac{\partial A_z}{\partial x} \qquad B_z = \frac{\partial A_y}{\partial x} - \frac{\partial A_x}{\partial y} \qquad (20.9)$$

If you look carefully, all the electric field and magnetic field formulas have a similar structure. The relationship between the vector potential A_μ and the components of the electric and magnetic fields can be shown using $F_{\mu\nu} = \partial_\mu A_\nu - \partial_\nu A_\mu$. For this, we need to start using the Einstein summation convention (see Box 20.4). It is the only way to keep track when things get more complicated.

Box 20.4   Einstein summation convention

The two key rules are:

(1) If a suffix appears twice, you sum over it
(2) If a suffix appears only once, it can take any value

Greek indices (μ, ν...) indicate spacetime and go from 0-3 ($t = 0$, $x = 1$...)
Roman indices (i, j...) indicate space only and go from 1-3 ($x = 1$, $y = 2$...)

For example, the expression $F_{\mu\nu} = \partial_\mu A_\nu - \partial_\nu A_\mu$ represents *16 separate* equations including: $F_{00} = ...$, $F_{01} = ...$, $F_{02} = ...$ which we summarise in one matrix. In contrast, if we write the expression $F_{\mu\mu}$ this is only one equation: $F_{00} + F_{11} + F_{22} + F_{33}$ because the μ is repeated.

The μ and ν indices follow the summation convention and stand for all the spacetime dimensions from 0 to 3 with 0 as t and 1-3 as x, y, z. The result for $F_{\mu\nu}$ can be summarised in a 4x4 matrix.

$$F_{\mu\nu} = \partial_\mu A_\nu - \partial_\nu A_\mu = \frac{\partial A_\nu}{\partial x^\mu} - \frac{\partial A_\mu}{\partial x^\nu} = \begin{bmatrix} 0 & +E_x & +E_y & +E_z \\ -E_x & 0 & -B_z & +B_y \\ -E_y & +B_z & 0 & -B_x \\ -E_z & -B_y & +B_x & 0 \end{bmatrix} \qquad (20.10)$$

Each value for the indices μ and ν produces a formula. Equation 20.11 shows the result for $\mu = 0$ (that is t) and $\nu = 2$ (that is y). Equation 20.12 shows an example when μ and ν are different spatial coordinates giving one of the magnetic field components. Switching the values of μ and ν gives the negative so $F_{\mu\nu} = -F_{\nu\mu}$ that is, the matrix is antisymmetric. It can be tricky to get the signs of the derivatives correct so, if you are not completely at ease with the summation convention, you should look at Box 20.5 that explains the meaning of upper and lower indices.

$$\mu = 0, \; \nu = 2 \qquad F_{02} = -\frac{\partial A_y}{\partial t} - \frac{\partial A_t}{\partial y} = E_y \qquad F_{01} = E_x, \; F_{02} = E_y, \; F_{03} = E_z \qquad (20.11)$$

$$\mu = 1, \; \nu = 3 \qquad F_{13} = \frac{\partial A_x}{\partial z} - \frac{\partial A_z}{\partial x} = B_y \qquad F_{13} = B_y, \; F_{21} = B_z, \; F_{32} = B_x \qquad (20.12)$$

The final step in our refresher is to tie A_μ and $F_{\mu\nu}$ back to the behaviour of the EM field by showing the link with Maxwell's equations. The *derivative* of $F_{\mu\nu}$ is related to charge. If we set $\partial_\mu F^{\mu\nu} = 0$, it means there is no charge in the region. This generates the relevant Maxwell equations shown earlier in Equation 20.6 (often referred to as Maxwell's equations in a vacuum).

I am not going to give all the proofs here, but as an example of how this works, the relationship $\nabla \cdot E = 0$ can be derived in the following way. If you set $\nu = t$ (this is the index $\nu = 0$) in $\partial_\mu F^{\mu\nu}$ it gives $-\nabla \cdot E$ as shown below, that is, $\partial_\mu F^{\mu t} = -\nabla \cdot E$.

The expression $\partial_\mu F^{\mu\nu} = 0$ means that the value is zero for all values of ν... so $\nabla \cdot E = 0$ in a region with no charge:

$$\nu = 0: \quad \partial_\mu F^{\mu\nu} = \partial_0 F^{00} + \partial_1 F^{10} + \partial_2 F^{20} + \partial_3 F^{30}$$

$$= 0 - \frac{\partial E_x}{\partial x} - \frac{\partial E_y}{\partial y} - \frac{\partial E_z}{\partial z} = -\nabla \cdot E = 0 \tag{20.13}$$

Box 20.5   Upper and lower indices (contravariant and covariant)

Things are complicated by Minkowski spacetime. I use the metric $< +--- >$ based on $d\tau^2 = dt^2 - dx^2 - dy^2 - dz^2$ so the *square* of time components is positive and of spatial components negative (note we are using $c = 1$). This requires two component types. *Covariant* components use a lower index such as V_μ and signature $(t, -x, -y, -z)$ matching the metric (the V stands for any typical 4-Vector). *Contravariant* components use an upper index V^μ and signature (t, x, y, z). Combining the two such as $V_\mu V^\mu$ gives a Lorentz invariant result by matching to the invariant interval of the Minkowski metric $(t^2, -x^2, -y^2, -z^2)$ because:

$$V_\mu V^\mu = V_t(+V_t) + V_x(-V_x) + V_y(-V_y) + V_z(-V_z) = V_t^2 - V_x^2 - V_y^2 - V_z^2 \tag{20.14}$$

Sometimes you need to think carefully. For example, with $F_{\mu\nu}$ and $F^{\mu\nu}$ you need to account for the signs of the components of A_μ and of the derivatives. The lower index derivative ∂_μ means $\frac{\partial}{\partial x^\mu}$ and the upper index ∂^μ means $\frac{\partial}{\partial x_\mu}$. As examples, F_{01} and F^{01} are shown below.

$$F_{01} = \partial_t A_x - \partial_x A_t = \frac{\partial(-A_x)}{\partial(+t)} - \frac{\partial(+A_t)}{\partial(+x)} \qquad = -\left(\frac{\partial A_x}{\partial t} + \frac{\partial A_t}{\partial x}\right) \tag{20.15}$$

$$F^{01} = \partial^t A^x - \partial^x A^t = \frac{\partial(+A_x)}{\partial(+t)} - \frac{\partial(+A_t)}{\partial(-x)} \qquad = +\left(\frac{\partial A_x}{\partial t} + \frac{\partial A_t}{\partial x}\right) \tag{20.16}$$

The combination means that the E components of $F_{\mu\nu}$ and $F^{\mu\nu}$ have opposite signs. The calculation $F_{\mu\nu}F^{\mu\nu}$ is also shown as an example because this will appear in the EM Lagrangian:

$$F_{\mu\nu} = \begin{bmatrix} 0 & +E_x & +E_y & +E_z \\ -E_x & 0 & -B_z & +B_y \\ -E_y & +B_z & 0 & -B_x \\ -E_z & -B_y & +B_x & 0 \end{bmatrix} \quad F^{\mu\nu} = \begin{bmatrix} 0 & -E_x & -E_y & -E_z \\ +E_x & 0 & -B_z & +B_y \\ +E_y & +B_z & 0 & -B_x \\ +E_z & -B_y & +B_x & 0 \end{bmatrix}$$

$$F_{\mu\nu}F^{\mu\nu} = F_{00}F^{00} + F_{01}F^{01} + F_{02}F^{02}... = -E_x^2 - E_y^2 ... = -2(E^2 - B^2)$$

Note that the electric field components each become $\frac{E}{c}$ if the speed of light is included. And please do not forget that the E here is the electric field, not energy.

– and watch out for the metric!

There is another complication, and this one is a pain in the butt. Many courses use the metric $<-+++>$ based on $-d\tau^2 = -dt^2 + dx^2 + dy^2 + dz^2$, in which case the covariant components become $(-t, x, y, z)$ while the contravariant still are (t, x, y, z). If you do a calculation and find the signs wrong at the end, you can bet that you have made an index mistake!

20.4   The EM Quantum Field and Lagrangian

The electromagnetic (EM) quantum field is a vector field, A_μ, which has four real components (t, x, y, z). Imagine a little vector arrow at every point in spacetime with a direction in spacetime and a magnitude. This is very different from the Klein-Gordon quantum field that has one complex scalar value at each point.

The excitations of the EM quantum field are *photons*, of course. The energy-momentum relationship is $\mathbf{E} = \mathbf{p}$ for the photon (or $\mathbf{E} = \mathbf{p}c$ if you are not working in natural units) so the rules are different, but the underlying concept is the same. Similar to the electron field, the behaviour of the EM field is driven by the path integral summation across every possible development pathway. This is determined by the Action which we quantify with the EM Lagrangian. What is the EM Lagrangian? We will start with the answer and then show why. The Lagrangian for the standalone EM field (i.e. not interacting, so no charge in the region) is below. The calculation in terms of E and B can be found towards the bottom of Box 20.5:

$$\mathscr{L}_{EM} = -\frac{1}{4} F_{\mu\nu} F^{\mu\nu} = -\frac{1}{4} (\partial_\mu A_\nu - \partial_\nu A_\mu)(\partial^\mu A^\nu - \partial^\nu A^\mu) \tag{20.17}$$

$$= \frac{1}{2} (E^2 - B^2) \quad \text{E and B are the electric/magnetic fields}$$

We can use the Euler-Lagrange equation to derive the pathway of stationary Action for the standalone EM field. The result is $\partial_\mu F^{\mu\nu} = 0$ (see Box 20.6 for details) which gives Maxwell's equations. From those, we can derive the speed of light as we showed earlier in Box 20.3. Moving at light speed, the photon has the energy-momentum relationship of all massless particles: $\mathbf{E} = \mathbf{p}$. Hey presto! The EM Lagrangian quantises the field and gives us Maxwell's equations!

Box 20.6   Euler-Lagrange calculation for the standalone EM Lagrangian

It may help to compare Equation 20.18 with 19.20. The term on the right side in 20.18 is zero because $\mathscr{L}_{EM}$ (20.17) is composed entirely of derivatives.

$$\partial_\mu \left(\frac{\partial \mathscr{L}_{EM}}{\partial(\partial_\mu A_\nu)} \right) = \frac{\partial \mathscr{L}_{EM}}{\partial(A_\mu)} \implies 0 \quad \text{Euler-Lagrange for EM field} \tag{20.18}$$

The shortcut to calculating the term on the left side of 20.18 is that $F_{\mu\nu}$ is antisymmetric so $F_{\mu\nu} = -F_{\nu\mu}$ and $F^{\mu\nu} = -F^{\nu\mu}$. So, $F_{\mu\nu}F^{\mu\nu}$ can be re-expressed equally as either 20.19 and 20.20. If the the calculation confuses you, try with $\mu\nu = (0,1)$ and $\mu\nu = (1,0)$ and all should become clear.

$$F_{\mu\nu} F^{\mu\nu} = (\partial_\mu A_\nu - \partial_\nu A_\mu) F^{\mu\nu} = \partial_\mu A_\nu F^{\mu\nu} + \partial_\nu A_\mu F^{\nu\mu} \qquad = 2\partial_\mu A_\nu F^{\mu\nu} \tag{20.19}$$

$$= (\partial^\mu A^\nu - \partial^\nu A^\mu) F_{\mu\nu} = \partial^\mu A^\nu F_{\mu\nu} + \partial^\nu A^\mu F_{\nu\mu} \qquad = 2\partial^\mu A^\nu F_{\mu\nu} \tag{20.20}$$

We then differentiate by parts –

$$\frac{\partial(F_{\mu\nu} F^{\mu\nu})}{\partial(\partial_\mu A_\nu)} = \frac{\partial(2 \partial_\mu A_\nu)}{\partial(\partial_\mu A_\nu)} F^{\mu\nu} + \frac{\partial(2 \partial^\mu A^\nu)}{\partial(\partial_\mu A_\nu)} F_{\mu\nu} = 4 \frac{\partial(\partial_\mu A_\nu)}{\partial(\partial_\mu A_\nu)} F^{\mu\nu} = 4 F^{\mu\nu} \tag{20.21}$$

– and show that the Euler-Lagrange equation produces $\partial_\mu F^{\mu\nu} = 0$ which means the pathway of stationary Action in the standalone EM field conforms to Maxwell's equations in a vacuum.

$$\partial_\mu \left(\frac{\partial \mathscr{L}_{EM}}{\partial(\partial_\mu A_\nu)} \right) = -\frac{1}{4} \partial_\mu \left(\frac{\partial(F_{\mu\nu} F^{\mu\nu})}{\partial(\partial_\mu A_\nu)} \right) = -\frac{1}{4} \partial_\mu (4 F^{\mu\nu}) \implies \partial_\mu F^{\mu\nu} = 0$$

The Action of differing pathways dictates that $\mathbf{E} = \mathbf{p}$ (or $\mathbf{E} = \mathbf{p}c$ with the light factor) for stable excitations of the field. We call these *real* or *on-shell* photons. But what about the other pathways that are very close, those blue pathways in Figure 19.1 that also contribute to the overall path integral? Over even a tiny stretch of spacetime the Action of off-shell development pathways means they cancel out with surrounding pathways so only on-shell photons can be detected.

However, in interactions, the contribution of short-lived off-shell pathways is important. For these pathways $\mathbf{E} \neq \mathbf{p}c$. If these were stable enough to create an observable photon (which they are not), the photon would not be moving at light speed! As mentioned earlier, these pathways are described as *virtual photons* and play an essential role in the EM interaction as we will discuss further in Chapter 21. Having discussed Maxwell's work, you may also be interested in his history in Box 20.7.

20.5   EM Gauge Invariance

Now that we have worked our way through the legwork (a bit of a yawn), we can push ahead with the most exciting feature of the EM vector potential A_μ. The value of A_μ is *not* fully defined. We have discussed the global gauge invariance of the Klein-Gordon and Dirac fields, that is, you can make a global change to the phase of the fields without affecting the physics. It turns out that the EM vector field has its own different type of gauge invariance.

Take any function of time and space $f(t, x, y, z)$. We can change the EM vector potential A_μ to $A_\mu \pm \partial_\mu f$ (making it what I will call A'_μ), without changing anything measurable. Why? Equation 20.22 shows, as an example, the effect on B_x which is the x component of the magnetic field using the definition from Equation 20.9. The two additional terms introduced from the function f cancel out because the order of differentiation does not matter. The result is that B'_x is the same as B_x. Equation 20.23 shows the more general solution. *All* of the components of $F_{\mu\nu}$ remain the same, so what happens to the electric and magnetic fields? Nothing!

$$B'_x = \frac{\partial A_z}{\partial y} + \frac{\partial}{\partial y}\left(\frac{\partial f}{\partial z}\right) - \frac{\partial A_y}{\partial z} - \frac{\partial}{\partial z}\left(\frac{\partial f}{\partial y}\right) = \frac{\partial A_z}{\partial y} - \frac{\partial A_y}{\partial z} = B_x \tag{20.22}$$

$$F'_{\mu\nu} = \partial_\mu(A_\nu + \partial_\nu f_\nu) - \partial_\nu(A_\mu + \partial_\mu f_\mu) = \partial_\mu A_\nu - \partial_\nu A_\mu = F_{\mu\nu} \tag{20.23}$$

We now have two types of gauge invariance. Take note that the term *gauge* is used generally for these types of invariance, not just for phase. Fermion fields such as the electron field are invariant to phase change. Boson fields such as the EM field of the photon are invariant to adding derivatives of functions in the way that we have just shown. Let's see how they link up, how the EM vector field is like a balancing tank for the electron field's infinity swimming pool.

20.6   U(1) Local Gauge Invariance: Putting Together the Pieces

In this section, we will look at how the combination of the electron field and EM field is local gauge invariant to changes in phase. The correct Lagrangian for the electron field is the Dirac Lagrangian. However, in this section we initially will continue to model the electron field with the Klein-Gordon Lagrangian because the maths is easier to follow. Then, once we have the basics in place, we will switch to the Dirac Lagrangian in Section 20.7. This is a complicated topic so I have divided it into several short subsections to make it easier to absorb.

20.6.1   The Swimming Pool: The Electron Field

Let's review the challenge for a fermion field to achieve U(1) local gauge invariance. Using the Klein-Gordon Lagrangian as a model, additional terms appear in the derivatives that affect the value of the Lagrangian. When the Lagrangian changes, the Action changes. When the Action changes, interference patterns change and measurables change. A local phase change of $\theta(t, x, y, z)$ on Ψ (which I label $\Psi_{<\theta>}$) alters the value of the derivatives in the Klein-Gordon Lagrangian as shown below. I only am showing the t derivative as we saw earlier in Equation 20.3, but similar distortion applies to all derivatives (t, x, y, z):

$$\frac{\partial \Psi_{<\theta>}}{\partial t} = e^{i\theta} \left(\frac{\partial \Psi}{\partial t} + i \frac{\partial \theta}{\partial t} \Psi \right) \tag{20.24}$$

20.6.2 The Balancing Tank: The EM Field

The EM field is unaffected by the addition or subtraction of the gradient of a function $f(t, x, y, z)$ to the vector potential A_μ. This has no impact on the values of the electric or magnetic fields.

$$A_t \rightarrow A_t + \frac{\partial f}{\partial t} \quad \text{no change to measurables} \tag{20.25}$$

20.6.3 The Connection

The electron and EM fields are connected to create a system with overall U(1) local gauge symmetry. The fields are coupled so that a phase shift of θ in the electron field also changes the EM field A. If Ψ becomes $\Psi_{<\theta>}$ due to a phase shift of θ, then A becomes $A_{<\theta>}$ as shown in Equation 20.26. This has no effect on the EM electric and magnetic fields. The g is a constant called, unsurprisingly, the *coupling constant* which we will discuss later. For clarity, the t component of A is also shown as an example:

$$\text{If } \Psi \rightarrow \Psi_{<\theta>} \quad \text{then } A \rightarrow A_{<\theta>} = A_\mu - \frac{1}{g} \partial_\mu \theta \tag{20.26}$$

$$\text{For example: } A_t \rightarrow A_t - \frac{1}{g} \frac{\partial \theta}{\partial t}$$

20.6.4 The Interaction

Using the pool analogy, the final step to get the infinity pool working is for the water to flow correctly between pool and balancing tank. The electron field and the EM field need to interact in a specific way so that the change to $A_{<\theta>}$ offsets the impact of the change to $\partial \Psi_{<\theta>}$ in the electron field Lagrangian. To achieve this, an interaction term must be added to the derivative of the Klein-Gordon (which is acting as our model for the electron field). This new gauge invariant version of the derivative is typically labelled $D\Psi$ and it replaces $\partial \Psi$. To be clear, the use of $D\Psi$ instead of $\partial \Psi$ means that the derivative is unaffected by any local gauge changes in phase.

$$D_\mu \Psi = \partial_\mu \Psi + i g A_\mu \Psi \tag{20.27}$$

Let's show how this works. Equation 20.28 shows the impact of a local phase change of $\theta(t, x, y, z)$ on the derivative with respect to t including the new interaction term with the EM field.

$$D_t \Psi_{<\theta>} = \partial_t \Psi_{<\theta>} + (i g A_t \Psi)_{<\theta>} \tag{20.28}$$

$$= e^{i\theta} \left(\frac{\partial \Psi}{\partial t} + i \frac{\partial \theta}{\partial t} \Psi \right) + i g \left(A_t - \frac{1}{g} \frac{\partial \theta}{\partial t} \right) e^{i\theta} \Psi \quad \text{substitute using Equation 20.26}$$

$$= e^{i\theta} \left(\frac{\partial \Psi}{\partial t} + i g A_t \Psi + i \frac{\partial \theta}{\partial t} \Psi - i \frac{\partial \theta}{\partial t} \Psi \right)$$

$$= e^{i\theta} (\partial_t \Psi + i g A_t \Psi) = e^{i\theta} D_t \Psi \tag{20.29}$$

The key thing is that this means $(D \Psi_{<\theta>})^* D \Psi_{<\theta>} = (D \Psi)^* D \Psi$ as shown in Equation 20.30, so if we replace $\partial \Psi^* \partial \Psi$ with $(D \Psi)^* D \Psi$ in the Lagrangian, it will become U(1) local gauge invariant to changes in phase.

$$(D \Psi_{<\theta>})^* D \Psi_{<\theta>} = e^{-i\theta} (D \Psi)^* e^{i\theta} D \Psi = (D \Psi)^* D \Psi \tag{20.30}$$

20.6.5 The Infinity Pool: Combined Electron and EM Fields

We can now build the QFT equivalent of an infinity swimming pool. We convert the Klein-Gordon Lagrangian (shown as $\mathscr{L}_{KG}$) to a U(1) local gauge invariant form by replacing $\partial \Psi$ with the EM-connected version $D\Psi$. There is another thing missing. We must add a Lagrangian for the EM field. EM excitations are photons whose behaviour will also depend on the interference pattern between different possible pathways. The EM Lagrangian is $-\frac{1}{4}F_{\mu\nu}F^{\mu\nu} = \frac{1}{2}(E^2 - B^2)$. You can see at once that this is U(1) invariant because we have established that the change $A_\mu \rightarrow A_\mu + \partial_\mu f$ has no impact on the electric or magnetic fields (see Equations 20.22 and 20.23). Don't forget that E is the electric field, not energy. Adding this as shown in Equation 20.31 gives the full Klein-Gordon/EM combined Lagrangian:

$$\mathscr{L}_{KG/EM} = \frac{1}{2}(D\Psi)^* D\Psi - \frac{1}{2}m^2\Psi^*\Psi - \frac{1}{4}F_{\mu\nu}F^{\mu\nu} \tag{20.31}$$

Every one of the terms is U(1) local gauge invariant so the Lagrangian is *not* affected by a change in phase, not even by a local one. If you want to check back, this is shown for $(D\Psi)^* D\Psi$ in Equation 20.30, for $\Psi^*\Psi$ in Equation 20.2 and for $F_{\mu\nu}$ in Equation 20.23.

This U(1) local gauge symmetry requires a specific coupling interaction between the two fields. If you expand $(D\Psi)^* D\Psi$ as in Equation 20.32 you can see *interaction* terms that include Ψ, Ψ^* and A_μ. Ψ and Ψ^* are respectively the creation and annihilation operators for the electron field $\hat{\Psi}^-$ and $\hat{\Psi}^+$. A_μ is the creation *and* annihilation operator of the photon field (it can act as either). Therefore the interaction involves both creation and/or destruction of electrons in the electron field and creation or destruction of photons in the electromagnetic field. These interactions lurk within the combined Lagrangian:

$$\mathscr{L}_{KG/EM} = \frac{1}{2}(\partial^\mu\Psi^* - ig A^\mu\Psi^*)(\partial_\mu\Psi + ig A_\mu\Psi) - \frac{1}{2}m^2\Psi^*\Psi - \frac{1}{4}F_{\mu\nu}F^{\mu\nu} \tag{20.32}$$

$$= \frac{1}{2}\partial^\mu\Psi^*\partial_\mu\Psi + \frac{1}{2}g^2 A^\mu A_\mu \Psi^*\Psi... \quad \textit{(showing interaction term)} \tag{20.33}$$

At this stage, we wish a fond farewell to the Klein-Gordon Lagrangian. It is an excellent model to show what is going on, but the interaction term is not quite right for the electron.[1] Electrons are spin-half so they have a four-component wave function and their equation includes 4x4 matrices. Gird your loins! Here comes the Dirac Lagrangian. The maths is a little more complicated, but is necessary to derive the correct interaction term for the electron.

20.7 The Dirac Lagrangian

As a reminder the Dirac *equation* is shown in Equation 20.34. The equation applies for all spin-half fermions such as the electron. The γ notation refers to 4x4 matrices. Do not confuse it with the γ of time dilation. The matrices mix up the four elements of the Dirac wave function. I have slightly altered the order of the terms in the equation, but it is the same as Equation 18.68 where you can also find the full version showing all dimensions and the gamma matrices.

$$i\gamma^\mu \partial_\mu \Psi - m\Psi = 0 \quad \textit{which is...} \quad i\gamma^0 \frac{\partial\Psi}{\partial t} + i\gamma^1 \frac{\partial\Psi}{\partial x}... - m\Psi = 0 \tag{20.34}$$

The Dirac *Lagrangian* is shown in Equation 20.35. Each term is acted on by the Dirac adjoint operator $\overline{\Psi}$. This plays a similar role to the Ψ^* in the Klein-Gordon Lagrangian, but includes the γ^0 matrix as shown below. Don't worry too much about this subtle difference. The γ^0 matrix changes the sign of the Ψ_3 and Ψ_4 components that are associated with antiparticle states so that the Lagrangian is Lorentz invariant. The Dirac *Lagrangian* can be expressed as the Dirac *equation* acted on by the adjoint operator $\overline{\Psi}$ as shown in Equation 20.36.

$$\mathscr{L}_D = i\overline{\Psi}\gamma^\mu \partial_\mu \Psi - m\overline{\Psi}\Psi \quad \textit{where} \quad \overline{\Psi}\Psi = \Psi^*\gamma^0\Psi \tag{20.35}$$

$$= \overline{\Psi}(i\gamma^\mu \partial_\mu \Psi - m\Psi) \tag{20.36}$$

This gives us a neat shortcut to calculate the pathway of stationary Action for the Dirac Lagrangian. We apply the Euler-Lagrange equation with respect to the *adjoint operator* $\overline{\Psi}$. As the Lagrangian contains no derivatives with respect to $\overline{\Psi}$, the term on the right in Equation 20.37 is zero and the result is the Dirac equation (20.38).

$$\frac{\partial \mathcal{L}_D}{\partial \overline{\Psi}} = \partial_\mu \left(\frac{\partial \mathcal{L}_D}{\partial (\partial_\mu \overline{\Psi})} \right) \qquad \textit{Euler-Lagrange for Dirac field} \tag{20.37}$$

$$\frac{\partial \mathcal{L}_D}{\partial \overline{\Psi}} = i\gamma^\mu \partial_\mu \Psi - m\Psi \qquad\qquad \partial_\mu \left(\frac{\partial \mathcal{L}_D}{\partial (\partial_\mu \overline{\Psi})} \right) = 0$$

$$\implies i\gamma^\mu \partial_\mu \Psi - m\Psi = 0 \qquad \textit{Dirac equation} \tag{20.38}$$

Therefore, those development pathways of the Dirac field that have stationary Action are the pathways that conform to the Dirac equation which, of course, is the whole point of the Dirac Lagrangian. This means that they conform to the energy balance built into the Dirac equation which is in natural units: $m = \sqrt{\mathbf{E}^2 - \mathbf{p}^2}$. The Dirac Lagrangian holds the pathways on-shell. If the pathway drifts off-shell too far or for too long, its Action will mean the contribution of the pathway will be cancelled out by surrounding pathways. As discussed before, we cannot ignore the pathways close to those of stationary Action. The contribution of off-shell development pathways, albeit such pathways are of very limited duration, will spice up the interaction calculation. We will discuss this later.

The last piece of the jigsaw is to modify the Dirac Lagrangian to incorporate U(1) local gauge invariance with the EM field. We change the derivative terms in the same way that we did for the Klein-Gordon Lagrangian, switching $\partial_\mu \Psi$ into $D_\mu \Psi$. The modification is exactly the same as Equation 20.27. In the electron-photon interaction, the coupling constant g is electric charge e (in natural units)^2. Finally, to complete the quantum electrodynamic (QED) Lagrangian, we add the standalone EM Lagrangian term $-\frac{1}{4} F_{\mu\nu} F^{\mu\nu}$:

$$\partial_\mu \Psi \implies D_\mu \Psi = \partial_\mu \Psi + ie A_\mu \Psi \qquad \textit{U(1) local gauge invariance} \tag{20.39}$$

$$\mathcal{L}_{QED} = i\overline{\Psi}\gamma^\mu(\partial_\mu \Psi + ie A_\mu \Psi) - m\overline{\Psi}\Psi - \frac{1}{4} F_{\mu\nu} F^{\mu\nu}$$

$$= (i\overline{\Psi}\gamma^\mu \partial_\mu \Psi - m\overline{\Psi}\Psi) - e\overline{\Psi}\gamma^\mu A_\mu \Psi - \frac{1}{4} F_{\mu\nu} F^{\mu\nu} \tag{20.40}$$

We can expand the QED Lagrangian (just as we did earlier in Equation 20.32 for the Klein-Gordon/EM Lagrangian). The result is Equation 20.40. If you look carefully you can see that the two terms to the left are the Dirac Lagrangian without the U(1) modification and the term on the right is the standalone EM field. In the middle is the new interaction term that has to be included for U(1) local gauge invariance. This is the QED interaction between the electron field and the electromagnetic field.

20.8 Interaction and the Pathway of Stationary Action

In order to get U(1) local symmetry, the QED Lagrangian has an interaction term between the electron field and the EM field:

$$\mathcal{L}_{QED} = i\overline{\Psi}\gamma^\mu \partial_\mu \Psi - m\overline{\Psi}\Psi - e\overline{\Psi}\gamma^\mu A_\mu \Psi - \frac{1}{4} F_{\mu\nu} F^{\mu\nu} \tag{20.41}$$

You might be seduced into thinking: *Oh, the interaction term increases the Action so it will be disfavoured in the path integral...* No! We can recalculate the pathway of stationary Action for the electron field using the Euler-Lagrange equation. This is the same calculation that we showed in Equation 20.38 except we must not ignore the new interaction term in the QED Lagrangian that also has a factor of $\overline{\Psi}$:

$$\frac{\partial \mathcal{L}_{QED}}{\partial \overline{\Psi}} = \partial_\mu \left(\frac{\partial \mathcal{L}_{QED}}{\partial (\partial_\mu \overline{\Psi})} \right) \qquad \textit{Euler-Lagrange for QED Dirac field} \tag{20.42}$$

$$i\gamma^\mu \partial_\mu \Psi - m\Psi - e\gamma^\mu A_\mu \Psi = 0$$

$$\implies (i\gamma^\mu \partial_\mu - e\gamma^\mu A_\mu - m)\Psi = 0 \qquad \textit{Pathway of stationary Action} \tag{20.43}$$

The pathway of stationary Action (the Dirac equation) now includes an interaction term between the electron and the EM field. The level of interaction (the coupling) depends on the charge e.

We know that the motion of an electron is deflected in an EM field so this may strike you as obvious – and it is – but it is important that you understand how these things link together:

- The EM field gives the electron field U(1) local gauge symmetry.
- This requires a specific interaction term ($e\,\overline{\Psi}\,\gamma^\mu A_\mu\,\Psi$) in the QED Lagrangian.
- This interaction term affects the pathway of stationary Action for the electron field.
- The motion of the electron is affected by the EM field.

You might question this logic. It treats the U(1) local symmetry as the key factor that drives the whole thing. Why not the other way round? Could it be that the U(1) local symmetry is a coincidence? Might it be a side effect of the electron/EM interaction? You might think so, but you would be wrong. We know because there are two other local symmetries called SU(2) and SU(3) that appear in fermion fields. In the same way that U(1) leads to the electromagnetic force, SU(2) and SU(3) lead respectively to the weak force and strong force. The pattern is unmistakable and is a central topic in Chapters 23 to 25.

20.9   The Photon Must Be Massless

One important thing to point out is that the recipe we have analysed for QED's local gauge invariance to phase *requires* that the photon is *massless*. Consider how the Lagrangian would change if the photon had mass. The mass would manifest itself as an additional $\frac{1}{2}m^2A_\mu A^\mu$ term in the Lagrangian in the same way as mass affects the Klein-Gordon Lagrangian. You can see this in Equation 20.31 and/or you can follow how it runs through the Euler Lagrange equation to drive the equations of motion in Equation 19.19 (remember that the photon is spin-1 and therefore does not follow the Dirac equation).

This additional term would *not* be gauge invariant even to a global change in phase. This would be a serious problem because any quantum field near equilibrium resembles the QHO which means it should be invariant to such a phase change (see Section 19.8). Equation 20.44 shows how a change in phase affects A_μ (check back to Equation 20.26 if needed).

$$\text{If}\ \ \Psi \to \Psi_{<\theta>}\quad then\quad A \to A_{<\theta>} = A_\mu - \frac{1}{g}\partial_\mu\theta \tag{20.44}$$

The impact on the Lagrangian term that would exist for a massive photon is shown in Equation 20.45. It is clear that the term is *not* gauge invariant to any change in phase, not even a global phase change.

$$\frac{1}{2}m^2\,A_\mu A^\mu \to \frac{1}{2}m^2\,A_{<\theta>}A_{<\theta>} = \frac{1}{2}m^2\left(A_\mu - \frac{1}{g}\partial_\mu\theta\right)\left(A^\mu - \frac{1}{g}\partial^\mu\theta\right)$$

$$\neq \frac{1}{2}m^2\,A_\mu A^\mu \tag{20.45}$$

If such a term existed, it would torpedo the gauge invariance for the whole QED Lagrangian. Why is this important? The photon is an example of what is called more generally a *vector boson*. We will discuss the strong force in Chapter 23. It involves vector bosons which in many ways are comparable to the photon of QED. These are called *gluons* and they are also massless. However, the equivalent for the weak force are the W_+, W_- and Z bosons which are *massive*. This created a mystery that was one of the keys to unravelling the Higgs mechanism. Don't worry about this yet. I just want you to remember that there are significant difficulties in preserving gauge invariance if massive vector bosons are added to the mix.

20.10   Summary

In this chapter we have kissed a lot of mathematical frogs, we have floated on a quantum sea frothing with equations, we have dabbled with countless derivatives; well, I apologise for that, but this is very advanced stuff. The good news is that the whole thing can be summarised by Figure 20.3 so now we can sit back, relax and ponder the underlying themes.

Let's start with the standalone Dirac field which is the left hand term in the QED Lagrangian in Figure 20.3. The path of stationary Action obeys the Dirac equation so: $m = \sqrt{\mathbf{E}^2 - \mathbf{p}^2}$. These on-shell development pathways result in the stable excitations that we call electrons. Ψ always operates alongside its adjoint which, in the case of the Dirac field is $\overline{\Psi}$.

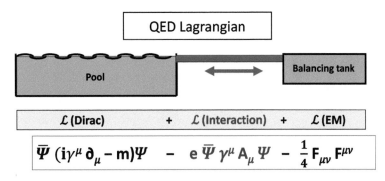

Figure 20.3 The QED Lagrangian is U(1) local gauge invariant.

This means the annihilation operator always operates alongside the creation operator. If an electron is annihilated, either another electron is created, or a positron is annihilated. This ensures *global* phase invariance.

Nature demands more. It requires *local* gauge invariance to phase. If the phase changes in only part of the quantum field, it means that the phase of the wave function varies with time and space:

$$\Psi \quad \rightarrow \quad e^{i\theta(t,x,y,z)}\,\Psi \tag{20.46}$$

When we say a quantum field is *local* gauge invariant to phase, we mean that any phase change of the type shown in Equation 20.46 has no effect on the physics. This requires a compensating interaction with another field, the EM field. This U(1) symmetry demands a specific interaction term between the two fields which is shown in the centre of Figure 20.3. It combines $\overline{\Psi}$, A_μ and Ψ. What is A_μ? It creates *or* annihilates a photon in the EM field.

Note that the photon does not experience time in the conventional sense (see Box 6.4) so there is no difference between a photon moving forwards and backwards in time. It is its own antiparticle. The creation and annihilation of a photon is mathematically identical. So what does $\overline{\Psi}A_\mu\Psi$ mean? It could be the destruction of an electron alongside the creation of a photon and a replacement electron. It could be the annihilation of a photon alongside the creation of a electron-positron pair. Every interaction is some combination of the three.

Let's end the chapter by comparing again U(1) symmetry with an infinity swimming pool. A local phase change affects the standalone Dirac Lagrangian (*the pool*). The phase change also affects the vector potential A_μ leading to an offsetting shift in the value of the interaction term (*in/outflow from the pool*). However, the vector potential A_μ changes in a way that has no effect on the electric and magnetic fields, so there is no change to the EM Lagrangian providing that the photon is massless (which it is). Effectively, the local phase change in the Dirac field is absorbed into the EM vector field (*the balancing tank*).

What *is* the electromagnetic force? It is what nature requires for fermions to be U(1) local gauge invariant (see Box 20.8). Similar mechanisms sit behind the strong and weak forces, but first we need to discuss the interactions of quantum electrodynamics (QED).

Box 20.8 Local gauge invariance – in a nutshell

Suppose we are working with a scientist, Jill. She does *part* of the calculations using a different phase for the electron. This local change of phase will affect her calculation of the Dirac component in the Lagrangian. However, the phase change also changes Jill's photon interaction term. Adding it all together we find that the Action of the QED Lagrangian is the same as it would have been without Jill's phase change. That is local gauge invariance to a change of phase.

Notes

1 Although wrong for the electron, this interaction term will prove important when we discuss the Higgs field in Chapters 24 and 25 as it is key to explaining the mass of the weak force bosons.

2 To switch to SI (International System) units you need to add some constant terms.

Chapter 21

QED and Feynman Diagrams

CHAPTER MENU

21.1 Feynman Diagrams
21.2 Example: Electron-positron Annihilation
21.3 Off-shell Drift and the QED Interaction
21.4 Feynman Rules
21.5 Resonance and the Search for New Particles
21.6 Do Virtual Particles Exist?

Box 21.1   Freeman Dyson recounts the following story of a meeting with Feynman:

Dick Feynman told me about his sum-over-histories version of quantum mechanics. 'The electron does anything it likes', he said. 'It just goes in any direction at any speed, forward or backward in time, however it likes, and then you add up the amplitudes and it gives you the wave-function'. I said to him, 'You're crazy' but he wasn't.   Freeman Dyson

---

21.1   Feynman Diagrams

Every interaction that can happen, does... sort of. Those of you who didn't snooze off during Chapter 6 should recognise this wording and know what is coming. If the behaviour of an electron is the path integral sum of every possible development pathway, then it follows that the interaction of an electron with the EM field must be the path integral sum of *every legitimate interaction pathway*. Yes, yes, it is that bad (see Freeman Dyson's reaction in Box 21.1).

Cast your mind back to the figure showing the path integral in our non-relativistic analysis (Figure 19.1). It shows a variety of pathways. They all contribute, but the outliers cancel out so you get a good estimate of the total path integral by adding up the contributions from the blue pathways. In that analysis the pathways are through spacetime. In QFT we ignore when and where things happen. Instead, we focus on how the *state* of the quantum field changes.

Compare the QFT path integral that I introduced in Equation 19.17 with the non-relativistic path integral shown below.

$$\text{Path integral in QFT:} \qquad \int_{\Psi_i}^{\Psi_f} D(\Psi)\, e^{\frac{i}{\hbar} \int \mathscr{L} d^4 x} \tag{21.1}$$

$$\text{Non-relativistic path integral:} \qquad \int_a^b D(x)\, e^{\frac{i}{\hbar} \int \mathscr{L} dt}$$

The non-relativistic path integral combines every possible pathway of an *object* or *particle* travelling from a to b, which are two points in spacetime. The QFT path integral combines every development pathway in the quantum field that goes from the initial state $|\Psi_i >$ to the final state $|\Psi_f >$.

Let's consider an example: Compton scattering. This is when two electrons repel each other. How many development pathways go from the initial state $|\Psi_i >$ (two electrons approaching each other) to the final state $|\Psi_f >$ (two electrons

Quantum Untangling: An Intuitive Approach to Quantum Mechanics from Einstein to Higgs, First Edition. Simon Sherwood.
© 2023 John Wiley & Sons Ltd. Published 2023 by John Wiley & Sons Ltd.

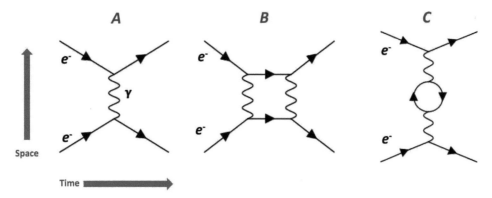

Figure 21.1   Some Feynman diagrams for Compton scattering.

moving apart)? The answer is that there are an *infinite* number. Figure 21.1 (a repeat of Figure 6.5) shows a few possibilities illustrated with *Feynman diagrams*. Each is a cartoon-style schematic of a pathway that contributes to the overall interaction path integral. Solid lines represent fermions (in this case electrons) and wavy lines represent bosons (in this case photons). Internal lines are described as *virtual*.

Diagram A in the figure shows one of the simplest development paths from $|\Psi_i>$ to $|\Psi_f>$. In layman's terms the electrons interact by the exchange of a virtual photon. It is tempting to think of a photon being fired from one electron at another and so repelling it. *Wrong!* That will confuse you. If the virtual photon were exchanged between an electron and a proton it would cause attraction. The virtual photon represents a transfer of energy and momentum through the EM field. It is shown as a photon because its behaviour is governed by the EM field Lagrangian.

Diagram B shows another pathway. The electrons exchange two virtual photons. Diagram C is even more complicated. The electrons exchange a virtual photon that turns into a virtual electron-positron pair. This pair then annihilates to form another virtual photon – you can make the Feynman diagrams as complicated as you like so there are an infinite number.

One source of confusion is that the Feynman diagram looks like a story in spacetime. Diagram C might be: 2 electrons $\longrightarrow$ photon emitted $\longrightarrow$ photon becomes electron-positron pair $\longrightarrow$ pair annihilates to become photon $\longrightarrow$ photon annihilates with energy transfer to second electron. This is not what is happening. All the interaction terms are treated as one indivisible event. For example, there is no difference in diagrams A or C between the virtual photon travelling from the top electron to the bottom or vice versa. The virtual photon and all the other interaction steps are just a figurative portrayal of potential pathways for the energy transfer between the electrons via the EM field. In QFT we assess changes to the state of the quantum field which can be summarised as: (1) initial field state (2) *BANG!* (3) final field state.

We use Feynman diagrams because they are easy to comprehend, but never forget that the so-called virtual photon in A is not really a photon. It is one of an infinite number of ways in which excitations could transfer energy and momentum between fields (both attraction and repulsion, see Box 21.2). The actual behaviour of particles depends on the combined effect of every possible interaction pathway. That's right, the whole lot – the total path integral including every option – wow!

Box 21.2   Virtual particles and attraction

It is fairly easy to visualise how the exchange of virtual particles can lead to a repulsive force, but somewhat harder to imagine how it might result in an *attractive* force. If this is perplexing you, please refer back to Subsection 16.4.2 which shows how the ability of electrons to flip-flop between atoms (quantum tunnelling) attracts atoms to each other to form covalent bonds.

I also find it helpful to use currency as an analogy. Imagine you can trade in the UK only for a whole numbers of pounds and, in the USA, only for a whole numbers of dollars. The different currencies would define what trades happen in each market. In the same way, virtual particles are the currency (quanta) in the exchange of energy and momentum through a quantum field. The attraction of an electron to a proton is a flow of energy and momentum through the electromagnetic field. That flow must be in the currency of photons.

21.2 Example: Electron-positron Annihilation

Let's imagine that after reading this magnificent book, you are hired to work at an electron-positron collider. The Large Electron-Positron collider (LEP) at CERN cost over one billion dollars! It operated for seven years and then was replaced by the Large Hadron Collider (LHC). An electron-positron collider does what it says on the tin. Electrons and positrons whizz round a magnetised tunnel in opposite directions because of their different charges. You get them close to light speed and then *BANG!* you smash them into each other.

Figure 21.2 is my clumsy attempt to illustrate what happens. The particles in the initial state (the electron and positron) enter the experiment from the left. Don't be confused by the arrow on the positron. It is not the direction of motion. It is a convention to remind you that a positron is the mathematical equivalent of an electron moving backwards in time. The right side of the figure shows the final state, that is, what you see in your detectors as a result of the collisions. Much depends on the momentum of the electron and positron going in, but for illustration let's consider how we calculate the probability of the output being two photons.

The centre of Figure 21.2 represents the development pathways through the quantum fields. There are an infinite number of pathways from the initial state (electron and positron) to the final state (two photons). In the figure, I have shown multiple pathways. For example, the electron and positron simply might annihilate or they could exchange one or more virtual photons and then annihilate. Some of these pathways constructively interfere (blue dotted paths) and others largely cancel out with surrounding paths (black dotted paths). Each pathway is a different sequence of interactions so a different Feynman diagram.

Theoretically we must add up the amplitude of every pathway leading to two photons and then square the total. That would be impossibly tiresome. Fortunately there is a shortcut. In QED we get a good approximation by adding up the key pathways that make the strongest contribution. This is the same concept as shown in Figure 19.1 where we can approximate the path integral by adding up the blue pathways. There is a simple rule for determining which are likely to be the most important pathways. In QED, it tends to be the ones with the fewest interactions because the coupling constant e is small. Let me give you a feel of why and how this works.

We will focus on the standalone Dirac component $\mathscr{L}_D$ and the interaction component $\mathscr{L}_I$ of the QED Lagrangian. Equation 21.2 takes the expression for the probability amplitude of a *single* pathway from Equation 21.1 and expresses it as a product of exponents using $e^{a+b} = e^a e^b$. This sounds horribly complicated, but is not. The integral is a long sum $e^{a+b+c+d...}$ that we can express as $e^a e^b e^c e^d$.... The $\prod$ sign means you multiply the components instead of adding them. With the same bit of maths[1] we can separate out the $\mathscr{L}_I$ term from the $\mathscr{L}_D$ term as shown in Equation 21.3.

$$e^{i\int(\mathscr{L}_D+\mathscr{L}_I)\, d^4x} = \prod e^{i(\mathscr{L}_D+\mathscr{L}_I)\, d^4x} \tag{21.2}$$

$$= \prod e^{i\mathscr{L}_I}\, e^{i\mathscr{L}_D}\, d^4x \tag{21.3}$$

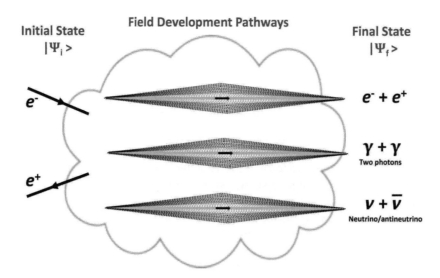

Figure 21.2 Illustrative quantum field development pathways.

Box 21.3   Taylor expansion of e^{ix}

$$e^{ix} = 1 + ix + \frac{(ix)^2}{2!} + \frac{(ix)^3}{3!} + \frac{(ix)^4}{4!} + \cdots$$

We then use a Taylor expansion (see Box 21.3). This gives the expression in Equation 21.4. An exclamation mark ! in maths means *factorial* so, for example, $4! = 4 \times 3 \times 2 \times 1$.

$$
\begin{aligned}
e^{i\mathscr{L}_I} &= 1 + i\mathscr{L}_I + \frac{(i\mathscr{L}_I)^2}{2!} + \frac{(i\mathscr{L}_I)^3}{3!} + \frac{(i\mathscr{L}_I)^4}{4!} + \cdots \\
&= 1 - ie\left(\overline{\Psi}\gamma^\mu A_\mu \Psi\right) + e^2 \frac{(\overline{\Psi}\gamma^\mu A_\mu \Psi)^2}{2!} - ie^3 \frac{(\overline{\Psi}\gamma^\mu A_\mu \Psi)^3}{3!} \cdots
\end{aligned}
\tag{21.4}
$$

Now comes the neat trick that makes QED a useful theory. As the coupling constant e is small, we can approximate the total path integral of all interactions by considering only the first few terms in Equation 21.4. A pathway with only one QED interaction includes the term $(\overline{\Psi}\gamma^\mu A_\mu \Psi)$ that has a weighting factor of e. A pathway that has two interactions includes $(\overline{\Psi}\gamma^\mu A_\mu \Psi)^2$ with a weighting factor of e^2 and so on. Because e is small, the amplitude of each term falls rapidly as the number of interactions increases. The probability of an outcome depends on the square of the amplitude, so this falls away even faster. This can give a finite result (see Box 21.4).

Box 21.4   Infinite sums can converge

How could the summed amplitudes of an infinite sequence of Feynman diagrams ever be finite? Consider the following series that *converges*. The sum gets closer and closer to 1 (try it)...

$$\frac{1}{2} + \frac{1}{4} + \frac{1}{8} + \frac{1}{16} + \frac{1}{32} + \frac{1}{64}\cdots \; < 1$$

Similarly, the QED coupling constant is small enough that the contribution of pathways decreases rapidly as the number of interactions increases.

What does this mean for our task at the LEP collider? It means that we can get a good estimate of the probability of producing two photons by considering only those pathways that involve few interactions. We sum the amplitude of the pathways and square the result. And if we want a more accurate estimate, we simply take into account more complicated pathways. Job done!

If you study QFT seriously, you will see this type of analysis again and again. It is called *perturbation theory*. The idea is to break down an impossible problem (in this case an infinite number of interaction pathways) into bite-size chunks that can be analysed to give an approximate result. The deeper you delve, the more calculations you make, the better the approximation.

But our analysis so far has ignored something very important. We have focussed on the $\mathscr{L}_I$ interaction term in the Lagrangian of the electron field. What about the $\mathscr{L}_D$ term? We know the standalone Dirac Lagrangian $\mathscr{L}_D$ holds pathways on-shell. The more a pathway drifts off-shell, the greater the Action and the less the pathway is likely to contribute to the path integral. Do QED interaction pathways drift off-shell sometimes? Yes, they do. Let's take a look at that.

21.3   Off-shell Drift and the QED Interaction

Consider what the QED interaction term would imply for on-shell particles. The interaction term is shown again in Equation 21.5. In QFT, Ψ, $\overline{\Psi}$ and A_μ are creation or annihilation operators. Ψ and $\overline{\Psi}$ create or annihilate an electron or positron. A_μ creates or annihilates a photon.

$$e\,\overline{\Psi}\,\gamma^{\mu}\,A_{\mu}\,\Psi \qquad \textit{QED interaction term} \tag{21.5}$$

The interactions that Ψ, $\overline{\Psi}$ and A_{μ} can represent are shown below:

(a) Annihilate electron – create electron and photon   *Electron emits photon*
(b) Annihilate electron and photon – create electron   *Electron absorbs photon*
(c) Annihilate positron – create positron and photon   *Positron emits photon*
(d) Annihilate positron and photon – create positron   *Positron absorbs photon*
(e) Annihilate electron and positron – create photon   *Electron-positron pair annihilation*
(f) Annihilate photon – create electron and positron   *Electron-positron pair creation*

Now for the punch line. All the processes listed above require something to drift off-shell. If they were on-shell, they would break the conservation of energy or momentum. Here's why.

Interaction (a): imagine yourself in the rest frame with respect to the electron *before* it emits a photon. You measure total energy as $m_e c^2$, the rest mass of the electron. After emission you measure energy again. Now there is the rest mass energy of the electron plus the energy of the photon and the electron's kinetic energy (the electron must move in the opposite direction of the emitted photon if momentum is conserved): $\Longrightarrow$ *energy conservation breach. Oops!* The same logic applies for (c).

Interaction (b): put yourself in the rest frame of the electron *after* the photon is absorbed. You measure energy after the absorption as $m_e c^2$, the rest mass of the electron, but before absorption there was an electron plus a photon, so where did the extra energy go? $\Longrightarrow$ *energy conservation breach. Oops!* The same logic applies for (d).

Interaction (e): imagine you observe things from the frame of reference that is the centre of mass of the electron and the positron as they approach each other from either side of you to annihilate. You measure net momentum as zero. After the annihilation a photon emerges but a normal photon moves at light speed and must have momentum $\Longrightarrow$ *momentum conservation breach. Oops!* The same logic applies in reverse for (f).

The upshot is that any process with, in totality, only *one* QED interaction term can be discounted. For example an electron and positron cannot annihilate to produce only one photon. *One* interaction term would leave something off-shell – a photon, electron or positron.

You may think: *Aha, what if the electron is constrained in an atom? The electron can absorb or emit a photon. One interaction works because the atom's momentum balances things out.* Wrong, for the atom to absorb momentum there must be an interaction between the electron and the atomic proton so there is more than one interaction.

Whichever way you cut things, the fundamental $\overline{\Psi}A_{\mu}\Psi$ interaction of QED involves some sort of off-shell drift. Small amounts of off-shell drift are permissible within the path integral as we discussed in Section 19.4, but not for long. Every Feynman QED diagram is a combination of $\overline{\Psi}A_{\mu}\Psi$ interactions. There must be more than one in order for the excitations to finish up back on-shell. To be sure this is all clear, Figure 21.3 shows three impossible QED processes for you to compare with those in Figure 21.1.

Diagram D is incorrect because the vertex (the point where particles meet) has two photons so it would be $\overline{\Psi}A_{\mu}A_{\mu}\Psi$ instead of $\overline{\Psi}A_{\mu}\Psi$. Diagram E is incorrect because the second vertex has two electrons emitted so it is $\Psi A_{\mu}\Psi$ which has two Ψ terms. Diagram F is incorrect because there is only one vertex so the photon, electron and positron cannot all be on-shell. Although the structure of the vertex is legitimate, there would have to be another vertex for F to represent a legitimate QED process.

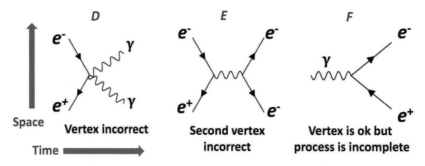

Figure 21.3 Some INCORRECT QED vertices and processes.

21.4 Feynman Rules

To get a good estimate of a particular outcome in QED (such as our example of producing two photons from an electron-positron collision), you need to work out the contribution of the simpler pathways one by one. You then sum these amplitudes and square the total. The more complicated the pathways you include in your calculation, the better the estimate of total probability. How do you calculate the contribution of a particular pathway? The maths is a bit tricky, but the concept is simple. The amplitude of a pathway is affected by two things:

- The number of interactions in the pathway – calculated using the *coupling constant*.
- The amount of off-shell drift in the pathway – calculated using the *propagator*.

In the following discussion of Feynman's rules, I will be using the language of particles and virtual particles so, I want to remind you that the lines in Feynman diagrams are not actual particles. They are *contributing development pathways* in the quantum field.

21.4.1 The Vertex and the Coupling Constant

Figure 21.4 shows the fundamental $\overline{\Psi} A_\mu \Psi$ interaction of QED. This involves the meeting of three lines at a vertex: two solid lines (electron or positron) and a wavy line (photon). Let's briefly run through the examples in the figure. **A** shows an electron emitting or absorbing a photon. **B** shows an electron annihilating with a positron to produce a photon. The top line is the electron. The bottom line is the positron (by convention, distinguished from the electron using an arrow pointing backwards in time). **C** shows a photon converting into an electron-positron pair. Please note carefully that Figure 21.4 shows Feynman diagrams of *vertices* not of legitimate processes. Legitimate QED processes require at least two vertices so that the resulting particles are back on-shell.

Feynman noticed something that made his diagrams much easier to handle. The $\overline{\Psi} A_\mu \Psi$ interaction term contains no derivatives with respect to time or space. Different orientations of the vertex represent very different interactions, but the maths at the vertex remains the same. Each vertex is an interaction and so brings a factor of the coupling constant e to the amplitude of the pathway. Often you will see the symbol g used more generally for the coupling constant of all the forces. In the case of QED the value of the coupling constant is about 0.3. Typically this is expressed using α, the fine structure constant, which is approximately $1/137$:

$$\text{QED coupling constant:} \quad g_e = \sqrt{4\pi\alpha} \approx 0.30 \tag{21.6}$$

Consider what this means for Feynman diagrams A and B in Figure 21.1. Pathway A includes two vertices whereas pathway B contains four. The extra pair of vertices reduces the amplitude of B to about 0.1 of A (0.3×0.3). The probability contribution of the pathway is the square of the amplitude so it is reduced to about 0.01 (see Box 21.5). I have simplified things by ignoring phase.[2]

Box 21.5   For every vertex, multiply by a factor of: g_e   **Effect of coupling constant**

Rule of thumb: Every extra pair of vertices reduces the probability to 1% of what it was. In most QED calculations, you don't have to worry about anything but the simplest Feynman diagrams.

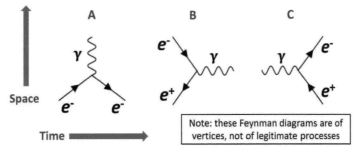

Figure 21.4   Different vertex orientations represent different interaction terms.

21.4.2 The Propagator

The contribution of a pathway to the total probability amplitude of an interaction also depends on off-shell drift. Let's start with an example. Consider the off-shell drift of pathway C in Figure 21.1. The excitation in the EM field (virtual photon) passes back into the electron field (virtual electron-positron pair) on its development path. It takes a lot of energy to create an on-shell electron-positron pair because of the rest mass, so this excitation is likely to be way off-shell. This creates Action in the C pathway and its contribution to the path integral falls.

The effect of off-shell drift is encapsulated in the *Feynman propagator*. There is a propagator associated with every virtual particle (internal line) in a Feynman diagram. The mathematical derivation is a bore, but the result is simple and intuitive: the amplitude of the pathway goes down in proportion to the extent to which the excitation is off-shell (Box 21.6).

Box 21.6   For every internal line, multiply by the propagator to account for off-shell drift

Particle	On-shell formula	Propagator	Summary notation	
photon	$\mathbf{E}^2 - \mathbf{p}^2 = 0$	$\dfrac{1}{\mathbf{E}^2 - \mathbf{p}^2}$	$\dfrac{1}{p_\mu p^\mu}$	(21.7)
electron	$\gamma\mathbf{E} - \gamma\mathbf{p} - m = 0$	$\dfrac{1}{\gamma\mathbf{E} - \gamma\mathbf{p} - m}$	$\dfrac{1}{\gamma_\mu p^\mu - m}$	(21.8)
massive boson	$\mathbf{E}^2 - \mathbf{p}^2 - m^2 = 0$	$\dfrac{1}{\mathbf{E}^2 - \mathbf{p}^2 - m^2}$	$\dfrac{1}{p_\mu p^\mu - m^2}$	(21.9)

Let's look at the virtual photon as an example (Equation 21.7). The first column shows you the energy relation for a real photon. The QED reaction throws the internal lines in Feynman diagrams off-shell so a virtual photon is an off-shell excitation pathway in the EM field. The propagator is inversely proportional to the amount that the excitation is off-shell. QFT boffins should note that I have ignored phase and the infinitesimal to keep things simple.[3]

Note the summary notation in the final column. In this form, energy and momentum are handled as a 4-vector $(\mathbf{E}, \mathbf{p}_x, \mathbf{p}_y, \mathbf{p}_z)$ with relativistic energy $\mathbf{E}$ treated as p_0, the time component of the momentum 4-vector. As a reminder of Einstein summation, $p_\mu p^\mu$ is shown expanded below. If you are confused, please refer back to Box 20.5. On most occasions, it is further simplified to just p^2 (or k^2). For the rest of this book, I am going to use this notation and assume you remember that it includes momentum in all three dimensions and also energy:

$$p^2 \;=\; p_\mu p^\mu \;=\; \mathbf{E}^2 - \mathbf{p}_x^2 - \mathbf{p}_y^2 - \mathbf{p}_z^2 \tag{21.10}$$

The propagator of the electron (Equation 21.8) is slightly different. It is spin-half so the definition of on-shell is the energy-momentum relationship of the Dirac equation, but the concept is the same: the propagator is inversely proportional to the off-shell drift. The γ is the Dirac matrix mixing (not to be confused with time dilation).

One moment! You cry. *Don't we have to integrate the off-shell drift over time and space as set out in Equation 21.3?* That would be hard, so Feynman built his propagator in energy-momentum coordinates. In earlier sections on non-relativistic quantum mechanics (e.g. Section 10.1), we showed that energy-momentum coordinates are the Fourier transform of space-time coordinates. The one fully defines the other. This allows us to integrate the Feynman propagator over all values of energy and momentum $\int d^4 p$ instead of time and space $\int d^4 x$. That may not strike you as a big simplification but it is. In simple Feynman diagrams (called *tree* diagrams that have no loops) the value of the virtual particle's energy and momentum is fully determined by the particles going in and out of the interaction so you do not have to integrate; you can just pop the value into the propagator.

All of this stuff about Feynman rules is best demonstrated with an example. I should warn again that I have drastically simplified things by ignoring phase changes, the four-component structure of the electron wave function and the polarisation of the virtual photon. In spite of this, it should give you a good feel of how the Feynman rules work.

21.4.3 Illustrative QED Calculation (Simplified)

For this illustration, I will return to electron-positron collisions. Figure 21.5 shows the simplest pathways for the scattering interaction between an electron and positron, which goes by the rather charming name Bhabha scattering (see Box 21.7 on Bhabha's untimely death). Time flows from left to right so in both cases you start with an electron and a positron and end with an electron and a positron.

The left of the figure shows what is called the T-channel. In layman language the electron and positron interact (attraction) by exchanging a virtual photon. The incoming and outgoing electron (e_a^- and e_c^-) has energy and momentum values p_a and p_c. The incoming and outgoing positron has momentum values p_b and p_d. Remember that each of these is a 4-vector. Check back to the explanation of this notation in Equation 21.10 if you have doubts.

To calculate the amplitude, which is typically labelled M, we apply the Feynman rules for the two vertices and the propagator as shown in Equation 21.11. We must integrate over all possible values for the virtual photon's energy and momentum which I will label p_q. This is a 4-vector with the time component p_0 being energy. In this case we know from conservation of energy and momentum that the values of p_q can be derived from the corresponding change in the electron or positron that is, $p_q = p_a - p_c$ or $p_q = p_b - p_d$ (we will use the former).

$$M \propto g^2 \int_{-\infty}^{+\infty} \frac{1}{p_q^2} \, d^4 p \tag{21.11}$$

$$M_T \propto \frac{g^2}{(p_a - p_c)^2} = \frac{4\pi\alpha}{(p_a - p_c)^2} \qquad \alpha \text{ is the fine structure constant} \tag{21.12}$$

Don't forget that the p terms in Equation 21.12 contain four components of energy and momentum:

$$
\begin{aligned}
(p_a - p_c)^2 &= (p_a - p_c)_\mu (p_a - p_c)^\mu \\
&= (E_a - E_c)^2 - (\mathbf{p}_a - \mathbf{p}_c)_x^2 - (\mathbf{p}_a - \mathbf{p}_c)_y^2 - (\mathbf{p}_a - \mathbf{p}_c)_z^2
\end{aligned}
\tag{21.13}
$$

Box 21.7   Conspiracy theory: was Bhabha (1906–1966) murdered by the CIA?

Homi Bhabha (as in Bhabha scattering) was one of the leading experts in nuclear fission. In October 1965, he publicly announced that, having received approval from the government of India, he would rapidly develop an atomic bomb for them. Three months later he died when an Air-India Boeing 707 crashed into Mont Blanc on its way to Vienna. Reportedly a former CIA operative, Robert Crowley admitted to a journalist that the CIA was responsible. Conspiracy theory or fact? I have no idea.

Turning to the S-channel on the right of Figure 21.5, by a similar logic it is easy to calculate its value of p_q. In layman's terms the electron and positron annihilate each other creating a virtual photon which in turn becomes an electron-positron pair. Clearly in this case $p_q = p_a + p_b$.

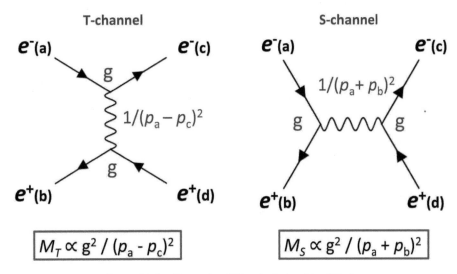

Figure 21.5   Illustrative QED calculation (simplified).

$$M_S \ \propto \ \frac{g^2}{(p_a + p_b)^2} \ = \ \frac{4\pi\alpha}{(p_a + p_b)^2} \tag{21.14}$$

With more complicated Feynman diagrams, you simply multiply together all the coupling constants and all the propagators. Fortunately there are computer programs to help with the maths. Loops in the diagrams create a special challenge that we will address later.

21.4.4 From Amplitude to Cross Section

We now need to combine the results for the T-channel and S-channel. The rule is fairly simple. If the outcome of Feynman diagrams is the same (i.e. the resulting electron and positron of one Feynman diagram are indistinguishable from those of another) then the *amplitudes* add: $M_{tot} = M_1 + M_2....$ On the other hand, if the outcomes can be distinguished (i.e. there is some difference such as spin) then the *probabilities* add $M_{tot}^2 = M_1^2 + M_2^2...$

Theoretically, to calculate the total probability of Bhabha scattering, you need to combine the result of every possible one of the infinite number of Feynman diagrams that result in an electron and positron. However, we have shown that adding up the simpler ones gives a good approximation. In this case, in principle, we add and then square the result for the two simplest routes: $|M|^2 = |M_T + M_S|^2$. Needless to say, things are actually more complicated. These amplitudes are complex numbers and I am ignoring phase differences that are very important in how the amplitudes combine.

QED experts will howl *if only it were so easy...* and they are right. Look at Bhabha scattering solutions on the web (Wikipedia has one) and you will see it is much more complicated to account for phase changes, the four-component nature of the fermion wave functions and the spin of the fermions and of the photons. If you want to work at the Large Hadron Collider (LHC) you are going to have to dig deep into that mathematical nightmare and, for that, you need somebody better informed and more patient than I am. The simplified example that we have done covers the concept. For me, enough is enough.

To tie $|M|^2$ to the probability of Bhabha scattering in the collider takes a couple more steps. You use *Fermi's Golden Rule* (see Box 21.8 if you are interested in Fermi's life) to account for the density of electrons and positrons that you are smashing into each other. With this you can calculate the experimental *cross section* which is the measure of probability in a collider particle experiment. Why cross section? Basically, each electron and positron is treated mathematically like a little circular target, a bullseye which if it is hit, results in the outcome under measurement. The larger the cross section of the target, the more likely the experimental outcome. The maths linking $|M|^2$ to cross section is mechanical and dull so I leave it up to you to learn more if you want although, amusingly, the cross section is measured in *barns* from the expression *as big as a barn door*.

Here is a conceptual question for you to ponder. Imagine you calculate M_T and it is much higher than M_S so $|M_T|^2$ is much higher than $|M_S|^2$. Does this mean that Bhabha scattering is more likely to be through the interchange of a virtual photon than electron-positron annihilation? *No! No! No! Absolutely not!* Bhabha scattering is the path integral of *all* contributing pathways. Extinguish from your mind any thought of one interaction following the T-channel and the next the S-channel. Neither channel is a standalone interaction. Look back to Equation 21.4 to remind yourself that the T-channel and S-channel are just terms in a mathematical expansion of the interaction. The terminology of virtual particles is seductive, so be careful.

Box 21.8   Fermi (1901–1954): one of the architects of the atomic bomb

Enrico Fermi (the Fermi of fermion) quickly understood the explosive potential of energy-mass conversion through $E = mc^2$, and is reported as concluding in 1923: *The first effect would be an explosion so terrible that it would tear the physicist who tried it to pieces.*

He won the Nobel Prize in 1938, but abandoned his Italian home because of Mussolini's race laws that banned Jews from public office, higher education and even from writing books. Fermi emigrated with his wife to the USA where he built the first experimental nuclear reactor (under one of Chicago University's football fields) and then became one of the leading lights in the Manhattan Project that constructed the first nuclear bomb, proving that, if you have a Nobel-winning expert in nuclear fission and you want to win a war, don't piss him off and drive him out! Towards the end of his life, Fermi became uneasy about the potential he had helped unleash:

What is less certain, and what we all fervently hope, is that man will soon grow sufficiently adult to make good use of the powers that he acquires over nature.

21.5 Resonance and the Search for New Particles

It is time for some fun. You may wonder how colliders such as the Large Electron-Positron collider (LEP) and the Large Hadron Collider (LHC) find new particles such as the Z boson, a massive neutral (zero electric charge) boson associated with the weak force.[4] How do you use the LEP to prove it exists? Believe it or not, you have the basics at your fingertips. It all comes down to the *propagator*.

Consider the Feynman diagram on the upper left of Figure 21.6 that shows the annihilation of an electron and positron to produce a virtual photon that then decays into hadrons (these are relatively heavy composites of quarks such as mesons and occasionally protons and neutrons). The interaction is most easily analysed by working in the centre of mass frame. In this frame the electron and positron approach with equal and opposite momentum. Added together, they have *zero* momentum. Therefore the virtual photon will have their combined energy but *zero* momentum (remember that as a virtual particle, it can be off-shell). If we call the combined relativistic energy of the electron and positron E then the propagator of the virtual photon which is $\frac{1}{E^2 - p^2}$ must be $\frac{1}{E^2}$.

What happens as you ramp up the energy in the collider? Your gut instinct might be that the harder you smash the electron and positron together, the more hadrons you will get. Wrong! The higher the energy E, the further the virtual photon is off-shell so the lower the propagator and the lower the contribution of the pathway to the overall interaction path integral. The graph on the right of Figure 21.6 shows how the likelihood of producing hadrons, measured in terms of cross section σ, varies with energy E. As E starts to rise, σ falls.

But what happens if something like the neutral Z boson exists? The energy from the electron-positron annihilation can travel through the field of the weak force as a virtual Z boson. It turns out that the rest mass of the Z boson is about 90 GeV (giga-electron volts). As we are working in the centre of mass frame, the combined momentum of the electron and positron is zero. The same must be true for the Z boson so the propagator becomes:

$$Z \text{ boson with zero momentum:} \quad \frac{1}{E^2 - p^2 - m^2} = \frac{1}{E^2 - 90^2} \tag{21.15}$$

When the collider energy approaches the rest mass of the Z boson, the propagator value increases rapidly[5] and there is a marked peak in the cross section of hadron production. This is called a *resonance*. In fact, the scale of σ on the graph is logarithmic so the increase is about 150-fold. Eureka! This is how the LEP proved the existence of the Z boson in 1983.

The same principle was used to spot the Higgs boson in 2012 by smashing protons together in the LHC. At about 125 GeV there is an increase in the *4-lepton channel*. The proton energy becomes an excitation in the Higgs field (a virtual Higgs boson) which then decays to four leptons (electrons and muons). There is also an increase in the *diphoton channel* which is when the virtual Higgs decays into two high energy photons. Nobel Prize! We will discuss this in more detail in Chapter 25.

21.6 Do Virtual Particles Exist?

If you search on the web you will find countless comments and articles on this topic. For some reason it excites physicists to a level of pugnacious enthusiasm that few other subjects can match. Much comes down to semantics, but it provides a useful backdrop to summarise the topic:

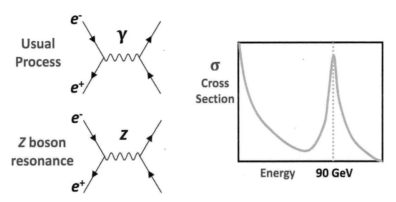

Figure 21.6 The search for the Z boson.

- *Do virtual particles physically exist?* - No
- *Do virtual particles reflect characteristics of real particles?* - Yes

Let me remind you what virtual particles are. They enter QED because path integrals in QFT include development pathways with excitations in the fields that are off-shell (see Section 19.4). Without this, there could be no QED. The fundamental QED interaction $\overline{\Psi}A_\mu\Psi$ by necessity involves off-shell excitations (Section 21.3). A pathway can stray off-shell for only a short period before its Action means it cancels out with surrounding pathways and its contribution to the path integral becomes negligible. This is why there must be more than one vertex in a QED process. The particles that exit a Feynman diagram must be back on-shell. What physicists call a virtual particle is a step in one of these off-shell *contributing development pathways*.

We showed that we can express the probability amplitude for a certain interaction as a path integral sum of all possible ways that the interaction could occur (Equation 21.4). Fortunately, the coupling constant in QED is small so the simpler paths contribute the lion's share of the probability amplitude and we do not have to delve into too many Feynman diagrams to generate a decent estimate of the probability of a particular outcome.

A simple way to understand the existence (or rather non-existence) of virtual particles is to compare this with the non-relativistic path integral. Feynman showed that the journey of a particle between two points is the path integral sum of every possible route, even to the moon and back. Does this mean that a physical particle travels along every route? Does a particle physically go to the moon and back? Of course not. If you want to label something as *existing* then what exists is the overall result of the path integral. Similarly, no virtual particle physically exists. It is just one of many contributing pathways, like the journey to the moon and back. What exists is the overall interaction that is the sum of all the pathways. *Virtual particles do not physically exist.*

However, virtual particles are more than just a mathematical trick. The development of the quantum field is the composite of every possible way it might have developed, each weighted according to the Action of the field Lagrangian. What we call *real* particles are the outcome of combining the contribution of all these pathways. While a particle does not physically go to the moon and back, that pathway reflects how a particle would have behaved if it had. The same holds true for virtual particles. For example, a photon moving at light speed has only two possible spin states (see Section 17.3). However off-shell virtual photons do not have zero mass because $\mathbf{E} \neq \mathbf{p}$. Therefore we must account for all three spin states that a massive spin-1 particle would have. *Virtual particles reflect characteristics of real particles.*

To calculate the probability amplitude of a Feynman diagram we account for interaction vertices (the coupling constant), off-shell drift (the propagator) *and* the characteristics of particles associated with that particular field. QED is not easy (see Box 21.9). And there is yet another complication – loops in Feynman diagrams. The loop problem bogged down QED for almost 40 years so this saga deserves its own chapter.

Box 21.9   QED is not an easy subject

If you have understood even a little of this chapter, you should be proud. Every calculation rests on distilling an approximate answer from an infinite number of possible pathways. Richard Feynman, the architect of QED, recognised what a very challenging subject it is. In his words:

Hell, if I could explain it to the average person, it wouldn't have been worth the Nobel Prize.

Notes

1 Technical point: this is not always true if a and b are operators that don't commute, but breaking the path into smaller steps progressively reduces any commutator terms towards zero (look up the Trotter Product and Baker-Campbell-Hausdorff formulas if you need to know more).
2 Technical point: each extra vertex brings a phase change of i. This can be important when summing across different pathways.
3 Technical point: propagators carry an i phase factor. Also, the denominator typically includes an infinitesimal (very small) factor, $i\varepsilon$, to facilitate integrating around the pole such as where the photon is on-shell and $p^\mu p_\mu = 0$.
4 We discuss the Z boson and weak force in Chapters 24 and 25.
5 Technical point: Equation 21.15 becomes infinite when $E = 90$. This creates some mathematical challenges in performing the calculation. However, from a practical perspective, the energy will never be *exactly* 90. There is always some spread because of the uncertainty principle.

Chapter 22

Renormalisation and EFT

CHAPTER MENU

22.1 Troublesome Loops
22.2 The Dressed Electron
22.3 Using Feynman Diagrams
22.4 Renormalisation
22.5 Ken Wilson's Effective Field Theory (EFT)
22.6 Summary

There was a serious problem with QED that troubled physicists for a long time: untamed loop infinities. Feynman came up with his diagrams 10 years after Dirac laid the foundations of QED, but it took a further 30 years to sort out the troublesome loops. When we analysed Bhabha scattering, we limited ourselves to two Feynman *tree* diagrams (i.e. loop-free as in Figure 21.5). In a tree diagram, the energy and momentum of the virtual photon is fully defined by the values of the ingoing and outgoing particles, so we simply pop the value into our calculation. Things are more complicated when a Feynman diagram contains a loop because there is no constraint on the energy and momentum of some virtual particles. To solve this, we use the process of *renormalisation* and treat QED as an Effective Field Theory (EFT), so read on.

22.1 Troublesome Loops

Box A in Figure 22.1 is a loop Feynman diagram for two electrons repelling each other through the exchange of two virtual photons. The two electrons approach each other with energy E_1 and E_3. After the exchange they have energy E_2 and E_4. The virtual photons have energy E_a and E_b. I have given each energy a direction so we can keep track of it, but E_a and E_b are as likely to be negative as positive (the transfer of energy could be from the top electron to the bottom or vice versa). We can balance energy in and out to derive relationships for $(E_a - E_b)$.

$$E_1 + E_b \ = \ E_2 + E_a \qquad\qquad \Longrightarrow \qquad\qquad E_1 - E_2 \ = E_a - E_b \qquad\qquad (22.1)$$

$$E_3 + E_a \ = \ E_4 + E_b \qquad\qquad \Longrightarrow \qquad\qquad E_4 - E_3 \ = E_a - E_b \qquad\qquad (22.2)$$

The problem with the loop is that E_a can take *any* value. Imagine that $E_1 - E_2$ is 3 eV (choosing a number at random). E_a could be 4 eV and E_b 1 eV or E_a could be 1,000,000 eV and E_b 999,997 eV or E_a could be −50,000 eV and E_b −50,003 eV. The same is true for the momentum of the virtual photon. Indeed, if the basic relationships in Equation 22.1 and 22.2 are observed, the value of the energy and momentum of all four internal lines connecting the vertices in diagram A can be anything from −∞ to +∞.

This is very troublesome. In QED, an interaction is the path integral sum of all development pathways, so we should integrate over all possible values up to infinitely high momentum (from now on I will use momentum to refer to the 4-vector that includes both momentum *and* energy). The sum usually heads off to infinity so QED appears to fall apart, but there is an extraordinary twist. If you simply *ignore* higher values of momentum, QED works fairly well! What this means

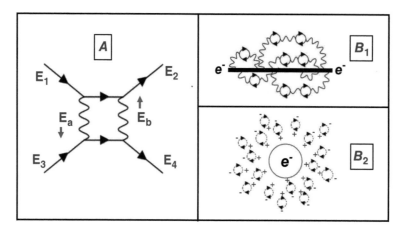

Figure 22.1 Loops, untamed infinities and dressed electrons.

is that you select an arbitrary momentum cut-off that is generally labelled Λ. It should, of course, be well above the range of the particles you are experimenting with. Then, rather than integrating from $-\infty$ to $+\infty$ you integrate over the defined range from $-\Lambda$ to $+\Lambda$. The infinities disappear and QED gives a good fit with your results. This led to much anxiety, teeth-grinding and hair-pulling by physicists. I have added a joke in Box 22.1 to ease the strain on your brain. QED works, but what the heck is going on?

Box 22.1 Just for laughs:

I want to include an infinite loop joke... but you would never hear the end of it.

22.2 The Dressed Electron

Let's examine how virtual pathways with very high momentum differ from those with low momentum. High momentum means shorter wavelength so the pathway reflects conditions closer to the electron. This is similar to the difference between an electron-microscope and a light microscope. The electron-microscope has higher resolution because it operates with a shorter wavelength beam.

Boxes B_1 and B_2 in Figure 22.1 are schematic representations for QED conditions close to an electron. Aside from the interaction with other electrons, an electron has *self-interaction* pathways. Box B_1 shows a Feynman diagram for an electron showing a host of virtual photons and virtual electron-positron pairs surrounding it. Box B_2 shows the same thing at a snapshot in time. The electron is enveloped by a cloud of electron-positron pairs. The negative charge of the electron creates an asymmetry. It attracts each virtual positron a little closer to it and repels each virtual electron a little further away. This creates what is called a *dielectric effect*. Each electron-positron pair acts like a tiny dipole with an electric field that lines up in the opposite direction, cancelling out some of the underlying electron field. The cloud of electron-positron pairs screens the electron charge. An observer outside the cloud sees the electron to have less charge. The observer sees the electron *dressed*.

On the other hand, an observation made inside this cloud experiences less screening so the electron is perceived to have a higher charge. This creates a difference between high momentum and low momentum virtual pathways. The charge of the electron is the coupling between the electron field and the EM field so, if you include very high momentum pathways in your QED calculation, the coupling constant grows. It turns out that if you include infinite momentum pathways, you observe the electron from zero distance and the perceived charge (the *bare coupling constant*) becomes infinite. Oops!

This means that we cannot base any calculations on the bare electron charge. The best we can do is make calculations based on how the charge appears from a distance. Essentially this is what renormalisation is. But before we get into that, we need to dig a bit deeper.

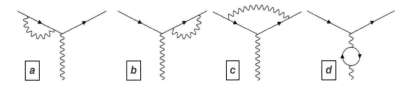

Figure 22.2 Vertices with loops. Note that these are vertices, not diagrams of legitimate processes.

22.3 Using Feynman Diagrams

The description of the dressed electron is intuitive and helpful, but I imagine some purists among you muttering: *tut tut... we are back to talking about electric fields – aren't we meant to do this with Feynman diagrams and QED?* Quite right.

Figure 22.2 shows Feynman diagrams of the basic vertex for the QED interaction. As drawn they all show an electron emitting or absorbing a photon plus options for the next possible level of complexity at the vertex. In diagram *a* the electron emits and reabsorbs a virtual photon prior to the vertex. In diagram *b* it does so after the vertex. In diagram *c* the virtual photon is emitted by the electron prior to the vertex and reabsorbed after the vertex. Diagram *d* also has three vertices, but in this case the photon becomes a virtual electron-positron pair that then decays back into a photon.

How does this relate to the dressed electron? In more complicated Feynman diagrams, the virtual photons in diagrams *a*, *b* and *c* can become virtual electron-positron pairs. To model an interaction we must add up all Feynman diagrams and there is no limit to the possible number of virtual photons and electron-positron pairs. When you account for all these options, the electron is dressed as shown schematically in B_1 of Figure 22.1.

Effects based on diagrams *a* and *b* of Figure 22.2 create what is called the *self energy* of the electron. The electron's energy is increased by the virtual photons and electron-positron pairs that surround it. How big is this self-energy? Well, these are loops, so if you integrate over all values of momentum up to infinity, then you are including pathways where the electron self-energy, and therefore its rest mass, is infinite. However, the existence of these off-shell pathways is constrained by their Action. We showed back in Equation 19.16 that a virtual pathway that is off-shell by energy ΔE can exist only for time Δt where: $\Delta E \, \Delta t \approx \hbar$. Therefore, if a virtual pathway has infinite momentum, it exists for an infinitely short time so it is infinitely close to the electron. The infinite self-energy and rest mass is the result of measuring things infinitely close to the electron.

Box 22.2 George Green (1793–1841): a mathemagician, a mystery and a windmill

The propagators in QFT are Green's functions that help to solve tricky differential equations. George Green's family were millers. The young Green had one year of formal secondary school at the age of eight. Beyond that, his education was probably limited to Sunday school at the local church. After 20 years tending his father's mill, Green suddenly published a tract on mathematics. With no formal education he did not approach any scientific journal. He published it himself and sold 51 copies mostly to friends.

Although his ability was recognised in part and he attended Cambridge university at the ripe old age of 40, he came to fame many decades after his death. Nobody knows where his ideas came from. In his treatise he referred to, and expanded on, complicated theorems that should have been completely inaccessible to him. Where did he find them? How could he understand them? Green's windmill in Nottingham has been restored to working condition and is now open as a science centre; but at the end of any visit the question remains: *How on earth did he do it?*

The good news is that the impact on the coupling constant of loops such as in diagrams *a*, *b* and *c* cancels out so we need only to focus on *d*-type diagrams where electron-positron (and other particle-antiparticle) loops appear in the photon propagator (see Box 22.2 on the astonishing George Green who developed the mathematical functions used to calculate propagators). For a virtual photon with low momentum, the contribution of *d* is insignificant because the formation of an electron-positron pair is way off-shell. The propagator for electrons and positrons, which is $\frac{1}{\gamma_\mu p^\mu - m}$ as shown in Box 21.6, will be minuscule. The contribution from pathways with more loops rapidly fades away so the total contribution is small and finite.

Things change as the momentum of the virtual photon grows. The propagator is inversely proportional to the off-shell drift. Once the virtual photon has energy-momentum of the order of the electron-positron pair's rest mass, the pathway opens up (see discussion of resonance in Section 21.5). Ramp up the energy further and all sorts of pathways open, such as the formation of multiple electron-positron pairs and quark-antiquark pairs. What about proton-antiproton pairs at 10^6 eV? What happens if the energy of the virtual photon is equivalent to the rest mass energy of the moon at 10^{60} eV? What if it is bigger than the rest mass energy of the entire universe? Such things seem ridiculous and laughable, but that is the problem with infinities – there is no place to stop.

As we account for virtual pathways with higher and higher momentum, more and more pathways open up. The propagators for loop pathways increase so the probability grows of the electron and photon interacting. Rather than altering an infinite number of propagators, we account for this as an increase in the overall QED coupling constant $\sqrt{4\pi\alpha}$ where α is the fine structure constant (see Box 22.3 for some more weird history). We can take this approach because each propagator is part of a larger diagram and is coupled at both ends. It turns out that we can account mathematically for changes in the myriad of possible pathways by a single adjustment to the coupling constant that is applied to each end.

For a calculation involving only lower-momentum virtual pathways, α is the usual value of about $1/137$. However, once virtual pathways are included that are close to the electron-positron rest mass (about 1 MeV), the value of α will start to rise. And if the path integral covers *all* possible virtual pathways up to *infinite* momentum, α becomes *infinitely* large.

What does this mean for QED? The whole theory is built around the path integral that combines every possible development pathway in the quantum field. We must add the contribution of every Feynman diagram, but all QED interactions include Feynman diagrams with loops, so we have infinities all over the place. The path integral includes *infinite* momentum pathways that are *infinitely* close to the particles and vertex. This gives *infinite* rest mass and *infinite* α which gives an *infinite* probability for every interaction. Abandon ship! All is lost – or is it?

Box 22.3   Eddington and the fine structure constant

Arthur Eddington (1882–1944) was the first to speculate that hydrogen fusion fuelled the stars. More controversially he believed numerology could provide insights into physics such as:

I believe there are 15 747 724 136 275 002 577 605 653 961 181 555 468 044 717 914 527 116 709 366 231 425 076 185 631 031 296 protons in the universe and the same number of electrons.

He *proved* using numerology that the fine structure constant α is exactly $1/136$. When subsequent experiments differed, he changed his *proof* to match a new value of $1/137$.

22.4   Renormalisation

As noted above, the effects of the a, b and c interactions in Figure 22.2 cancel out, so we only have to take account of d-type diagrams. Every QED vertex includes a photon. Therefore, the way that the coupling increases because of d is the same for *every* vertex in *every* Feynman diagram and the impact on self-energy and rest mass is the same for every electron.

It turns out that we can solve the loop problem by applying a cut-off that eliminates the highest momentum pathways. This is called *regularisation*. Suppose we run an experiment at, say, 10 MeV in a collider. We want to compare the results with the predictions of QED. We know that if we integrate the loops in Feynman diagrams up to infinite momentum, the predictive power of QED crumbles because the probability of any interaction is infinite. Therefore, we apply a cut-off of, say, $\Lambda = 1,000$ MeV to our calculation so we integrate the possible values for momentum of virtual particles from $-1,000$ MeV to $+1,000$ MeV instead of $-\infty$ to $+\infty$. As an example, this changes the propagator term for a virtual electron as follows:

$$Propagator \quad \propto \int_{-\infty}^{+\infty} \frac{1}{\gamma_\mu p^\mu - m} \; d^4p \quad \longrightarrow \quad \int_{-1,000\,MeV}^{+1,000\,MeV} \frac{1}{\gamma_\mu p^\mu - m} \; d^4p \tag{22.3}$$

Next, we do our QED calculation by adding up Feynman diagrams. As a first step, we use the base value for α of $1/137$ which I will now label α_0. We have excluded all the very high momentum pathways so the contribution of more complicated Feynman diagrams fades away as the number of vertices increases. Therefore, our QED calculation gives a finite total probability amplitude for the interaction. We compare this with the results of the collider experiment and notice a discrepancy.

We know why there is a discrepancy. Our integration includes pathways up to 1,000 MeV which is $1,000\times$ the rest mass energy of an electron-positron pair. This affects the calculated probability for coupling. It changes α from α_0 to a new higher value for the cut-off Λ which I will call α_Λ. Importantly, we have excluded the infinite momentum pathways so α_Λ is still finite. We can use the results of our collider experiment to give a corrected value for α_Λ.

This all seems circular, but we now have a value for α_Λ that works with the cut-off $\Lambda = 1,000$ MeV. Because of the symmetries, this value of α_Λ applies for *all* the vertices in *all* the Feynman diagrams in *all* experiments. The process of applying this adjusted value is *renormalisation*. In summary, we choose a cut-off safely above the working energy of the experiment and apply the correct (experimentally determined) value of α_Λ for that cut-off. We have banished the infinities. We have brought QED back from the dead.

In fact, QED rose like a phoenix from the ashes to deliver spectacular results. Probably the most quoted calculation is that of the magnetic moment of the electron that has been verified up to one part in 10^{13} making it, according to some, the most accurately verified prediction in the history of physics. QED works.

Figure 22.3 summarises the process. It looks horribly complicated, but the underlying concept is fairly simple. Integrating over all possible values of momentum gives us the equation for the QED Lagrangian labelled *Total*. However, as we have shown, the rest mass and the coupling constant associated with this equation are both infinite. I have labelled the coupling constant e_∞ (or in terms of the fine structure constant α, this would be $\sqrt{4\pi\alpha_\infty}$). This is the effect of including infinite momentum pathways in the integral. It makes the Lagrangian unusable for calculations.

In renormalisation, we divide the field into two parts. This is shown in the figure as a *UV* pool and a *Bounded* pool. The *UV* pool is called *ultraviolet* because it is higher energy (nothing to do with UV light). It includes all the pathways with momentum above the Λ cut-off, that is, from $+\Lambda$ all the way up to $+\infty$ and from $-\Lambda$ down to $-\infty$. The presence of infinite momentum pathways means that e_{uv} must still be infinite. The *Bounded* pool contains those pathways within the momentum cut-off from $-\Lambda$ to $+\Lambda$. In this case there are no infinite momentum pathways so the coupling constant e_Λ is finite. We generally refer to this in terms of the fine structure constant α_Λ. This is finite as is the value for rest mass m_Λ.

The key point is that we separate out the lower momentum pathways. The structure of the Lagrangian is not changed by the Λ cut-off so we can use experimental results to provide values of the coupling constant and rest mass for use in other QED calculations. In fact, once you have one value α_Λ established by experiment for cut-off Λ, it is possible to *calculate* the value of α for other values of Λ by taking into account all significant particle-antiparticle loops.[1] As α varies depending on the value you choose for Λ, it is called a *running constant*.

Renormalisation made QED usable, but many physicists were left with a feeling of discomfort – a bad smell. As you can see in Figure 22.3, renormalisation relies on subtracting one infinity from another to give a finite result. Mathematicians don't like this sort of thing. Dirac complained: *sensible mathematics involves neglecting a quantity when it is small... not neglecting it just because it is infinitely great and you do not want it.* Even Feynman, who helped to develop the maths behind renormalisation, described it as: *sweeping the infinities under the rug.*

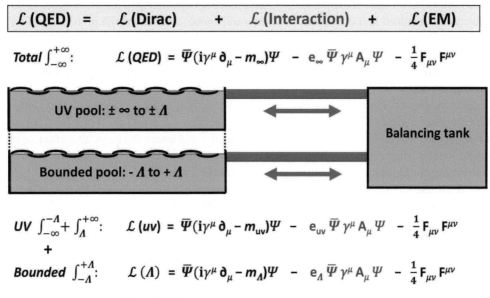

Figure 22.3   Local gauge invariance and UV cut-offs.

22.5   Ken Wilson's Effective Field Theory (EFT)

Physicists struggled with this for decades until Ken Wilson, twice winner of the impossibly difficult Putnam mathematics competition (Box 22.4), brought some perspective to things. His work challenged the prevailing wisdom that we should worry about infinities that are generated by infinite momentum pathways at zero distance from the electron. In the words of *Gone With the Wind – Frankly my dear, I don't give a damn.*

Many theories work for a certain scale or distance, but break down at another. That does not make them bad theories. Do we really expect QED to work up to infinite energies and down to infinitely small lengths of space and time? While it would be super-duper to have a Theory of Everything, is that a realistic goal? Max Planck proposed a natural scale for the universe by setting 1 as the value for several fundamental constants including c, $\hbar$ and G (the gravitational constant). On this basis, he hypothesised that a natural unit of energy is about 10^{28} eV (the *Planck energy*). The maximum energy of the LHC is about 10^{12} eV so we are less than a million billionth of the way there. The approach of EFT is to focus on the vast range of energies where QED works, rather than worry about infinite energy scenarios that are way beyond our experimental capability.

Let me give an analogy. In Chapter 15 we discussed the Coulomb potential. The potential energy of the electron in the hydrogen atom varies with distance from the proton by $\frac{1}{r}$. We say the electron's potential energy is zero at an infinite distance from the proton and grows progressively more negative as the electron approaches the proton (see Section 15.2). Using this scale the ground state potential energy of the electron in a hydrogen atom has a defined value of -14 eV.

Suppose I do the reverse and set the Coulomb potential energy to zero at the *centre* of the proton where $r = 0$. What then is the ground state energy of the electron in the hydrogen atom? Oops! It is infinitely more than at $r = 0$. We have an undefined mess. To keep the Coulomb potential workable, we avoid the infinite pole. The last thing we want to do is to take it head on. And what if we did manage to take the Coulomb potential and somehow tame the infinity at $r = 0$? So what? It would be a complete waste of time because we know now that at scales less than 10^{-15} metres you have to deal with conditions inside the proton: quarks, gluons and the strong force.

Wilson approached QED with the same mindset. We have no idea what happens at $r = 0$ distance from an electron, so why worry? Just accept that QED is not the ultimate whizz-bang Theory of Everything. Accept that there is stuff that we do not know. Celebrate that QED works at all energies experimentally available to us (currently) and does so with fabulous accuracy.

22.6   Summary

To handle the problem of infinite loops, we apply a high energy cut-off Λ and treat the fine structure constant α and particle rest mass m as variables that depend on this cut-off.

Why? Because without this, the sum of the probability amplitude for Feynman loop diagrams heads to infinity as higher momentum pathways are included in the integral. It becomes impossible to calculate the probability of an interaction by adding up successive Feynman diagrams.

How? The high momentum pathways have the same effect on every vertex and particle. This symmetry allows us to account for their exclusion by adjusting the value of α and m in the lower momentum QED Lagrangian: α and m become α_Λ and m_Λ... *running constants.*

This has no significant impact at low energies. For example, chemistry is based on the interaction between electrons in the outer shells of atoms at energies of the order of 0-10 eV. This energy scale is well short of 1 MeV (1,000,000 eV) when loops start to matter. For most purposes, we can use the base value of α and ignore the loops altogether, as classical physics has happily done for so long. This creates only very slight anomalies. Perhaps the most famous is the Lamb shift where loop pathways affect the energy levels in hydrogen, but only by a minuscule 4×10^{-6} eV.

At colliders, we must use higher energy cut-offs for experimental calculations.[2] How does this affect α? At 100 GeV (100,000,000,000 eV) α increases from 1/137 to about 1/127. At 1 TeV (1,000,000,000,000 eV) which is the energy scale of the LHC, it is about 1/125.

Viewed as an effective field theory, QED is exquisitely accurate for everything up to and including the energy range of modern colliders. However, renormalisation comes at a cost. Physicists had hoped QED might give some insight, some logic, for the values of α and the electron's mass. But QED is silent on this. Both must be determined by experiment. The fact that α becomes infinite at high energy tells us something else: that at some energy scale, we must expect QED to break down and no longer match experimental results. When it does, it will be an indication of new physics. At what energy level will this happen? We don't know.

Box 22.4   The Putnam mathematics competition

Calling all mathematical prodigies! The Putnam competition is an annual challenge consisting of 12 maths questions worth up to 10 points each. It is so hard that the median score for contestants is typically 0 or 1 out of 120. The top five contestants each year are named Putnam Fellows. Richard Feynman was a Fellow in 1939. Ken Wilson achieved the feat in 1954 and again in 1956. If you fancy your chances you can find past papers on line, go on – give it a go.

Notes

1 Technical point: this is easier said than done. At 100 GeV this includes the electron, muon, tau, and five of the six quarks.

2 Technical point: choosing the right cut-off is not so simple. It changes the relative contributions of different Feynman diagrams, some of which may cancel out. Sometimes a variety of cut-offs is used to generate a margin of error in calculations. And sometimes the cut-off involves adjusting the dimensions in the integral (dimensional regularisation). Tricky, tricky.

Chapter 23

The Strong Force

CHAPTER MENU
23.1 The Elementary Particles
23.2 The Strong Force: An Overview
23.3 QCD Local Gauge Invariance
23.4 The Residual Strong Force
23.5 Oh No! Here Comes Jill Again!

In this chapter, after a short introduction to the leptons, quarks and bosons that play the leading roles in what is called the *Standard Model* of particle physics, we will discuss the *strong force* that binds together the components in the atomic nucleus. In the following two chapters we will delve into the *weak force* which is harder to fathom as it is scrambled by *spontaneous symmetry breaking*, and the *Higgs field*. I will continue to use natural units ($\hbar = c = 1$).

23.1 The Elementary Particles

Before quantum mechanics and QFT, the concept of an *elementary* particle was straightforward: a particle with no internal components that exists in perpetuity. Thus, the atom is not elementary as it is made of electrons, protons and neutrons (which are in turn made of quarks), but the electron is elementary as it is not believed to have any internal components.

Reality is a bit more complicated. You can smash together an electron and positron in a collider and produce a host of quark–antiquark pairs, photons, neutrinos and anti-neutrinos, but nobody is suggesting that these particles come from inside the elementary electron. Another possibility is that the electron-positron pair annihilate into a couple of high energy photons, so elementary particles are not perpetual. A similar example is that an elementary particle called the muon decays rapidly into an electron, a muon-neutrino and an anti-electron neutrino, but that lot was never inside the muon. As for the quarks, they are elementary, but you will learn that a quark can *never* exist alone. Please bear all of this in mind when I refer to particles as *elementary* or *fundamental*. Don't forget that these particles are actually excitations of fields.

This warning aside, the Standard Model is clear on what are considered to be the elementary particles. As far as we know (currently) they are pointlike with no sub-structure. Table 23.1 shows the fermions. These are the spin-half elementary particles. They follow Pauli's exclusion principle and never occupy the same quantum state. There are three *families* or *generations* (we have no idea why). The *electron* has two more massive cousins which are the *muon* and the *tau* particles. Similarly there are three generations of each quark: *charm* and *top* for the *up quark*, *strange* and *bottom* for the *down quark*. These more massive particles decay rapidly, so we will focus on the stable lighter generation (shown in capitals in the table). Things work differently for the three generations of *neutrino*, all of which have minuscule mass. We will discuss that later (I jokingly call it the *nutty* neutrino).

At the end of Chapter 16, I briefly described protons and neutrons in the atomic nucleus. Each is composed of three quarks. Particles made of quarks are called *hadrons* (from the Greek *hadros* meaning bulky) versus the lighter electrons and neutrinos which are called *leptons* (from the Greek *leptos* meaning thin or slight).

Quantum Untangling: An Intuitive Approach to Quantum Mechanics from Einstein to Higgs, First Edition. Simon Sherwood.
© 2023 John Wiley & Sons Ltd. Published 2023 by John Wiley & Sons Ltd.

Table 23.1 The elementary particles: fermions.

Leptons		Charge	Quarks		Charge
ELECTRON	e	-1	**DOWN QUARK**	d	$-1/3$
NEUTRINO (e)	ν_e	0	**UP QUARK**	u	$+2/3$
Muon	μ	-1	Strange	s	$-1/3$
Neutrino (μ)	ν_μ	0	Charm	c	$+2/3$
Tau	τ	-1	Bottom	b	$-1/3$
Neutrino (τ)	ν_τ	0	Top	t	$+2/3$

Table 23.2 The elementary particles: bosons.

Interaction/Force	Bosons	Number	Spin	Symmetry
Electromagnetic	**PHOTON** (γ)	1	1	U(1)
Weak	**W+, W-, Z**	3	1	SU(2) (*massive*)
Strong	**GLUON** (g)	8	1	SU(3)
Mass	**HIGGS** (H)	1	0	*Scalar* (*massive*)
Gravity	*Graviton*	1	2	*Outside of model*

The proton is two up quarks and a down quark with a positive charge $[(2 \times \frac{2}{3}) - \frac{1}{3} = 1]$. The neutron is two down quarks and an up quark $[(2 \times -\frac{1}{3}) + \frac{2}{3} = 0]$ and carries no net charge. You are familiar with the electron, which has been the main topic of this book. The electron-neutrino ν_e is a featherweight (even the electron is over 100,000 times more massive). It is produced in nuclear reactions including fusion in the sun. Carrying no electric charge, it is hard to detect. Each second, about 100 trillion pass through your body unnoticed.

Table 23.2 lists the bosons. You should think of these as the force carriers. The photon carries the electromagnetic force. You have seen in the last few chapters how this gives U(1) local gauge invariance to the electron (invariance to phase change). The W and Z bosons and the 8 *gluons* play a similar role with the weak and strong force providing SU(2) and SU(3) symmetry to fermions (we will discuss these symmetries in the next few chapters). Taken together, these 12 particles are the spin-1 *vector bosons*. There are good grounds to expect them to be massless but, as always, things turn out trickier. The three weak force bosons have mass. This is one of the clues that led to the discovery of the Higgs field.

Rather embarrassingly, gravity and the predicted spin-2 *graviton* are not yet incorporated into the Standard Model. This conspicuous shortcoming has tormented physicists for over 60 years and arguably is the greatest challenge facing them. Not something to be laughed at, but I cannot help myself (Box 23.1).

Box 23.1 Just for laughs:

What does a sub-atomic duck say?   *Quark!*

23.2 The Strong Force: An Overview

It is now time to discuss the strong force that binds together the quarks forming the protons and neutrons of the atomic nucleus. Physicists knew that there had to be an additional force at work in the nucleus: if there wasn't, electromagnetic repulsion would push the protons apart. I will start by describing the basic properties of the strong force, the quarks and the gluons. This will highlight the differences between the strong force and the electromagnetic force. Following that, we will look at the local gauge symmetry of the strong force and show that, in spite of the differences, the mathematical structure of the strong force actually has much in common with that of the electromagnetic force.

23.2.1 Colour Charge

The quarks that make up protons and neutrons are fermions (spin-half). This means they are described by the Dirac equation and have antiparticles called antiquarks (see Subsection 18.5.1 if you need a reminder on the Dirac equation and antiparticles). The up quark has an electric charge of $+\frac{2}{3}$ and the down quark has $-\frac{1}{3}$. Antiquarks have the opposite electric charge of the quarks in the same way that the electron and positron have opposite charges. For example, an anti-up quark has an electric charge of $-\frac{2}{3}$.

Each quark also has a charge associated with the strong force. The strong force charge is very different from that of electromagnetism (EM). In EM there is only one type of charge, the electric charge, which when positive (+) for a particle (such as the up quark) will be negative (−) for its antiparticle. In the case of the strong force there are *three* different forms of charge that are given the colourful names red, green and blue. Quarks can have positive charge of red (R), green (G) or blue (B). Antiquarks carry the negative form of the charge and so can be anti-red ($\overline{R}$), anti-green ($\overline{G}$) or anti-blue ($\overline{B}$).

These so-called colour charges attract each other *very* strongly (hence *strong force*). In fact they interact so strongly that composite particles cannot have a net colour charge. They must be *colourless*. One way to achieve this is with a quark–antiquark combination, called a *meson*, giving an overall colourless combination such as $R\overline{R}$. You know what happens when particles get together with antiparticles – so, it will not surprise you that mesons are unstable and short lived.

There is a more stable colourless recipe. The combination red-green-blue RGB is colourless. The colour names originated from this. They come from the parallel that red-green-blue light adds to white light. That is the only link between the strong force and the language of colours. The quarks have three different *charges*, not three colours. An example of a colourless combination for a proton is a red up quark, a green up quark and a blue down quark, u_R, u_G, d_B. A neutron might be u_B, d_R, d_G. It turns out that the specific colour charge of each quark does not matter. It changes as the quarks interact. The important thing is how the colour charge of each quark *differs* in relation to the colour charge of the others so that the three quarks always form a colourless combination.

23.2.2 QCD, Gluons and Confinement

Let me try to illustrate this with the highly schematic Feynman diagram on the left in Figure 23.1. The three black dots and the solid lines represent the quarks in a proton or neutron and are labelled red R, green G, and blue B. Time is moving to the right. The curly lines represent their interactions which involve *gluons* that are *massless*. Well done Murray Gell-Mann for coming up with the fabulous quark and gluon names (see Box 23.2 for an interesting anecdote from his life).

Following the quarks in the figure over time, the first quark starts red R and then interacts with a neighbouring green G quark. They interchange a gluon with the result that the first quark becomes green G and the second red R. I want to highlight a very important difference between this and the electromagnetic interaction. In the case of quantum electrodynamics (QED) the interacting photon connects two charged particles, but it does *not* change the charges of the particles it interacts with. The photon does *not* carry any charge. In sharp contrast, in the study of the strong force which is quaintly called quantum chromodynamics (QCD), the gluon *changes* the colour charge of the quarks it interacts with. Gluons *carry* colour charge.

The easiest way to picture this is to remember that colour charge is conserved. For example, the gluon at the top left of the figure takes red R over to the second quark while cancelling out the green G of the second quark, so the gluon is coloured red-antigreen $R\overline{G}$. If the gluon exchange was in the opposite direction time-wise it would be $G\overline{R}$ (green-antired). Overall, the three quark combination remains RGB colourless, starting as red-green-blue and ending as blue-green-red.

By now, you should be familiar with the fundamentals of quantum field theory and fully understand that the sequence shown on the left side of the figure is just a comic-strip version of events. The behaviour of the quarks comes from combining the Action of all possible development pathways. There is no way to specify a quark as red or blue or green at any moment in time. Indeed you will learn later that each specific colour is undefined in the same way that the absolute phase of the wave function is undefined in QED. All we can say is that the three quarks attract strongly and form a colourless combination.

You will be alarmed to learn that the full picture is much more complicated. As mentioned earlier, gluons are not colourless. They carry colour charge. That means that they attract each other. This is called gluon *self-interaction* and is a major

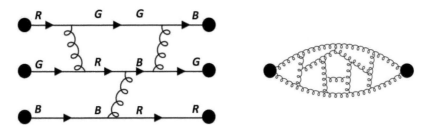

Figure 23.1 Schematic illustrations for QCD. The left side shows three quarks interacting through gluon exchange. The right side highlights gluon–gluon interactions.

difference between QED and QCD. The box on the right of Figure 23.1 is a schematic snapshot of the next level of complexity. Between the quarks (shown as black dots) there is a tangle of gluon-on-gluon interactions.

Box 23.2 Gell-Mann: the non-commutative mathematics of survival

Murray Gell-Mann (1929–2019), who named the quarks, was determined to work at Harvard or Princeton, but he lacked the necessary funds and in a fit of depression contemplated suicide. Upon being offered a place at the Massachusetts Institute of Technology (MIT), he decided to postpone any precipitous action concluding that he could attend MIT and then commit suicide (if he did not like it), but that he could not commit suicide and then attend MIT. In his words as a true mathematician, *the two did not commute*. Twenty years later, he had developed the quark model and won the 1969 Nobel Prize: a triumph for physics, for history, and for non-commutative maths.

In a high energy collider, you can smash a particle into one of a proton's quarks and force it away from the other two. As the quarks separate, the gluon–gluon interactions grow linearly with the distance apart. As a result, potential energy increases linearly. Smash the quark harder and you push it further away, but when the potential energy is high enough, a new quark–antiquark pair is created. The new quark restores the original three quark balance of the proton and the new antiquark forms a quark–antiquark colourless pair with the departing quark. With enough energy, that quark–antiquark pair will in turn be stretched apart and split into two quark–antiquark pairs, then four, then eight and so forth, creating a shower of quark–antiquark pairs (mesons) discharged in a narrow cone that decay into what is called a *hadron jet*.

It means that you cannot and never will detect a lone quark or any lone particle with a net colour charge. This is *confinement*. Quarks and gluons exist only within colourless composite particles such as protons and neutrons (three quarks each) and mesons (quark–antiquark pairs). Because of the linear growth of gluon–gluon interactions, the attractive force between quarks does not diminish with distance. As you force them further apart, the attractive force remains of the order of 100,000 Newtons, the equivalent of lifting about 10 tons. That is the weight of a large elephant or the anchor of a cruise ship; whichever way you add it up, it is a mind-blowing force resisting the separation to distance of these subatomic particles.

23.2.3 Strong Force Coupling Constant

Protons, neutrons and mesons are about 1 femtometre ($10^{-15}m$) in diameter. Forcing a lone quark or antiquark further away than this requires more energy than the creation of a new quark–antiquark pair. However, when the quarks are close together the attractive force drops to a much lower level and the quarks react more like free particles; this is called *asymptotic freedom*. The strong force looks very different depending on how closely you look at the quarks.

If we probe a proton with a collider at energies of say 1 GeV (10^9 eV), the probing wavelength is about 1 femtometre, which is similar to the diameter of the proton. It will not detect the quarks' asymptotic freedom within the proton so, at this sort of energy, the coupling constant of the strong force is very large (α_S is about 0.5 or higher). This creates a huge analytic headache for QCD. In QED the coupling constant is low (α_{QED} is 1/137), so we can ignore more complicated Feynman diagrams that involve multiple interactions. With QCD, the more complicated interactions cannot be ignored. A quick glance at Figure 23.1 highlights the challenge. QCD involves multiple quark–quark and gluon–gluon interactions. Feynman analysis becomes impractical.

Things become slightly more manageable at higher energies. At say 100 GeV, equivalent to a wavelength of about 0.01 femtometre, we are probing well inside the proton. The interaction between quarks at this range reflects asymptotic freedom. This is reflected in the strong force coupling constant α_S which decreases to about 0.1. Note that the QCD coupling constant *decreases* with energy whereas in QED it *increases* (see Section 22.6). Even with this decrease, α_S is still over 10 times as high as α_{QED}. Analysing and predicting results for QCD remains an enormous challenge.

Confinement means that the strong force operates over a short range. The strong force is so strong that it pulls quarks together in colourless combinations – it is always cloaked within a colourless shell. Gluons are massless like the photon so, without confinement, we would expect the strong force to operate over long distances just as the EM force does. It is the very strength of the strong force that reduces its range.

23.3   QCD Local Gauge Invariance

In the last few pages we have covered some of the basic features of the strong force (QCD) and highlighted how different it is from the electromagnetic force (QED). It is time now to examine what they have in common. Both forces are a manifestation of local gauge invariance. For the electromagnetic force it is local gauge invariance to changes in phase. For the strong force it is local gauge invariance to changes in colour.

Before we get into the maths, let me give a simple example. We analyse the quarks of a proton with their three colour charges. We catalogue the first colour charge as red, the second green and the third blue. Another scientist, whom I will call Jill, uses a different colour catalogue calling them blue, green and red that is, swapping the red and blue. Who is right? I hope it is obvious that it makes no difference. To switch to Jill's colour catalogue, I change all the red tags in my calculations to blue and vice versa. Providing I am consistent and follow the new scheme for *all* my results, nothing changes. This is because it does not change the Lagrangian of the quark. It is a fermion so it follows the Dirac Lagrangian (Equation 20.35). Switching the red and blue tags *globally* changes $\overline{\Psi}$ and Ψ. The new tags match so there is no change to $\overline{\Psi}\Psi$ or the derivatives. There is *global gauge invariance* to this change in how colour is categorised.

I need you to cast your mind back to the discussion of QED. The QED Lagrangian is not altered by a global phase change (see Section 19.8 using the Klein-Gordon Lagrangian as a model). It is also unaffected by a *local* phase change. A local phase change alters the derivatives of the Lagrangian, but is offset by a simultaneous change in the EM interaction (see Figure 20.3). Overall, the QED Lagrangian is unchanged. The photon gives local gauge invariance to phase.

QCD is similar. The gluons give the quarks *local gauge invariance to the colour catalogue*. Imagine we are studying a proton and we switch to Jill's colour catalogue for only one of the quarks. Perhaps that turns it from red to blue. That gives two blue quarks and one green. Oops! However, we will show that switching to Jill's colour catalogue simultaneously changes the gluon interaction terms in the QCD Lagrangian. Add it all up and you find the other quark colours have changed and the overall combination remains colourless in spite of the local colour change.

There is a further twist. The gluons have colour. Imagine if we switch to Jill's colour catalogue for one gluon changing it from perhaps $R\overline{G}$ to $B\overline{G}$. Again, this creates a mismatch in our calculations. However, this is rectified by changes in the gluon's interaction with surrounding gluons. Local gauge invariance cannot work without these gluon–gluon interactions.

23.3.1   SU(3) Symmetry and Colour

Let's get to work. Imagine that Jill decides on a radically different colour catalogue for the three quarks in a colourless proton. She calls them lemon (L), mandarin (M) and nectarine (N). We compare her colour catalogue with ours (let's assume there is a way to do this) and conclude that our colour systems are related in the following complicated way. The difference between Jill's basis and ours can be identified by the matrix shown as Equation 23.1.

$$L = \frac{1}{2}(R + iG + B + iB) \qquad M = \frac{1}{2}(-iR + G - B + iB) \qquad N = \frac{1}{2}(-R + iR + G + iG)$$

$$\frac{1}{2}\begin{bmatrix} 1 & -i & -1+i \\ i & 1 & 1+i \\ 1+i & -1+i & 0 \end{bmatrix}\begin{bmatrix} L \\ M \\ N \end{bmatrix} = \begin{bmatrix} R \\ G \\ B \end{bmatrix} \tag{23.1}$$

Equation 23.2 gives an example. It translates Jill's *lemon* quark into our [*RGB*] basis and you can check that it does indeed match with Jill's colour catalogue.

$$\begin{bmatrix} L \\ M \\ N \end{bmatrix} = \begin{bmatrix} 1 \\ 0 \\ 0 \end{bmatrix} \qquad \frac{1}{2}\begin{bmatrix} 1 & -i & -1+i \\ i & 1 & 1+i \\ 1+i & -1+i & 0 \end{bmatrix}\begin{bmatrix} 1 \\ 0 \\ 0 \end{bmatrix} = \frac{1}{2}\begin{bmatrix} 1 \\ i \\ 1+i \end{bmatrix} \tag{23.2}$$

We can use this (multiplying with complex conjugate) to calculate the probability of Jill's lemon quark being red, green or blue. Similar calculations can be done for her mandarin and nectarine quarks:

$$\text{LEMON:} \quad Red: \frac{1}{2}\frac{1}{2} = \frac{1}{4} \quad Green: \frac{i}{2}\frac{-i}{2} = \frac{1}{4} \quad Blue: \frac{(1+i)}{2}\frac{(1-i)}{2} = \frac{1}{2} \tag{23.3}$$

$$\text{MANDARIN:} \quad Red: \frac{-i}{2}\frac{i}{2} = \frac{1}{4} \quad Green: \frac{1}{2}\frac{1}{2} = \frac{1}{4} \quad Blue: \frac{(-1+i)}{2}\frac{(-1-i)}{2} = \frac{1}{2}$$

$$\text{NECTARINE:} \quad Red: \frac{(-1+i)}{2}\frac{(-1-i)}{2} = \frac{1}{2} \quad Green: \frac{(1+i)}{2}\frac{(1-i)}{2} = \frac{1}{2} \quad Blue: 0$$

If you add up the probabilities you can see that Jill's colourless combination of 1x*L*, 1x*M*, 1x*N* is also a colourless combination of 1x*R*, 1x*G*, 1x*B*. Based on this, we tell Jill that it really doesn't matter because we agree on what is colourless. We can use either her [*LMN*] or our [*RGB*] catalogue providing we are consistent. This is QCD global gauge invariance to colour.

We may point out to Jill that what she calls a lemon quark is actually bits of red, green and blue quarks. Her response is that we are the ones using bits of quarks. Our red quark is actually bits of lemon, mandarin and nectarine quarks. Who is right? Both of us. Quarks are excitations in a field. You cannot pin them down. They can be classified either way.

How many systems exist for cataloguing the colours while agreeing on what is colourless? There are an infinite number. Each can be represented by a 3x3 matrix such as that in Equation 23.1. The key thing is that the Dirac Lagrangian does not change. The value of $\Psi^*\Psi$ must stay the same. $\overline{\Psi}\Psi$ is $\Psi^*\gamma^0\Psi$ but the γ^0 matrix can be ignored because it mixes the components of the bi-spinor without affecting the overall colour mix. This group of matrices is called U(3). The U is for unitary, and the 3 is for 3x3. The definition of a unitary matrix is that multiplying it by its *conjugate transpose* gives the identity matrix (which I label **1**) so $U^*U = \mathbf{1}$ (see Box 23.3).

Box 23.3 Unitary matrices and the conjugate transpose   (2x2 illustration)

Consider a wave function Ψ with two types of charge, x of the first and y of the second (both are complex numbers). We change the catalogue of charges using a 2x2 matrix A with components a, b, c, d (all complex). This changes Ψ as shown below:

$$\Psi = \begin{bmatrix} x \\ y \end{bmatrix} \qquad A\Psi = \begin{bmatrix} a & b \\ c & d \end{bmatrix}\begin{bmatrix} x \\ y \end{bmatrix} = \begin{bmatrix} ax+by \\ cx+dy \end{bmatrix}$$

Ψ^* changes to Ψ^*A^* where A^* is the *conjugate transpose* of matrix A, created by taking the complex conjugate of every entry and transposing them across the lead $a : d$ diagonal. This matches the terms of Ψ^* to the new Ψ as shown below:

$$\Psi^*A^* = [x^*, y^*]\begin{bmatrix} a^* & c^* \\ b^* & d^* \end{bmatrix} = [a^*x^* + b^*y^*, c^*x^* + d^*y^*] = [(ax+by)^*, (cx+dy)^*]$$

The value of $\Psi^*\Psi$ does not change if A^*A equals the identity matrix **1** as shown in the example below. In this case, A is a *unitary* matrix, shown usually as U and defined as $U^*U = \mathbf{1}$.

$$\Psi^*\Psi \quad \rightarrow \quad (\Psi^*A^*)(A\Psi) = \Psi^*(A^*A)\Psi = \Psi^*\Psi \qquad if \quad A^*A = \begin{bmatrix} 1 & 0 \\ 0 & 1 \end{bmatrix} \tag{23.4}$$

Let's investigate further. Consider the U(3) matrix in Equation 23.1. You can multiply it by i to form another U(3) matrix because its conjugate transpose will be multiplied by $-i$ so $U^*U = \mathbf{1}$. However, this does *not* shuffle any of the colours. In fact, you can multiply it by any value on the complex unit circle (see Figure 9.1) because $e^{i\theta}e^{-i\theta} = 1$ for any value of θ. If we define each colour catalogue as a different shuffling of colours, then the U(3) group includes an infinite number of versions of each catalogue.

We can exclude multiple copies by limiting ourselves to U(3) matrices with determinant one which limits the $e^{i\theta}$ multiple to only one value. This new more limited group of matrices is called SU(3). The S stands for *special*. It contains all the distinct colour catalogues (in terms of distinctly shuffled colours). There are still an infinite number so you may not feel we have made much progress, but we have a cunning trick to simplify things. Every one of these SU(3) matrices can be expressed as a combination of eight 3x3 *generator* matrices. This means that every possible catalogue for the three colours can be expressed as follows where α_1 to α_8 are real numbers and T_1 to T_8 are the eight generator matrices (see note on matrix exponentials in Box 23.4).

$$SU(3) = e^{i(\alpha_1 T_1 + \alpha_2 T_2 + \alpha_3 T_3 + \alpha_4 T_4 + \alpha_5 T_5 + \alpha_6 T_6 + \alpha_7 T_7 + \alpha_8 T_8)}$$

(23.5)

Box 23.4   Matrix exponential

Taking the exponential of a matrix may seem a weird idea. The mathematician Edmund Laguerre introduced it back in the 1860s. The definition is shown below. $[M]$ is an $n \times n$ real or complex matrix, $\mathbf{1}$ is the identity matrix and $k!$ is k factorial. The result *always* converges (i.e. does not head off to infinity):

$$e^{[M]} = \sum_{k=0}^{k=\infty} \frac{1}{k!}[M]^k = \mathbf{1} + [M] + \frac{1}{2}[M][M] + \frac{1}{6}[M][M][M] + \frac{1}{24}[M][M][M][M]...$$

Fortunately, we do not need the actual result to make use of the relationship in Equation 23.5.

23.3.2   A Short Detour into Group Theory

This may all seem very obscure. You may ask how physicists came up with these ideas. However, this is (yet again) well trodden ground for mathematicians. This sort of multi-component symmetry with its accompanying groups of matrices has been studied by mathematicians such as Lagrange in the 1700s, followed in the 1800s by Lie, Abel and, a personal favourite, the tragically short-lived Evariste Galois (see Box 23.5). Most textbooks on group theory are horribly difficult to read so, I think it worth spending a few moments on the subject if only to make it less intimidating.

Let's start with U(1) which is the group of 1x1 unitary complex matrices. Of course, a 1x1 matrix is just one complex number. In QCD, we consider the different ways to catalogue three-colour components without altering their total combined probabilities. We can ask the same for one component. The answer is simple. The Lagrangian is based on $\Psi^*\Psi$. You can multiply a single complex component by $e^{i\theta}$ for any value of θ without changing the probability. This is why we describe the phase invariance of QED as a U(1) symmetry. If you were to require the determinant to equal one, that would be SU(1) which is a bit dull because it has only one possible value: SU(1)=1.

$$U(1) = e^{i\theta}$$

(23.6)

What if we want to categorise two components such as black and white, instead of the three colours of QCD? Any 2x2 unitary matrix would leave $\Psi^*\Psi$ unchanged. This is U(2). If we want only distinct catalogues, we cut out the multiple copies in U(2) by setting the matrix determinant to 1 taking us to SU(2). Again, there is a cunning trick. All the SU(2) matrices can be expressed using three generator matrices. In this case, they are essentially the 2x2 σ Pauli matrices just as in Subsection 17.2.4. By convention, we use the label T for generator matrices, but clearly these 2×2 matrices are *not* the same as the 3×3 ones of SU(3). SU(2) will be relevant when we discuss the weak force in the next chapter.

$$SU(2) = e^{i(\alpha_1 T_1 + \alpha_2 T_2 + \alpha_3 T_3)}$$

(23.7)

The eight T generators of SU(3) are based on what are called *Gell-Mann* matrices (listed later in Equation 23.13). If you ever want to study SU(4) (I am not sure why you would), there are fifteen 4x4 T generator matrices. If you kept going up and up, you can express SU(n) with ($n^2 - 1$) generator matrices. I guess mathematicians enjoy this sort of thing. What odd creatures they are! That is enough group theory for now.

Box 23.5   Evariste's duel

Evariste Galois (1811–1832) would have described himself as a revolutionary with an interest in maths. While in his teens, he built Galois theory, a new way to link group theory to field theory, as well as solving polynomial problems that had plagued mathematicians for 350 years. At 18, he was furious that his academic director locked him in school so he could not take part in France's July revolution of 1830. Active at several armed protests, he was imprisoned twice.

Sadly, this same volatility led him into a duel, most likely over a woman. The night before the duel, he pulled together his work in a letter that the mathematician Hermann Weyl (Box 19.10) described as *if judged by the novelty and profundity of ideas... perhaps the most substantial piece of writing in the whole literature of mankind.*

Galois, shot in the abdomen, died in his brother's arms only 20 years old. His last words were: *Don't cry any more, Alfred. I need all my courage to die at 20.*

23.3.3   The QCD Lagrangian

Summarising the story so far, QCD involves three charges (colour). The actual colours do not matter. What is important is that they combine to a colourless combination. There are an infinite number of ways to catalogue the colours, each of which is equally valid providing we consistently stick to that colour catalogue in all our calculations (QCD *global* gauge invariance).

Each distinct colour catalogue can be represented by a SU(3) matrix: U stands for unitary, S for special (determinant 1). Every SU(3) matrix can be expressed as a combination of eight generator matrices as shown in Equation 23.5. This means that you can switch from one colour catalogue to any other catalogue by applying an appropriate combination of the generators:

$$\Psi \longrightarrow e^{i\alpha_1 T_1} e^{i\alpha_2 T_2} e^{i\alpha_3 T_3} e^{i\alpha_4 T_4} e^{i\alpha_5 T_5} e^{i\alpha_6 T_6} e^{i\alpha_7 T_7} e^{i\alpha_8 T_8} \Psi \tag{23.8}$$

You now have all the background needed for us to address *QCD local gauge invariance*. Let's imagine that, instead of using the [RGB] catalogue, Jill uses [GRB] (switching green and red), but she applies this new system to only part of her calculations. To keep things simple, let's assume that this change in colour catalogue can be achieved by applying only one generator, T_1 as shown on the left of Equation 23.9. The fact that Jill has used this catalogue for only part of her calculations means that α_1 varies with space and time (t, x, y, z). On the right of Equation 23.9 is the change wrought by a local gauge change to phase in QED (this is from Equation 20.46). In the case of a QED local gauge change, it is θ that varies with space and time (t, x, y, z).

$$QCD\ (Jill):\ \Psi \longrightarrow e^{(i\alpha_1 T_1)}\Psi \qquad\qquad QED:\ \Psi \longrightarrow e^{(i\theta)}\Psi \tag{23.9}$$

The challenge is very similar and so is the solution. The QED Lagrangian is local gauge invariant to phase because it uses a *covariant* form of the derivative (labelled D instead of ∂). This adds to the QED Lagrangian an interaction term with the photon (EM) field that balances out the effect of the local gauge change (see Equation 20.40 for a reminder). QCD has exactly the same mechanism. This is shown in Equation 23.10. The QCD Lagrangian also uses a *covariant* form of the derivative which introduces a similar interaction term, this time with a gluon field labelled below as A_μ^1. The strong force coupling constant is shown as g. In the case of QCD, the term must also include the generator matrix T_1, which tells you that the interaction alters the colour of Ψ. This must be the case to rebalance the colours after Jill's local change to the colour catalogue.

$$QCD\ (Jill):\ D_\mu \Psi = \partial_\mu \Psi + ig A_\mu^1 T_1 \Psi \qquad\qquad QED:\ D_\mu \Psi = \partial_\mu \Psi + ie A_\mu \Psi \tag{23.10}$$

In this example we are catering for a local gauge change that involves only one of the generator matrices. Gauge changes can involve all eight so, to achieve QCD local gauge invariance, there must be eight interaction terms with eight distinct gluon fields (Equation 23.11). These are combined into one term using the Einstein summation convention.

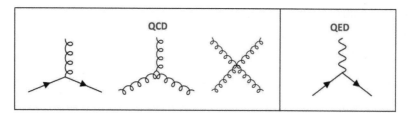

Figure 23.2 QCD has three possible vertices whereas QED has one. Note that these Feynman diagrams are of QCD and QED vertices, not of legitimate processes.

$$D_\mu \Psi = \partial_\mu \Psi + ig A_\mu^1 T_1 \Psi + ig A_\mu^2 T_2 \Psi + ig A_\mu^3 T_3 \Psi... = \partial_\mu \Psi + ig A_\mu^a T_a \Psi \qquad (23.11)$$

We have to include a term for the Action of each gluon field. In QED this is of the form $F_{\mu\nu}F^{\mu\nu}$. In QCD similar terms combine with the Einstein summation and are written as $G_{\mu\nu}^a G_a^{\mu\nu}$ making the final QCD Lagrangian (Equation 23.12). You can compare this with the QED Lagrangian below it. The QED Lagrangian is from Equation 20.41 if you want to refer back. The gluons are massless as required for this to be local gauge invariant (see Section 20.9).

$$\mathscr{L}_{QCD} = i\overline{\Psi}\gamma^\mu \partial_\mu \Psi - m\overline{\Psi}\Psi - g\overline{\Psi}\gamma^\mu A_\mu^a T_a \Psi - \frac{1}{4}G_{\mu\nu}^a G_a^{\mu\nu} \qquad (23.12)$$

$$\mathscr{L}_{QED} = i\overline{\Psi}\gamma^\mu \partial_\mu \Psi - m\overline{\Psi}\Psi - e\overline{\Psi}\gamma^\mu A_\mu \Psi - \frac{1}{4}F_{\mu\nu}F^{\mu\nu}$$

The final twist is that lurking inside each $G_{\mu\nu}G^{\mu\nu}$ term are gluon–gluon interaction terms. These appear because the T generator matrices do not commute (QCD is called *non-Abelian* because of this). This should not surprise you. The colours of the gluons also are affected by the local gauge change, so offsetting interaction terms are needed to restore the overall colour balance. The gluon–gluon interaction terms mean that QCD Feynman diagrams have three possible vertices as shown in Figure 23.2. The solid lines represent fermions (quarks in the case of QCD) and the curly lines are gluons.

23.3.4 Gluons and the Generators

This all looks horribly complicated. We have eight separate gluon fields all interacting with each quark. You can get a more intuitive feel of what is happening by examining what the interaction term $T_a \Psi$ does to the colour of the quark. The T generator matrices are derived from the Gell-Mann matrices ($T_1 = \frac{\lambda_1}{2}, T_2 = \frac{\lambda_2}{2}...$). The Gell-Mann matrices are labelled λ and shown below.

$$\lambda_1 = \begin{bmatrix} 0 & 1 & 0 \\ 1 & 0 & 0 \\ 0 & 0 & 0 \end{bmatrix} \quad \lambda_2 = \begin{bmatrix} 0 & -i & 0 \\ i & 0 & 0 \\ 0 & 0 & 0 \end{bmatrix} \quad \lambda_3 = \begin{bmatrix} 1 & 0 & 0 \\ 0 & -1 & 0 \\ 0 & 0 & 0 \end{bmatrix} \quad \lambda_4 = \begin{bmatrix} 0 & 0 & 1 \\ 0 & 0 & 0 \\ 1 & 0 & 0 \end{bmatrix} \qquad (23.13)$$

$$\lambda_5 = \begin{bmatrix} 0 & 0 & -i \\ 0 & 0 & 0 \\ i & 0 & 0 \end{bmatrix} \quad \lambda_6 = \begin{bmatrix} 0 & 0 & 0 \\ 0 & 0 & 1 \\ 0 & 1 & 0 \end{bmatrix} \quad \lambda_7 = \begin{bmatrix} 0 & 0 & 0 \\ 0 & 0 & -i \\ 0 & i & 0 \end{bmatrix} \quad \lambda_8 = \frac{1}{\sqrt{3}}\begin{bmatrix} 1 & 0 & 0 \\ 0 & 1 & 0 \\ 0 & 0 & -2 \end{bmatrix}$$

As an example of the interaction, Equation 23.14 shows the effect of λ_1 on a red quark as defined by the $[RGB]$ colour catalogue. If it starts red $[1,0,0]$, it is changed to green $[0,1,0]$ which removes the red and adds green so it is: $G\overline{R}$. If it starts green $[0,1,0]$ it is changed to red $[1,0,0]$ which is $R\overline{G}$.

$$\lambda_1 [RED] = \begin{bmatrix} 0 & 1 & 0 \\ 1 & 0 & 0 \\ 0 & 0 & 0 \end{bmatrix}\begin{bmatrix} 1 \\ 0 \\ 0 \end{bmatrix} = \begin{bmatrix} 0 \\ 1 \\ 0 \end{bmatrix} \qquad \lambda_1 [GREEN] = \begin{bmatrix} 0 & 1 & 0 \\ 1 & 0 & 0 \\ 0 & 0 & 0 \end{bmatrix}\begin{bmatrix} 0 \\ 1 \\ 0 \end{bmatrix} = \begin{bmatrix} 1 \\ 0 \\ 0 \end{bmatrix} \qquad (23.14)$$

This describes the first gluon exchange in Figure 23.1 (we cannot say whether the gluon went from the red quark to the green quark or vice versa). We can characterise each gluon interaction in terms of its colour exchange. There are many ways to write out the eight gluons (not nine, see Box 23.6) depending on the colour catalogue you use; this is just one option.

$$\frac{1}{\sqrt{2}}\left(R\overline{G}+G\overline{R}\right) \qquad \frac{i}{\sqrt{2}}\left(G\overline{R}-R\overline{G}\right) \qquad \frac{1}{\sqrt{2}}\left(R\overline{R}-G\overline{G}\right) \qquad \frac{1}{\sqrt{2}}\left(R\overline{B}+B\overline{R}\right) \qquad (23.15)$$

$$\frac{i}{\sqrt{2}}\left(B\overline{R}-R\overline{B}\right) \qquad \frac{1}{\sqrt{2}}\left(G\overline{B}+B\overline{G}\right) \qquad \frac{i}{\sqrt{2}}\left(B\overline{G}-G\overline{B}\right) \qquad \frac{1}{\sqrt{6}}\left(R\overline{R}+G\overline{G}-2B\overline{B}\right)$$

Box 23.6   Might there have been a ninth gluon?

We might wonder why QCD does not include local gauge invariance for all U(3) rather than just the SU(3) colour catalogues. This would require the identity matrix as a ninth generator to change phase *without* shuffling the colours. Therefore, the ninth gluon interaction would be $(R\overline{R}+G\overline{G}+B\overline{B})$. Being colourless, this gluon would *not* be confined and the strong force would act between colourless particles over long distances. This does not happen so we know there is no ninth gluon.

What does all of this mean for a lepton such as the electron that is colourless? You can think of it as having no colour $[RGB] = [000]$ or having equal amounts of each colour along the lines of $[RGB] = [111]$. Either way, changes to the colour catalogue (global or local) have absolutely no effect. It has no gluon interactions, so it does not feel the strong force.

23.3.5   Summary: QCD As an Infinity Swimming Pool

Stepping back from the detailed maths, we can picture QCD as a (very complicated) infinity swimming pool. In the case of QED there was only one balancing tank (the EM field) and one inflow/outflow (the massless photon) as shown earlier in Figure 20.3. With QCD the SU(3) symmetry requires eight balancing tanks and eight inflow/outflows (the eight massless gluons). Furthermore, these gluons interact with each other. This is illustrated in a highly schematic way in Figure 23.3 by showing connections between the inflow/outflow pipes.

Suppose our scientist Jill decides to change colour catalogue for one quark, calling it red, not green. In her calculation, this *local* change in colour gauge will affect the derivatives in the quark's Dirac Lagrangian. However, her change to the colour catalogue will set off a sequence of other changes. It will alter the interaction terms between the quark and the gluon fields. It will alter the gluon–gluon interaction terms. Add it all up, and we find that the Action of the QCD Lagrangian is exactly the same as we would have calculated without Jill's change.

The interaction terms in the QCD Lagrangian feed through into the equations of motion in a similar way to QED (see Section 20.8). The result is that the quarks and gluons attract each other. This and the strength of the coupling constant result in the tight interaction patterns schematically portrayed on the right of Figure 23.1 that lead in turn to confinement. This is why quarks and antiquarks always combine in colourless combinations.

It is fairly obvious why QCD involves such tough calculations and is much harder than QED. There are a host of moving parts so the Feynman diagrams can be very complicated. Furthermore, the strong force coupling constant is large enough to make the more complicated diagrams significant to the development of the quantum fields. Aarrgghh!

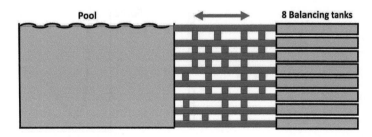

Figure 23.3   Schematic of QCD Lagrangian as an infinity swimming pool.

23.4 The Residual Strong Force

The protons and neutrons in the nucleus (collectively called *nucleons*) are colourless which means that there are no gluon interactions between them. What then is it that binds the nucleons together? I addressed this question back in Subsection 16.4.2 where I compared it with the attractive force of molecular covalent bonds. The attraction is due to the ability of quarks to quantum tunnel between nucleons. However, the interchange must be colourless, so the virtual particles are *mesons* (quark–antiquark pairs).

The attraction is called the *residual strong force*. Let's go into a bit more detail. The left of Figure 23.4 shows a schematic illustration of a proton and neutron. Time flows from left to right. From the left, the top nucleon (which I will call nucleon-*A*) starts as a proton (*d*, *u*, *u* quarks). The bottom one in the figure (nucleon-*B*) starts as a neutron (*d*, *d*, *u*). They exchange a π^+ meson which is the combination of an up quark and down antiquark ($u\bar{d}$) with the result that the proton becomes a neutron and the neutron becomes a proton.

The pathway works as follows. A $d\bar{d}$ quark–antiquark pair is created in the proton (nucleon-*A*). The $\bar{d}$ antiquark forms the π^+ pi-meson ($u\bar{d}$) with one of the nucleon-*A* u quarks. This meson travels across to nucleon-*B*. Nucleon-*A* loses that *u* quark which is replaced in the quark triplet by the new *d* quark. Nucleon-*A* now has two down quarks and an up quark so it is a neutron. The π^+ meson ($u\bar{d}$) arrives at nucleon-*B*. Its $\bar{d}$ antiquark annihilates with one of the *d* quarks of nucleon-*B*. This is replaced by the *u* quark of the meson, so nucleon-*B* becomes a proton.

There are lots of variations on the pathway. It can be a π^- meson ($d\bar{u}$) from neutron to proton, delivering a *d* and annihilating its *u*. It can be a π^0 meson ($u\bar{u}$ or $d\bar{d}$) between two neutrons or two protons; don't forget each meson is a quark–antiquark pair so it is integer spin. It is a boson.

The process repeats with proton/neutron → neutron/proton → proton/neutron... and, as usual, let me remind you that the schematic is a comic-strip version of events. Reality is the combination of all possible pathways. The result is a force of attraction between nucleons. This is shown on the right of Figure 23.4. The force is at its maximum of about 20,000 Newtons (2 tons!) when the nucleons are 1 femtometre ($10^{-15}m$) apart.

The strongest attraction is between spin-aligned pairs of nucleons which, given the Pauli exclusion principle for fermions (nucleons are spin-half), makes an equal number of protons and neutrons a particularly stable combination in the atomic nucleus. This is the combination for many of the commonest isotopes of elements. This is sometimes called the *Valley of Stability* for isotopes. Take for example helium-4 (protons:neutrons 2 : 2), carbon-12 (6 : 6), nitrogen-14 (7 : 7), oxygen-16 (8 : 8) and many more, although much larger atomic nuclei (over atomic number 30) tend to have more neutrons than protons (1.7x is a rule of thumb).

The residual nuclear force is very short range and falls off to negligible levels beyond a few femtometres, which is why that is the diameter of the atomic nucleus. In sharp contrast, the strong force between quarks does not decrease with separation (see Subsection 23.2.2). Why the difference? The strong force is mediated by massless gluons whereas the residual nuclear force is mediated by mesons that each have a mass about 100 times more than the tiny difference in mass between the proton and neutron. In the language of QFT, the gluon and meson *propagators* are very different (Subsection 21.4.2). The mesons that mediate the residual nuclear force are a long way *off-shell*.

Hideki Yukawa (1907–1981), a Japanese physicist, realised the short range of the residual nuclear force is indicative of a massive off-shell mediator. We can follow in his footsteps using the rule of thumb that relates how long virtual particles can exist off-shell while remaining significant to the development of the quantum field. The relevant equation is $\Delta E\, \Delta t \approx \hbar$,

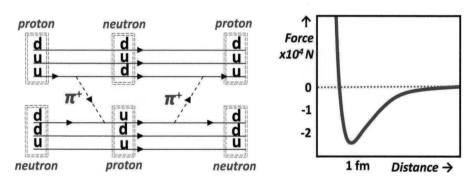

Figure 23.4 Nuclear force mediated by π^+ meson ($u\bar{d}$ quark–antiquark pair).

where $\Delta \mathbf{E}$ is off-shell drift, and Δt is the time off-shell (refer back to Equation 19.16 if you need to). Yukawa knew from scattering studies that the atomic nucleus is about $2 \times 10^{-15} m$. He assumed that the mediator moves at close to light speed so $\Delta t \approx \frac{2\times10^{-15}}{c}$. Also, the difference in mass between proton and neutron is so small that we can assume the mediator is off-shell by its rest mass energy: $\Delta \mathbf{E} \approx mc^2$.

$$\Delta \mathbf{E} \, \Delta t \approx mc^2 \, \frac{2 \times 10^{-15}}{c} \approx \hbar \qquad\qquad \hbar \approx 10^{-34} Js \qquad c \approx 3 \times 10^8 ms^{-1}$$

$$\implies \quad m \approx \frac{10^{-34}}{(2 \times 10^{-15})(3 \times 10^8)} \approx 2 \times 10^{-28} \, kg \tag{23.16}$$

This is a mass of about 100 MeV. Yukawa published his prediction in 1935. Twelve years later the pi-meson was discovered; its mass was about 140 MeV, and won Yukawa the 1949 Nobel Prize.

23.5 Oh No! Here Comes Jill Again!

Our colour categorist Jill is still puzzling over our $[RGB]$ versus her $[LMN]$ colour system when Figure 23.4 catches her eye. When the proton and neutron are close together you cannot tell them apart. The residual strong force treats them the same and their masses are very similar. Looking within them, the difference in quarks is d, u, u versus d, u, d so it comes down to the third being u versus d. The strong force treats u and d quarks identically. Is this a coincidence? They have different electric charge, but still, might *up* and *down* reflect some subtle sort of symmetry in the way that the red, green and blue colour charges do? This brings us to the main topic of the next chapter – the *weak force*.

Chapter 24

The Weak Force and Higgs Field (1)

CHAPTER MENU
24.1 Idealised Weak Force and SU(2) Symmetry
24.2 The Real Weak Force
24.3 What About SU(2) Gauge Symmetry?
24.4 Mass, Chirality and the Higgs Field
24.5 The Story So Far

Just when you think that you have a grip on the basics of QFT and gauge invariance, the weak force jumps up and bites you in the rear. You will learn that it is based on SU(2) symmetry and involves three bosons as expected (W_+, W_-, and Z), *but* the symmetry is very distorted. The weak force interacts with all fermions the same way, *but* is affected by the orientation of their spin (specifically it interacts only with left-chiral states). Its gauge bosons should be massless, *but* they have mass. Overall, that is a whole lot of *buts*. What a mess! To piece it together takes the Higgs field and plenty of mathematical and mental magic (see Higgs' quote in Box 24.1).

Let me warn in advance that this is complicated stuff. I have divided the topic into two chapters simply to give you a pause for breath. In the first chapter, we start by considering how the weak force would appear if its SU(2) symmetry were perfect and in what ways it differs from this idealised model, including the surprising left-handed bias of the weak force. This leads to the idea of mass being an interaction with some scalar field (the *Higgs field*). The second chapter continues the story. We analyse how the Higgs field leads to massive weak force bosons (the *Higgs mechanism*) and then discuss how including U(1) symmetry leads to the unification of the weak force and the EM force into the combined *electroweak force*.

Box 24.1   Peter Higgs upon the discovery of the Higgs boson
It is very nice to be right sometimes

24.1   Idealised Weak Force and SU(2) Symmetry

Let me take a moment to introduce the concept of the *flavour* of a fermion. If we focus again on the most stable light family of fermions, a quark can be a down quark or an up quark (you may want to refer back to the list of fermions in Table 23.1). This down/up distinction is referred to as the flavour of the quark. Similarly the electron/neutrino distinction is described as the flavour of the lepton (remember that lepton is the term for the lighter fermions).

I want to start with an overly simplistic description of the weak force. Things will turn out to be more complicated, but we can get into those details later. The strong force works between different *colours* of fermion, specifically red/green/blue for the quarks. The weak force works between different *flavours* of fermion, for example, down/up for quarks and electron/neutrino for leptons. In the same way that a quark can change say red↔green through a strong force interaction, it can change down(d)↔up(u) through a weak force interaction. Similarly a lepton can change electron(e)↔neutrino(ν_e). The same applies for the heavier generations of fermion such as strange(s)↔charm(c) and muon(μ)↔muon-neutrino(ν_μ).

Let us suppose for a moment that the universe were designed for the convenience of physics students and that the weak force exhibits a pure SU(2) symmetry in the same way that the strong force exhibits SU(3) symmetry. In this imaginary universe, the *only* difference between a down quark and an up quark would be its flavour (for example, we ignore the difference in mass). The same would be true for the electron and neutrino. We could look on up and down quarks as having different *flavour charge* in the same way as they can have different *colour charge*.

We can study flavour symmetry in the same way that we studied colour symmetry in the last chapter. Let's reintroduce the fictional scientist Jill, from the last chapter. In terms of colour, the quarks are colour *triplets*. She proposed the lemon/mandarin/nectarine [LMN] triplet colour catalogue for QCD versus our red/green/blue [RGB]. In terms of flavour, a fermion is a flavour *doublet*. She takes a look at our up/down [ud] doublet for quarks and proposes instead that we use a salt/pepper [sp] flavour catalogue (actually, those would be rather more sensible names for flavour). I have chosen her flavour catalogue at random to illustrate. It is compared with ours in Equation 24.1 and summarised as a 2 × 2 matrix on the left side of Equation 24.2. The right side of 24.2 shows how this restates Jill's salt quark ([s, p] = [1,0]) in terms of our up and down ([u, d] = $\frac{1}{\sqrt{2}}[i, -1]$).

$$s = \frac{1}{\sqrt{2}}(iu - d) \qquad p = \frac{1}{\sqrt{2}}(u - id) \tag{24.1}$$

$$\frac{1}{\sqrt{2}}\begin{bmatrix} i & 1 \\ -1 & -i \end{bmatrix}\begin{bmatrix} s \\ p \end{bmatrix} = \begin{bmatrix} u \\ d \end{bmatrix} \qquad\qquad \frac{1}{\sqrt{2}}\begin{bmatrix} i & 1 \\ -1 & -i \end{bmatrix}\begin{bmatrix} 1 \\ 0 \end{bmatrix} = \frac{1}{\sqrt{2}}\begin{bmatrix} i \\ -1 \end{bmatrix} \tag{24.2}$$

Continuing our fantasy physics, it is simple to show that switching from our [ud] to Jill's [sp] makes no difference providing we are consistent. A global change of catalogue will not affect $\Psi^*\Psi$, and therefore $\overline{\Psi}\Psi$, providing the 2 × 2 matrix is unitary (see Equation 23.4 if you need a reminder). Jill's matrix is indeed unitary ($U^*U = 1$) as shown in Equation 24.3.

If the only difference is the flavour, this means that there is no change to $m\overline{\Psi}\Psi$ or to the derivative terms in the quark's Dirac Lagrangian. The Lagrangian is *globally* gauge invariant to the change in flavour catalogue from [ud] to [sp] in our fantasy scenario.

$$\frac{1}{\sqrt{2}}\begin{bmatrix} -i & -1 \\ 1 & i \end{bmatrix} \frac{1}{\sqrt{2}}\begin{bmatrix} i & 1 \\ -1 & -i \end{bmatrix} = \frac{1}{2}\begin{bmatrix} 2 & 0 \\ 0 & 2 \end{bmatrix} = \begin{bmatrix} 1 & 0 \\ 0 & 1 \end{bmatrix} \tag{24.3}$$

We can step through exactly the same logic that we used for QCD in Subsection 23.3.1 so please refer back if needed. There are an infinite number of potential flavour catalogues which can each be represented by a unitary 2 × 2 matrix, collectively called U(2). If we define a flavour catalogue as a distinct shuffling of flavours, U(2) contains an infinite number of copies of each catalogue, but we can eliminate the multiple copies by requiring the determinant of the matrix to be 1. This reduces us to SU(2) which is the set of *special* (determinant = 1), *unitary* ($U^*U = 1$), 2 × 2 matrices. There are still an infinite number.

In this fantasy universe, the consistent (global) use of any of these SU(2) matrices as a flavour catalogue produces the same result. The Dirac Lagrangian does not change, which means the Action of the quark quantum field does not change – which means the physics does not change. This is SU(2) *global* gauge invariance.

Assuming that you did not fall asleep in the last chapter, you will know what is coming: SU(2) *local* gauge invariance. If you find this section confusing, please refer to the material following Subsection 23.3.1 where the argument is covered in more detail for the strong force. All SU(2) matrices can be expressed using three generator matrices which we label T_1, T_2 and T_3 (essentially, the three Pauli matrices) as shown in Equation 24.4 where α_1, α_2 and α_3 are real numbers. More detail on matrix exponentials can be found back in Box 23.4.

$$SU(2) = e^{i(\alpha_1 T_1 + \alpha_2 T_2 + \alpha_3 T_3)} \tag{24.4}$$

$$T_1 = \frac{1}{2}\begin{bmatrix} 0 & 1 \\ 1 & 0 \end{bmatrix} \qquad T_2 = \frac{1}{2}\begin{bmatrix} 0 & -i \\ i & 0 \end{bmatrix} \qquad T_3 = \frac{1}{2}\begin{bmatrix} 1 & 0 \\ 0 & -1 \end{bmatrix}$$

A *local* change to the flavour catalogue means that the factors α_1, α_2 and α_3 vary over time or space (or both). Such a change does not affect $\overline{\Psi}\Psi$ in the Lagrangian, but *does* affect the derivative terms (you can check back to Equation 20.3 for a reminder of why this happens). This can be offset in exactly the same way as we saw in QED and QCD by switching to a covariant form of the derivative that reflects interaction with three gauge boson fields labelled below A_μ, one for each of the three generators. This is the same structure as in QCD except for QCD there are eight gauge boson fields (you can compare with Equation 23.11). The term on the right of Equation 24.5 uses the Einstein summation convention to show all three gauge boson fields in one mathematical term.

$$D_\mu \Psi = \partial_\mu \Psi + ig A_\mu^1 T_1 \Psi + ig A_\mu^2 T_2 \Psi + ig A_\mu^3 T_3 \Psi = \partial_\mu \Psi + ig A_\mu^a T_a \Psi \tag{24.5}$$

We can now construct the Lagrangian for our fantasy universe *idealised* SU(2) weak force as Equation 24.6. The required adjustment to the Dirac Lagrangian is compared with that for SU(3) QCD and U(1) QED. The coupling constant is shown as g_W. For ease of comparison, I have used $F_{\mu\nu} F^{\mu\nu}$ for all the vector fields and I have used the same A_μ symbol for all the gauge bosons, but don't forget that they are distinct: one boson in QED (the photon), eight in QCD (the gluons) and three idealised massless bosons for the weak force which I will call W_1, W_2 and W_3.

The three W bosons can change the flavour of a fermion, so they themselves can carry flavour charge. This means that they self-interact. Mathematically, these self-interaction terms appear in the vector field Lagrangian because the SU(2) T generator matrices do not commute. This is described in group theory as *non-Abelian*. The same occurs in QCD because the gluons carry colour charge (see Equation 23.12 and the paragraph that follows it).

$$\text{Idealised} \quad \mathscr{L}_{weak} = i\overline{\Psi}\gamma^\mu \partial_\mu \Psi - m\overline{\Psi}\Psi - g_W \overline{\Psi}\gamma^\mu A_\mu^a T_a \Psi - \frac{1}{4} F_{\mu\nu}^a F_a^{\mu\nu} \tag{24.6}$$

$$\mathscr{L}_{QCD} = i\overline{\Psi}\gamma^\mu \partial_\mu \Psi - m\overline{\Psi}\Psi - g_S \overline{\Psi}\gamma^\mu A_\mu^a T_a \Psi - \frac{1}{4} F_{\mu\nu}^a F_a^{\mu\nu}$$

$$\mathscr{L}_{QED} = i\overline{\Psi}\gamma^\mu \partial_\mu \Psi - m\overline{\Psi}\Psi - e\overline{\Psi}\gamma^\mu A_\mu \Psi - \frac{1}{4} F_{\mu\nu} F^{\mu\nu}$$

If only things were so simple! We have a tantalisingly pretty model for the weak force. The only problem is that it does *not* match experimental facts. As a starting point, this idealised model is very important for you to understand, but the complete Lagrangian associated with the weak force will turn out to be much more complicated.

24.2   The Real Weak Force

You will soon learn that the SU(2) symmetry of the weak force is hidden or obscured or distorted (take your pick). There are some major differences between the SU(2) symmetry of the weak force and the SU(3) symmetry of QCD. Changing the colour of a quark does not affect its mass, but changing its flavour does. Up and down quarks have different mass. As for different flavours of lepton, the masses of an electron and a neutrino differ by a factor of 10^5. The gluons of QCD are massless, but the bosons of the weak force are massive (the W_+, W_- and Z bosons). The QCD coupling constant g_S is the same for all fermions. The coupling constant g_W of the weak force is the same for all left-chiral fermions, but is *zero* for right-chiral. The term *chiral* may be new to you. It is related to spin as I will explain later.

At first glance these factors appear to drive a dagger through the heart of SU(2) local gauge invariance. On the other hand, weak reactions often involve a change in flavour and the weak force involves three bosons as expected. This reflects SU(2). How do we reconcile things?

24.2.1   Weak Isospin

Let's review the basic features of the weak force starting with some nomenclature. To distinguish between the flavours of quark, we assign a flavour charge of $-\frac{1}{2}$ for the down quark and $+\frac{1}{2}$ for the up quark. This is just a convention to make it easy to remember, as the fermions with negative electric charge are labelled with negative flavour charge. The same is true for the other quark generations so, for example, the strange quark is $-\frac{1}{2}$ and the charm quark is $+\frac{1}{2}$. Applying this same logic to the leptons, the flavour charge is $-\frac{1}{2}$ for the electron and therefore $+\frac{1}{2}$ for the neutrino.

The flavour charge goes by the extremely unfortunate name T_3 *weak isospin*. For those interested, the T_3 is because we use eigenstates of the T_3 generator matrix (see Equation 24.4) to define the flavour catalogue. An example of how this works for an up quark is shown in Equation 24.7. The isospin name is misleading because it is *not* a form of spin or angular momentum. That is physicists for you, needlessly confusing things.

$$T_3 \begin{bmatrix} u \\ d \end{bmatrix} \qquad T_3 \begin{bmatrix} 1 \\ 0 \end{bmatrix} = \frac{1}{2} \begin{bmatrix} 1 & 0 \\ 0 & -1 \end{bmatrix} \begin{bmatrix} 1 \\ 0 \end{bmatrix} = +\frac{1}{2} \begin{bmatrix} 1 \\ 0 \end{bmatrix} \tag{24.7}$$

Weak interactions can affect T_3 weak isospin in three different ways. The T_3 can *increase* (+1) from $-\frac{1}{2}$ to $+\frac{1}{2}$ such as a quark changing from down to up, or a lepton changing from electron to neutrino. It can *decrease* (−1) with a flavour change in the opposite direction. Or, T_3 can be unchanged (0). The last happens because the flavour catalogue can be shuffled without changing the up/down status. For example, if salt and pepper were switched in Jill's flavour catalogue it would not alter particles in terms of their up-ness and down-ness (see Equation 24.1).

24.2.2 Weak Interactions

Examples of the weak force vertices and processes are shown in Figure 24.1. Box *A* on the left of the figure shows examples of what are called *charged current* interactions. These are vertices where the value of T_3 changes. The top one shows an electron changing flavour into a neutrino with the emission of a W_- boson which carries away −1 of T_3 weak isospin *and* −1 of electric charge. The bottom shows an up quark changing flavour into a down quark with the emission of a W_+ boson that carries away +1 of T_3 weak isospin and +1 of electric charge. Please note that this Box *A* shows vertices, not legitimate processes. The W bosons are off-shell, so they must connect to another vertex for the full process to end up back on-shell (if needed, please refer to Section 21.3 that covers the same topic in QED).

Box *B* shows how the Z boson (often called the *neutral current*) can play a role even though there is no change to flavour. The Feynman diagram shows the annihilation of an electron-positron pair to create a Z boson that decays into a quark–antiquark pair. The resonance of this pathway led to the discovery of the Z boson in electron-positron colliders (as illustrated earlier in Figure 21.6).

Box *C* of Figure 24.1 shows a particularly important weak interaction. One of the quarks in a neutron changes flavour from down to up with the emission of a W_- boson. This changes the neutron (*ddu*) into a proton (*udu*). The W_- boson rapidly decays into an electron and anti-neutrino.[1] The reverse can also happen with a proton becoming a neutron by one of its quarks emitting a W_+ boson that decays subsequently into a positron and a neutrino.

Outside of the nucleus, the energy balance favours neutron→proton because the neutron is slightly more massive. In fact, a lone neutron will decay into a proton in about 10 minutes. However, inside the nucleus the energy balance is dominated by the strong force. The weak interactions mean that the proton/neutron mix in the nucleus of an atom can change to become more energetically favourable. This is what happens in nuclear fusion. Four hydrogen nuclei (each a proton) fuse in a sequence of steps to form a helium nucleus (two protons, two neutrons). The energy release is substantial so why is fusion so hard to achieve?

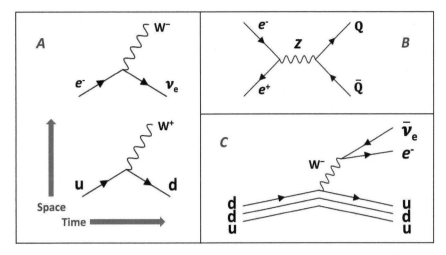

Figure 24.1   Some weak force vertices and interaction processes.

24.2.3 Massive Weak Bosons

The reason is that the W_+, W_- and Z weak force bosons are hugely massive, weighing in at circa 10^{-25} kg (about 80 GeV) rest mass. To put this in context, consider the top of Box A in Figure 24.1. An electron changes flavour. Its rest mass is about 0.5 MeV, so the rest mass of the emitted W_- boson is $100,000\times$ greater. In the bottom diagram the difference in mass between the up and down quark is about 3 MeV. The W_+ boson is about $25,000\times$ more massive. The weak bosons are a long way off-shell which gives a very low value to their propagators, dramatically reducing the probability of a weak interaction occurring (for a reminder on propagators see Subsection 21.4.2).

In the last chapter, we discussed how the range of the residual strong force (the force between colourless nucleons) is limited in range because the mediating π meson is massive. Well, these weak force bosons are about $1,000\times$ more massive than the π meson. The weak bosons are truly humongous! The same calculation that we did for the π meson (back in Equation 23.16) shows that the weak force is limited in range to about 10^{-28} metres because of the mass of the bosons. This is one thousandth the range of the strong residual force and is why the weak force is so *weak*.

I find it fascinating that both the strong and weak force are short range, but for completely different reasons. The short range of the *strong* force has nothing to do with the mediating gluons, which are massless. It is driven by the strong coupling constant that causes confinement by locking quarks into colourless combinations. In contrast, the short range of the *weak* force is caused by the mass of the weak bosons and has very little to do with its coupling constant. How intriguing that completely different causes can lead to similar results.

24.2.4 Wu and the Weak Left-handed Bias

The physics community got a huge shock in 1957 when Chien-Shiung Wu showed that *parity* is not conserved in weak force interactions (see the Nobel controversy surrounding this discovery in Box 24.2). Parity conservation is the assumption that the physics of any system is not changed by taking its mirror-image. Imagine the footballer David Beckham kicking a free kick with his right foot. The spin he puts on the ball means that it bends to the left into the goal. In the mirror image, he kicks with his left foot, which generates spin in the opposite sense and bends the ball to the right. The physical laws linking boot-to-spin-to-bend are the same in the mirror image.

Every interaction ever examined that involves gravity, the electromagnetic force or the strong force had been found to conserve parity. Nobody had bothered to check that it also held for weak interactions such as radioactive decay. However in 1956, two young theoretical physicists, Tsung-Dao Lee and Chen-Ning Yang, analysed some decay patterns in K-mesons (strange quark–antiquark pairs) that suggested parity was not conserved. This hint led to a rush of experimental activity in 1957. The most famous is the Wu experiment.

Box 24.2 No thanks to Wu

Chien-Shiung Wu (1912–1997) was nicknamed the *Chinese Madame Curie*. She designed and ran her ground-breaking experiment that led directly to the award of the 1957 Nobel Prize – to Lee and Yang – not to her. Although her results came too late to be eligible for that year's prize, she subsequently was nominated again and again – in all, eleven times. The lack of recognition seems inexcusable and carries more than a whiff of sexism.

In Wu's own words: *I wonder whether the tiny atoms and nuclei, or the mathematical symbols, or the DNA molecules have any preference for either masculine or feminine treatment.*

The experiment and its results are summarised in Figure 24.2. Wu and her team focused on the *beta* decay of the radioactive isotope Cobalt-60 to Nickel-60. The cobalt was cooled to an extremely low temperature and placed in an intense magnetic field to align the spin of the cobalt atoms. The decay process is shown on the left of Figure 24.2. Cobalt decays into nickel, an electron and an antineutrino. The blue arrows in the figure show the *spin* of the particles as vectors using the right-hand convention. If you curl the fingers of your right hand in the direction of the spin, your thumb will give you the direction of the spin vector.

The cobalt spin is +5 and the nickel produced is +4 so the spin of the emitted electron and antineutrino must also be aligned with the magnetic field. The particles are not physically spinning, but their intrinsic spin must obey the conservation of angular momentum (see Section 17.1 for a reminder).

The right of Figure 24.2 shows possible outcomes for the decay. To conserve angular momentum all the spins (blue arrows) must remain aligned with the magnetic field. To conserve linear momentum, the electron and antineutrino must

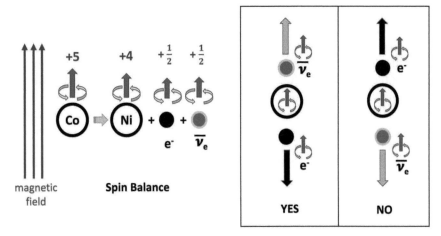

Figure 24.2 The Wu experiment.

be emitted in opposite directions (black and grey arrows). The experimenters could detect the electron and thereby infer the direction of emission of the antineutrino. The decay also creates a pattern of gamma ray (photon) emission that the experimenters used to determine how well the cobalt atoms were aligned with the magnetic field.

By comparing the pattern of electron and photon emission, Wu was able to conclude that, when the cobalt atom was aligned with the magnetic field, the electron and antineutrino emission was always that labelled *YES* in the figure. The option labelled *NO* did not occur. This means that all the emitted antineutrinos were spinning the same way! Their spin vector was always aligned with their motion. This is described as *positive* or *right-handed helicity*. An easy way to remember right-handed from left-handed helicity is to imagine that the spinning object is a corkscrew. Twist it clockwise and it moves into the cork. If that matches the movement of the particle it is right-handed helicity as is the case for the antineutrinos in the *YES* scenario. In the *NO* scenario the antineutrino would have left-handed helicity. None were emitted.

Box 24.3   Wolfgang Pauli's initial reaction to Wu's result
That's total nonsense!

Similar experiments have been conducted analysing decay paths that involve the emission of a positron and a neutrino. In each case the neutrino produced is left-handed helicity. Let me emphasise how shocking this is. The emitted neutrinos *must* have spin *anticlockwise* in their direction of motion (left-handed helicity). For antineutrinos, it *must* be *clockwise* (right-handed helicity). Now, that is weird! See Pauli's reaction in Box 24.3 (and if you are unfamiliar with neutrinos, you may want to glance at Box 24.4 plus Box 24.5 about Majorana).

Subsequent research including high energy electron and quark scattering experiments confirm that the weak force interacts only with *left-chiral* particles and *right-chiral* antiparticles. I will explain chirality in a moment, but for neutrinos and antineutrinos, it is almost the same as helicity. For something as light as a neutrino, left-handed helicity is left-chiral and right-handed helicity is right-chiral. Therefore the antineutrino in the *YES* box of Figure 24.2 is right-chiral and can be produced by the weak force. The antineutrino in the *NO* box would be left-chiral and cannot. This is the reason for the asymmetric result in the Wu experiment.

24.3 What About SU(2) Gauge Symmetry?

The weak force reflects SU(2) symmetry. The fact that it applies *only* to left-chiral particles and right-chiral antiparticles complicates the Dirac Lagrangian (to avoid my having to mention the antiparticles every time, please just remember that everything we say of left-chiral particles also applies to right-chiral antiparticles). Let's label the left-chiral part of the wave function as Ψ_L and the right-chiral as Ψ_R. Then Ψ_L is a flavour *doublet* such as $[ud]$ for a quark and the derivative terms for Ψ_L in the Lagrangian will be covariant along the lines of our idealised model (see Equation 24.6) with the three W vector fields balancing any local change in the flavour catalogue. On the other hand, Ψ_R is *not* a flavour doublet and is unaffected by any change in the flavour catalogue. Furthermore, it turns out that you cannot simply pull apart the Ψ_L and Ψ_R components of a fermion. Aarrgghh!

At first glance, SU(2) local gauge symmetry has become a complete mess. Changing the flavour of a particle changes its mass (Oops!). The bosons are massive which appear to break the boson field Lagrangian's gauge symmetry (Oops!). Only the left-chiral component of the particle wave function is affected by the weak force and so must reflect SU(2) symmetry differently from the right-chiral component (Oops!).

One might be tempted to abandon the whole idea of SU(2). However, the similarity to the U(1) of QED and the SU(3) of QCD is compelling. Some physicists started to wonder whether something was missing. Perhaps all these problems with SU(2) are interrelated. If you think back, the Dirac Lagrangian became U(1) gauge invariant through its interaction with the EM field. It became SU(3) invariant through its interaction with the gluon fields. In the same way, SU(2) local gauge invariance could come from additional interaction terms in the Lagrangian. If something were indeed missing, it would mean that the weak force SU(2) symmetry appears mixed up because we are not seeing the whole picture.

There was a clue. All of the challenges to SU(2) local gauge symmetry are in some way related to *mass*. The differing *mass* of different flavoured fermions. The *mass* of the *W* bosons. And you will learn that the left-chiral and right-chiral components of fermions are linked by – you guessed it – *mass*. Could these mass terms actually be interaction terms with a new unknown field?

Box 24.4   The nutty neutrino

There are three generations of neutrino: the electron neutrino ν_e, the muon neutrino ν_μ and the tau neutrino ν_τ (see Table 23.1). For many years they were believed to be massless. We now know the neutrino oscillates between generations over time: $\nu_e \rightarrow \nu_\mu \rightarrow \nu_\tau$... and such a change in state over particle time means that it must have rest mass. However, this mass is some 500,000 times lighter than the electron. The neutrino is the only elementary fermion with no electric charge and its mass is so small and out of line with the others that many wonder if it actually follows the Dirac equation or might be a *Majorana particle*. The Majorana equation is a radically different solution which could apply in theory for a fermion that is the same as its antiparticle (so electrically neutral). One consequence is that left and right-chiral forms of a Majorana particle can have different rest mass!

A further outstanding question is whether the right-chiral neutrino even exists. If it does, it would not interact with the EM force (it has no electric charge), the strong force (it is not a quark) or the weak force (being right-chiral). This is described as *sterile* because only gravity would affect it. If the neutrino is a Majorana particle (and that is a big *if*), the right-chiral neutrino might be massive and numerous enough to account for *dark matter*. This is the missing invisible 85% of matter that seems to be needed to account for the gravitational behaviour of galaxies.

For these reasons, the neutrino is the subject of much active research.

24.4   Mass, Chirality and the Higgs Field

Before we discuss the rest mass of elementary fermions, let me remind you what mass is. You might remember from Chapter 3 that the bulk of what we call rest mass energy is not resting at all (if needed, look back to Box 3.5). For example, the rest mass energy of the proton is about 900 MeV (million electron volts). This is its energy measured by an observer stationary relative to its centre of mass. The observer is *not* stationary relative to the proton's internal quark and gluon components that are thrashing around, interacting with each other. In fact, only about 1% of the 900 MeV comes from the rest mass energy of the proton's component quarks. That being said, the rest mass energy of the elementary fermions remains extremely important. If the electron and quarks were massless they would move at light speed and the universe would be a very different place.

24.4.1   Mass as an Interaction

It may seems strange to talk of mass as an interaction, but take a close look at how the rest mass of an electron fits into the QED Lagrangian (shown below as it appeared earlier in Equation 20.41). I have rewritten the terms in Equation 24.8 to show the similar construction of the mass and EM interaction terms. The mass m and the coupling constant e are numbers and so are unaffected by their position relative to $\overline{\Psi}$.

$$\mathscr{L}_{QED} = (i\overline{\Psi}\gamma^\mu\,\partial_\mu\Psi - m\,\overline{\Psi}\Psi) - e\,\overline{\Psi}\gamma^\mu A_\mu\,\Psi - \frac{1}{4}F_{\mu\nu}F^{\mu\nu}$$

$$= i\overline{\Psi}\gamma^\mu\,\partial_\mu\Psi - \frac{1}{4}F_{\mu\nu}F^{\mu\nu} - \overline{\Psi}\,m\,\Psi - \overline{\Psi}\,e\gamma^\mu A_\mu\,\Psi \tag{24.8}$$

The last term of Equation 24.8 is the interaction between the electron field and the electromagnetic field. In the language of QFT, an electron may be annihilated (Ψ) and another created ($\overline{\Psi}$) in this interaction. The mass term $\overline{\Psi}\,m\,\Psi$ has the same structure – an electron annihilated (Ψ) and another created ($\overline{\Psi}$) so, it is reasonable to think of the fermion's elementary mass as an interaction term.

There is one notable difference. In the EM interaction term, the value of $e\gamma^\mu A_\mu$ that sits between Ψ and $\overline{\Psi}$ *varies* with the strength and direction of the EM field. It is a *vector* quantity, which is why the photon is a spin-1 *vector boson*. In sharp contrast the mass interaction term is just m, a number. This means that the mass-creating field would have to have a *constant* non-zero *scalar* value (i.e. no direction) throughout time and space. And what of the excitations in this new field? The equivalent of the photon in the EM field would be a spin-0 *scalar boson* in this mass-creating field. It is almost time to say hello to the *Higgs* boson, but first, we must discuss chirality.

24.4.2 Chirality Versus Helicity

I will now try to explain the important difference between *chirality* and *helicity*. This is not easy so get your brains in top gear. I am going to resort to the shorthand of describing particles with spin as *spinning*. In reality, spin is the angular momentum that relates to the angular variation of the wave function. The particles are *not* actually spinning in the classical sense. Don't forget!

Helicity is the projection of spin on the axis of motion using the right-hand rule as shown in Figure 24.3. Helicity is said to be right-handed or positive if the spin axis projects in the direction of motion and is said to be left-handed or negative if not. For a given observer, helicity is conserved. If you measure a particle to have right-handed helicity (as on the left of the figure), then from your perspective that particle will maintain right-handed helicity as it continues on its path.

The bad thing about helicity is that it is not Lorentz invariant. Consider two particles A and B with opposite spin in the stationary frame (linear momentum $\mathbf{p} = 0$) as shown on the right of Figure 24.3. An observer moving to the right sees the particles moving to the left and concludes that A is right-handed helicity while B is left-handed. An observer moving to the left concludes that it is B that is right-handed helicity, not A. The measurement of helicity is not picking up anything fundamental about the particle state. It is *observer-dependent*.

Things are simpler with a massless particle like the photon. The spin axis is *always* either aligned or opposite to the direction of motion (as shown earlier in Figure 17.4). No observer can catch up with or pass a photon. All observers agree on its helicity. There is a fundamental Lorentz invariant difference between the two states. This is described as *chirality*. A massless particle with right-handed helicity is called *right-chiral*. If it has left-handed helicity, it is called *left-chiral*.

It turns out that chirality reflects a mathematical difference in the structure of the wave function. This difference between left-chiral and right-chiral is also true for massive particles such as the spin-half fermions. The best I can do to give you a feel of things is that where the right-chiral wave function rotates through values $+1 \to +i \to -1 \to -i \to +1$, the left-chiral goes $+1 \to -i \to -1 \to +i \to +1$. That seems a bit esoteric, but the important thing is that left-chiral is left-chiral and right-chiral is right-chiral. They are different states for *all* particles with spin, whether massless or massive.

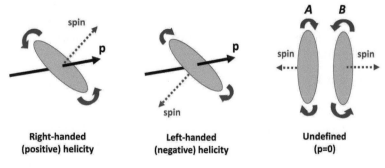

Figure 24.3 The helicity of a massive particle is observer-dependent.

This can be confusing. For a *massless* particle, helicity and chirality go hand in hand. The same is true for neutrinos and antineutrinos that have very small mass and move at close to light speed. In the Wu experiment, there is only about 10^{-12} chance of a mismatch between the chirality and helicity of a neutrino. However, knowing the helicity of a particle with significant rest mass, such as the electron, does not reveal its chirality. In fact, for a slow moving right-handed helicity electron, there is almost a 50:50 chance of it being right-chiral versus left-chiral. One intuitive way to think about chirality is that it is the helicity the particle would have if it were massless. I know that sounds odd, but hopefully it will make more sense shortly.

Box 24.5   The Majorana mystery

Ettore Majorana (1906–?) was an Italian physicist who worked on neutrinos. Widely regarded as a genius, he suffered from depression and isolated himself, living like a hermit for four years. In 1937, he discovered the *Majorana equation* which is a possible description of an electrically neutral fermion like the neutrino (see Box 24.4).

On the 25th of March 1938 he wrote a cryptic message to the director of his institute ending with a comment on his appreciation for all the people he had been working with:

I will keep a fond memory of them all at least until 11 pm tonight, possibly later too.

A letter followed stating that he had decided against suicide, but he disappeared and was never seen again. There are many theories – suicide, kidnapping, seclusion in a monastery. An official investigation in 2015 claimed to establish that he had lived a secluded life in Venezuela. We may never know what happened, and still don't know if the neutrino is a Majorana particle.

24.4.3   Chiral Dirac Equation

As discussed in Chapter 18, the Dirac equation requires a four-component wave function. The components are mixed using 4x4 γ matrices to reflect the energy balance in Equation 24.9 (see Equation 18.71 for a reminder). We need four γ matrices to allow Equation 24.9 to have a solution. Actually there are a total of 15 of these 4x4 γ matrices so there are many ways to express things. Each gives a slightly different way of looking at the same Dirac equation.

$$m = \sqrt{\mathbf{E}^2 - \mathbf{p}_x^2 - \mathbf{p}_y^2 - \mathbf{p}_z^2} = \gamma^0 \mathbf{E} + \gamma^1 \mathbf{p}_x + \gamma^2 \mathbf{p}_y + \gamma^3 \mathbf{p}_z \tag{24.9}$$

Of particular interest to us is the *chiral Dirac equation* that is organised to reflect chirality. A version is shown below. It is sometimes called the Weyl representation. I have not listed the precise γ matrices needed for the calculation, but you can easily find them online. Why is this representation interesting and what does it have to do with chirality? The telltale sign is what happens when you set mass $m = 0$.

$$m\,\Psi_1 = \hat{\mathbf{E}}\,\Psi_3 + \hat{\mathbf{p}}_x\,\Psi_4 - i\,\hat{\mathbf{p}}_y\,\Psi_4 + \hat{\mathbf{p}}_z\,\Psi_3 \tag{24.10}$$

$$m\,\Psi_2 = \hat{\mathbf{E}}\,\Psi_4 + \hat{\mathbf{p}}_x\,\Psi_3 + i\,\hat{\mathbf{p}}_y\,\Psi_3 - \hat{\mathbf{p}}_z\,\Psi_4 \tag{24.11}$$

$$m\,\Psi_3 = \hat{\mathbf{E}}\,\Psi_1 - \hat{\mathbf{p}}_x\,\Psi_2 + i\,\hat{\mathbf{p}}_y\,\Psi_2 - \hat{\mathbf{p}}_z\,\Psi_1 \tag{24.12}$$

$$m\,\Psi_4 = \hat{\mathbf{E}}\,\Psi_2 - \hat{\mathbf{p}}_x\,\Psi_1 - i\,\hat{\mathbf{p}}_y\,\Psi_1 + \hat{\mathbf{p}}_z\,\Psi_2 \tag{24.13}$$

When $m = 0$, Equations 24.10 and 24.11 involve only the Ψ_3 and Ψ_4 components of the wave function. Similarly, Equations 24.12 and 24.13 involve only the Ψ_1 and Ψ_2 components. Instead of one interlinked relationship involving four equations, we now have two independent relationships each involving two equations. For example, when $m = 0$, Equations 24.12 and 24.13 can be written as:

$$\hat{\mathbf{E}}\,\Psi_1 = \hat{\mathbf{p}}_x\,\Psi_2 - i\,\hat{\mathbf{p}}_y\,\Psi_2 + \hat{\mathbf{p}}_z\,\Psi_1 \tag{24.14}$$

$$\hat{\mathbf{E}}\,\Psi_2 = \hat{\mathbf{p}}_x\,\Psi_1 + i\,\hat{\mathbf{p}}_y\,\Psi_1 - \hat{\mathbf{p}}_z\,\Psi_2$$

The two equations in 24.14 are a massless form of the Dirac spinor called a *Weyl spinor* (him again!). $[\Psi_1, \Psi_2]$ is a *two component spinor* obeying[2] the energy balance of a massless particle which is: $\mathbf{E} = \mathbf{p}$. As it is moving at light speed, all observers agree on its chirality which is also its helicity. To make things clearer, let me relabel $[\Psi_1, \Psi_2]$ as $[l_1, l_2]$ which is *left* chiral and $[\Psi_3, \Psi_4]$ as $[r_1, r_2]$ which is *right* chiral (also see Box 24.6). The Dirac equation becomes:

$$m \begin{bmatrix} l_1 \\ l_2 \end{bmatrix} = \hat{\mathbf{E}} \begin{bmatrix} r_1 \\ r_2 \end{bmatrix} + \hat{\mathbf{p}}_x \begin{bmatrix} r_2 \\ r_1 \end{bmatrix} + i\hat{\mathbf{p}}_y \begin{bmatrix} -r_2 \\ r_1 \end{bmatrix} + \hat{\mathbf{p}}_z \begin{bmatrix} r_1 \\ -r_2 \end{bmatrix} \tag{24.15}$$

$$m \begin{bmatrix} r_1 \\ r_2 \end{bmatrix} = \hat{\mathbf{E}} \begin{bmatrix} l_1 \\ l_2 \end{bmatrix} - \hat{\mathbf{p}}_x \begin{bmatrix} l_2 \\ l_1 \end{bmatrix} - i\hat{\mathbf{p}}_y \begin{bmatrix} -l_2 \\ l_1 \end{bmatrix} - \hat{\mathbf{p}}_z \begin{bmatrix} l_1 \\ -l_2 \end{bmatrix}$$

Box 24.6   The chiral Dirac equation and the Pauli matrices

The mass linkage allows the chiral Dirac equation to be expressed using the three 2x2 Pauli matrices (these matrices are shown back in Equation 18.17). This is shown below using the labels $\Psi_L = [l_1, l_2]$ and $\Psi_R = [r_1, r_2]$

$$m\Psi_L = \hat{\mathbf{E}}\,\Psi_R + \sigma_x\,\hat{\mathbf{p}}_x\Psi_R + \sigma_y\,\hat{\mathbf{p}}_y\Psi_R + \sigma_z\,\hat{\mathbf{p}}_z\Psi_R$$

$$m\Psi_R = \hat{\mathbf{E}}\,\Psi_L - \sigma_x\,\hat{\mathbf{p}}_x\Psi_L - \sigma_y\,\hat{\mathbf{p}}_y\Psi_L - \sigma_z\,\hat{\mathbf{p}}_z\Psi_L$$

Let's put the maths to one side and think what this tell us. Without the mass term, the Dirac equation describes two distinct unconnected chiral particles: one left-chiral and the other right-chiral. The m mass term connects the two and pulls them together into one particle. A massive spin-half fermion is a connected mix of left-chiral and right-chiral; the strength of that connection is the fermion's *rest mass*.

Based on this, it clearly makes sense to think of the rest mass term in the fermion Lagrangian as an *interaction* term. The weak force interacts with left-chiral but not right-chiral, so they behave very differently. Effectively we have two distinct chiral quantum fields that are connected, which means that there is some interaction linking them together. This interaction is with the Higgs field and is shown in Equation 24.16. It is called *Yukawa coupling* after the Japanese physicist Hideki Yukawa.

$$y\,\overline{\Psi}\Psi\,\phi_H \qquad\qquad \textit{Yukawa coupling} \tag{24.16}$$

The coupling constant, typically labelled y for Yukawa, determines the strength for each fermion of the coupling between its left and right-chiral forms, and thus its rest mass. ϕ_H is the Higgs field.

24.5   The Story So Far

We are half way through the story of the weak force and the Higgs field so let's take a pause for breath and a moment to collect our thoughts. We know that the EM force reflects U(1) local gauge invariance to a change in phase. We know that the strong force reflects SU(3) local gauge invariance to a change in colour. Therefore, there is every reason to believe that the weak force is linked in some way with SU(2) local gauge invariance to a change in flavour.

Furthermore, there is a (somewhat distorted) symmetry between different flavours of particle and there are three weak force bosons which is exactly what you would expect. The overarching pattern is the same as for the EM and strong forces. Surely this cannot be a coincidence.

However, as we have seen, experimental evidence conflicts with an idealised model of the weak force as a standalone source of SU(2) local gauge invariance to a change in flavour. In the first place, different flavours of fermion, such as the up and down quarks, have different rest mass. In addition, the weak force bosons are massive which we know is inconsistent with a simple model of gauge invariance (not even *global* gauge invariance as discussed in Section 20.9).

This prompts the question: *are we missing something?* Physicists started to suspect that the Dirac Lagrangian of fermions *is* SU(2) local gauge invariant, but we are missing part of it – a part that would restore this gauge invariance. Bear in mind that the interaction with the EM field makes the fermion Lagrangian U(1) local gauge invariant (to phase change). If you did not know about that EM interaction term, you would have an incomplete picture of the Lagrangian and would not see the U(1) gauge invariance.

What might the missing interaction term be? The clues point to a connection with *mass*. Different flavours of fermion have different *mass*. The weak force bosons unexpectedly have *mass*. These suspicions were reinforced when Wu discovered that the weak force only interacts with left-chiral fermions. The chiral Dirac equation shows left-chiral and right-chiral fermions are coupled together by *mass*.

In the next chapter we will show that these suspicions were correct. The rest mass terms in the fermion Lagrangians are actually interaction terms with another field and that field has exactly the right configuration to restore SU(2) local gauge invariance... it is time to discuss in detail the Higgs field and the Higgs boson.

Notes

1 If this final step confuses you, bear in mind that the anti-neutrino is equivalent to a neutrino moving back in time so the vertex is analogous to a neutrino absorbing a W_- boson and changing flavour to an electron.

2 To check, use the first line of 24.14 to create an expression for Ψ_1 in terms of Ψ_2, then substitute it into the second.

Chapter 25

The Weak Force and Higgs Field (2)

CHAPTER MENU

25.1 The Higgs Interaction
25.2 The Higgs Field and Mechanism
25.3 The Maths of the Higgs Field
25.4 Visualising the Higgs Field
25.5 Spontaneous Symmetry Breaking
25.6 The Maths of the Higgs Mechanism
25.7 The Discovery of the Higgs Boson
25.8 Electroweak Unification
25.9 Summary

25.1 The Higgs Interaction

We are trying to understand the rest mass of an elementary fermion like the electron. In Section 24.4.1 we saw how the rest mass term in the Lagrangian has the structure of an interaction term. In Section 24.4.3 we learned that the fermion's rest mass m is the coupling between its left-chiral and right-chiral components. This coupling is the interaction with the Higgs field. Let's look at this in more detail.

It may help you to think of this as a step-by-step process. For example, for an electron we start with a left-chiral electron which interacts with the Higgs field to become a right-chiral electron and then interacts again to become left-chiral, and so on. The left side of Figure 25.1 gives a simple illustration. Each cross is an interaction with the Higgs field. The quark's greater rest mass means it is interacting more with the Higgs field. Its coupling constant with the Higgs field is higher. The effect of these interactions is to give the particles inertia, a little like wading through treacle.

By now, I am confident that you know enough not to take this step-by-step picture too literally. It is a helpful cartoon portrayal of the Higgs interaction, but in reality the behaviour of a fermion is a composite of every possible development pathway, not a sequence of individual left-chiral and right-chiral steps. Note in the figure that the massless photon does not interact with the Higgs field. Its chirality is the same as its helicity and does not change.

The changes in chirality of the fermions do *not* affect helicity which, being a form of angular momentum, is conserved. This means that helicity *is not* Lorentz invariant, but *is* conserved, while chirality *is* Lorentz invariant, but *is not* conserved (because it is constantly changing through the Higgs interaction). Confused? You are not alone, this is tricky stuff.

The continuous chiral switching via the Higgs field changes the Action of the fermion. Cast your mind back to the free particle wave function. Rest mass m is its rate of change over *particle* time (Section 5.9). In Chapter 6, I showed the link between the Action of the Lagrangian and Newton's equations of motion. Thus, the coupling with the Higgs field becomes the m in $F = ma$ for the elementary fermions.

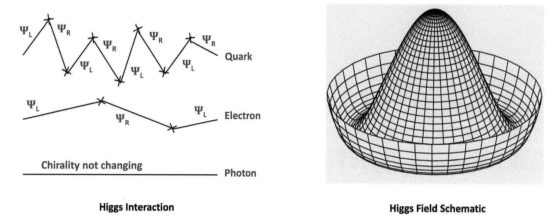

Figure 25.1 Schematics of Higgs interaction and field.

25.2 The Higgs Field and Mechanism

If rest mass is in reality the Higgs interaction term then the Higgs field must be a *scalar field* with an underlying minimum *non-zero* value. *Scalar* because, as discussed in Section 24.4.1, mass is a scalar quantity unaffected by direction. *Non-zero* because rest mass is ever-present, so the interaction must occur throughout space and time.

Before we dig into deeper mathematics, I think it will help if I give a road map of how the argument is going to unfold. Physicists were puzzled by the three massive weak force vector bosons. If there is SU(2) local gauge invariance in the Dirac Lagrangian, you would expect the bosons to be massless, so where does their mass come from? A scalar field can provide a natural solution because it follows the Klein-Gordon Lagrangian. For local gauge invariance, its derivatives must be covariant. This produces an interaction term that is different from that of the Dirac Lagrangian. The two are compared below.

$$\mathcal{L}_{Dirac} = i\overline{\Psi}\gamma^{\mu}(\partial_{\mu}\Psi + igA_{\mu}\Psi) - m\overline{\Psi}\Psi - \frac{1}{4}F_{\mu\nu}F^{\mu\nu}$$

$$= i\overline{\Psi}\gamma^{\mu}\partial_{\mu}\Psi - g\overline{\Psi}\gamma^{\mu}A_{\mu}\Psi + ... \tag{25.1}$$

$$\mathcal{L}_{KG} = \frac{1}{2}(\partial^{\mu}\Psi^{*} - igA^{\mu}\Psi^{*})(\partial_{\mu}\Psi + igA_{\mu}\Psi) - V_{KG} - \frac{1}{4}F_{\mu\nu}F^{\mu\nu}$$

$$= \frac{1}{2}\partial^{\mu}\Psi^{*}\partial_{\mu}\Psi + \frac{1}{2}g^{2}A^{\mu}A_{\mu}\Psi^{*}\Psi + ... \tag{25.2}$$

$$= \frac{1}{2}\partial^{\mu}\Psi^{*}\partial_{\mu}\Psi + \frac{1}{2}v^{2}g^{2}A^{\mu}A_{\mu} + ... \qquad if \ \Psi = v \ \ so \ \ \Psi^{*}\Psi = v^{2} \tag{25.3}$$

Equation 25.1 highlights the interaction term when the Dirac Lagrangian of a *fermion* field is combined with a vector field to give local gauge invariance as is the case for the QED Lagrangian. You should recognise the QED interaction term between fermion ($\overline{\Psi}\Psi$) and photon (A_{μ}). It is the same as Equation 20.40 (the coupling constant is labelled e in QED).

Equation 25.2 shows the interaction term for the Klein-Gordon Lagrangian of a *scalar* field to be local gauge invariant (we derived this in Chapter 20 as Equation 20.33). Note that I have generalised the potential of the scalar field as V_{KG} rather than assuming it to be a mass term. Equation 25.3 shows what happens if the scalar field has a non-zero constant base value, illustrated here as: $\Psi = v$ (I will explain this label later). As both the coupling constant g and the base value v are constants, the interaction term includes a *constant* $A^{\mu}A_{\mu}$ term. This is a mass term for the A_{μ} vector field. The $v^{2}g^{2}A^{\mu}A_{\mu}$ term means that the vector boson is *massive* in exactly the same way that a $\frac{1}{2}m^{2}\Psi^{*}\Psi$ term in the Klein-Gordon Lagrangian indicates a massive scalar particle.

The illustration above is for a single vector field (one boson), but the same argument holds for the three vector fields (three bosons) of the weak force. To reiterate, the covariant derivative of a Dirac Lagrangian creates interaction terms involving *massless* bosons. Massive vector bosons would destroy the gauge invariance (as discussed in Section 20.9). However, the covariant derivative of a Klein-Gordon Lagrangian for a scalar field with a constant non-zero minimum value involves *massive* vector bosons in order to be local gauge invariant. The Higgs is just such a scalar field and this is why the weak force bosons are massive.

25.3   The Maths of the Higgs Field

I will use the label ϕ_H for the Higgs field. A key feature of the field is that its base value in its ground state is not zero. This is called a non-zero *vacuum expectation value (v.e.v)* and it is typically labelled v (warning: an alternative convention is to label this $\frac{v}{\sqrt{2}}$). The use of a little v is a bit confusing because it is actually a *big* number – so, do not forget – little v is a BIG constant. I will label fluctuations in the value of the field around this base value as H. These fluctuations are small relative to v so $|\phi_H| = v + H$ with v *much larger* than H.

If the Higgs field is to play the overarching role behind SU(2) local gauge invariance, it must be a *doublet* field in the same way that the left-chiral fermion is a flavour doublet (as described in Section 24.1). Consider the equations in Box 24.6. On one side is a flavour doublet Ψ_L and on the other a flavour singlet Ψ_R. The Higgs interaction term (the mass m term) needs to act as a doublet for the equation to match. If you prefer, think about weak charge. Ψ_L carries weak charge. Ψ_R doesn't. In the $\Psi_L \rightarrow \Psi_R$, interaction, that weak charge must be absorbed by the Higgs field. If the Higgs force carries weak charge, it is affected by changes in the flavour catalogue, so it must have the doublet structure shown in Equation 25.4.

$$\phi_H = \begin{bmatrix} a + ib \\ c + id \end{bmatrix} = e^{i(\alpha_1 T_1 + \alpha_2 T_2 + \alpha_3 T_3)} \begin{bmatrix} 0 \\ v + H \end{bmatrix} \tag{25.4}$$

It is a scalar field, so the doublet is two complex numbers. It has four components or *degrees of freedom* shown above as real numbers a, b, c, d. The doublet looks tricky to handle, but there is a cunning simplification that will help us. The Higgs field is a doublet so we can use SU(2) symmetry. Without changing the magnitude $|\phi_H|$, we can select the flavour catalogue which leads to the top component being zero and the real part of the bottom component containing the value of the field so $c = v + H$ (Equation 25.4). T_1, T_2 and T_3 are the generator matrices for SU(2) as described back in Equation 24.4. Setting the right values of α_1, α_2 and α_3 gives every possible SU(2) matrix, so with the correct values, the expressions on the right and left in Equation 25.4 will match. The four degrees of freedom are now the four real numbers α_1, α_2, α_3 and $(v + H)$.

The ground state value of the Higgs field must be non-zero. Peter Higgs and others hypothesised that this is because the potential energy V_H of the Higgs field (not to be confused with the base value v of the field) is as shown in Equation 25.5 where μ^2 and λ are real constants (*warning*: there are different conventions so you may see slightly different versions). In the Higgs ground state, the potential energy is at its lowest and the ground state base value v of the Higgs field is such that $v^2 = \frac{\mu^2}{\lambda}$.

$$V_H = -\frac{\mu^2}{2}(\phi_H^* \phi_H) + \frac{\lambda}{4}(\phi_H^* \phi_H)^2 \tag{25.5}$$

Equation 25.6 shows how to calculate this. If you differentiate the potential energy with respect to the magnitude squared of the Higgs field $\phi_H^* \phi_H$, the derivative will be zero at the minimum. This gives the ground state value v^2 of $\phi_H^* \phi_H$ in terms of μ and λ.

$$\frac{d(V_H)}{d(\phi_H^* \phi_H)} = -\frac{\mu^2}{2} + \frac{\lambda}{2}(\phi_H^* \phi_H) = 0 \qquad at\ minimum\ for\ V_H$$

$$\implies \quad in\ ground\ state \quad (\phi_H^* \phi_H) = \frac{\mu^2}{\lambda} = v^2 \tag{25.6}$$

25.4   Visualising the Higgs Field

On the right of Figure 25.1 there is a simple schematic portrayal of the Higgs field. It is often described as the *Mexican Hat* potential because of its resemblance to a sombrero. The potential V_H of the Higgs field (shown vertically) initially drops as the value of the Higgs field (shown horizontally) increases from zero at the centre. This initial drop is driven by the negative μ^2 term in the potential (see Equation 25.5). However, at higher values of the Higgs field, the potential increases as the λ term dominates.

The potential is at a minimum when the magnitude of the Higgs field is the base value $|\phi_H| = v$. This is the gully in the schematic of Figure 25.1. Consider a change in the value of ϕ_H away from the centre of the schematic. From now on, I will refer to this as a *radial* fluctuation in the value of ϕ_H. Any radial shift from the gully will increase the potential, which will create a restoring force back to $|\phi_H| = v$. Therefore *radial* fluctuations will lead the value of the field to oscillate radially above and below the base value v in the same way that pushing a small ball bearing in the gully away from the centre would make it rock up and down the sides of the gully.

It is important to understand that the schematic in Figure 25.1 has nothing to do with space or time. It represents the possible value and potential of the Higgs field at one single point in space and time. Why then you might ask is the field potential shown as circular? It is to show schematically that for every absolute value $|\phi_H|$ there are an infinite number of possible values of ϕ_H because it is a *doublet*.

If this puzzles you, look back at Equation 25.4. Think of the Higgs field in its ground state $|\phi_H| = v$, so our imaginary ball bearing is in the gully. There is an infinite number of possible values for the a, b, c, d of the ϕ_H doublet for which $\phi_H^* \phi_H = v^2$ and so $|\phi_H| = v$. You can multiply it by any one of the infinite number of SU(2) matrices. As shown in Equation 25.4 we can express all possible options using the three SU(2) generator matrices and varying the values of α_1, α_2 and α_3.

The circular gully in Figure 25.1 represents all the possible doublet values for the Higgs field in its ground state that we can identify by varying *one* of the generators. Consider the T_1 generator matrix as an example, shown in Equation 25.7. Changing the value of α_1 is like the ball bearing rolling around the gully. From now on, I will refer to this as a *rotational* fluctuation in the value of ϕ_H. You can think of it as a rotation in SU(2) space. The value of the doublet ϕ_H changes, but the magnitude $|\phi_H|$ remains unchanged.

$$e^{i(\alpha_1 T_1)} \begin{bmatrix} 0 \\ v + H \end{bmatrix} = \phi_H \quad \Longrightarrow \quad \phi_H^* \phi_H = (v + H)^2 \qquad \textit{for any value of } \alpha_1 \tag{25.7}$$

This means that there are in reality three independent rotations through SU(2) space, one for each generator matrix (i.e. three gullies). All the possible values of ϕ_H that give the same $|\phi_H|$ can be expressed through a combination of these three rotations through SU(2) space, varying α_1, α_2 and α_3. The schematic in Figure 25.1 shows only one because there is no obvious way to picture all three in the same image.

25.5 Spontaneous Symmetry Breaking

The fact that the Higgs doublet ϕ_H has a particular value *matters*. In the ground state, ϕ_H can be any of an infinite number of doublets. But which? They are all equivalent in terms of potential. There is perfect SU(2) symmetry (the rotational symmetry of the schematic). However, ϕ_H must have a value and that value places it somewhere specific in the gully, and the circular SU(2) symmetry is broken.

You may read all sorts of comparative examples. A pencil standing on its tip has circular symmetry, but once it has fallen that symmetry is broken. My favourite is to compare the Higgs field potential to a roulette wheel. The roulette croupier spins the ball around the wheel. There is perfect symmetry and the betting continues. But as the energy of the ball dissipates, there is a shout of *les jeux sont faits* and the betting stops. Finally the ball settles on a number. The symmetry is gone. A perfectly symmetrical system has produced an asymmetric result. That is spontaneous symmetry breaking.

The fact that ϕ_H can change without changing $|\phi_H|$, and therefore with no change to the potential V_H, is important. These are *rotational* fluctuations in the Figure 25.1 schematic (such as swinging around the gully). They are excitations that involve a change in the value of the quantum field ϕ_H with no linked change in the potential V_H. This is the maths of massless particles. These problematic massless particles were called *Goldstone bosons*. They perplexed physicists for some time. A scalar field such as the Higgs solved so many issues, but if associated massless particles really existed, they would surely have been spotted in experiments.

The breakthrough is the Higgs mechanism or, more correctly, the Brout-Englert-Higgs mechanism as it was discovered by a number of physicists at the same time in the 1960s. The solution is simple and elegant. The rotational excitations that might lead to massless particles can be combined with those related to the weak force bosons to give a clearer view of what is happening. The result is massive weak force bosons and *no* massless Goldstone bosons. Let's look at how this works.

25.6 The Maths of the Higgs Mechanism

The aim of this section is to build SU(2) local gauge invariance into the Higgs field and show why this leads to three massive weak force bosons plus a massive scalar Higgs boson. To do this, we need to construct the Lagrangian of the Higgs field.

There are four steps to this. I have divided them into separate short subsections for clarity. Subsection 25.6.1 starts to build the SU(2) local gauge invariant form of the Higgs Lagrangian based on the Klein-Gordon Lagrangian. Subsection 25.6.2 quantifies the potential in the Higgs Lagrangian in terms of *radial* fluctuations of the Higgs field. Subsection 25.6.3 simplifies things by absorbing *rotational* fluctuations of the Higgs field into the weak force vector fields. This eliminates the terms that were feared to lead to Goldstone massless bosons. Subsection 25.6.4 pulls things together so that, hopefully, you end up with a clear idea of what is going on.

25.6.1 The Starting Point

We want to construct the Higgs Lagrangian including SU(2) local gauge invariance. The starting point is the Klein-Gordon Lagrangian from Equation 25.2 which by design is SU(2) local gauge invariant (note that texts using the convention of v.e.v. as $\frac{v}{\sqrt{2}}$, absorb the $\frac{1}{2}$ factor in the Lagrangian derivative into the value of the field). SU(2) local gauge invariance requires three vector fields, but for simplicity we will show only one, which I have labelled S_μ rather than A_μ (for no other reason than to distinguish it from the vector field in QED). We can substitute in the expression for the potential V_H of the Higgs field from Equation 25.5. We now have the Higgs Lagrangian in a form that is SU(2) local gauge invariant.

$$\mathscr{L}_H = \frac{1}{2}(\partial^\mu \phi_H^* - ig\,S^\mu \phi_H^*)(\partial_\mu \phi_H + ig\,S_\mu \phi_H) - V_H - \frac{1}{4}F_{\mu\nu}F^{\mu\nu} \tag{25.8}$$

$$= \frac{1}{2}\partial^\mu \phi_H^* \partial_\mu \phi_H + \frac{1}{2}g^2\,S^\mu S_\mu\,\phi_H^* \phi_H - \left(-\frac{\mu^2}{2}(\phi_H^* \phi_H) + \frac{\lambda}{4}(\phi_H^* \phi_H)^2\right) - \frac{1}{4}F_{\mu\nu}F^{\mu\nu}$$

25.6.2 The Potential of the Higgs Field

We can express the Higgs potential V_H in terms of *radial* fluctuations in the value of the Higgs field. If you look at Figure 25.1 you can see that the potential is affected only by changes in the radial value of the Higgs field. Rotational changes do not alter the potential.

We have set the radial value of the Higgs field as $|\phi_H| = (v + H)$. Remember that v is *big* and H is *small*. The calculation of $(\phi_H^* \phi_H)$ and $(\phi_H^* \phi_H)^2$ in terms of v and H is shown in Equations 25.9 and 25.10. As the fluctuations H are very small, we can ignore the H^3 and H^4 terms.[1]

$$\phi_H^* \phi_H = (v + H)^2 = v^2 + 2vH + H^2 \tag{25.9}$$

$$(\phi_H^* \phi_H)^2 = (v + H)^4 = v^4 + 4v^3H + 6v^2H^2 + 4vH^3 + H^4$$

$$\approx v^4 + 4v^3H + 6v^2H^2 \qquad \textit{for small } H \tag{25.10}$$

We can use this to calculate changes to the Higgs potential (as in Equation 25.5) in terms of H. The result is shown in Equation 25.11. We are interested in the variation in the potential, so we can ignore the constant C. As you can see, the variation in the potential V_H for a fluctuation H in the value of $|\phi_H|$ is $\mu^2 H^2$. This will turn out to be the mass term for the Higgs boson.

$$V_H = -\frac{\mu^2}{2}(\phi_H^* \phi_H) + \frac{\lambda}{4}(\phi_H^* \phi_H)^2 \qquad \textit{use } \lambda = \frac{\mu^2}{v^2} \textit{ from Equation 25.6}$$

$$= -\frac{\mu^2}{2}(\phi_H^* \phi_H) + \frac{\mu^2}{4v^2}(\phi_H^* \phi_H)^2 \qquad \textit{substitute from Equations 25.9 and 25.10}$$

$$= -\frac{\mu^2}{2}(v^2 + 2vH + H^2) + \frac{\mu^2}{4v^2}(v^4 + 4v^3H + 6v^2H^2)$$

$$= \mu^2\left(-\frac{1}{2}v^2 - vH - \frac{1}{2}H^2 + \frac{1}{4}v^2 + vH + \frac{3}{2}H^2\right)$$

$$= \mu^2 H^2 - \frac{1}{4}\mu^2 v^2 = \mu^2 H^2 - C \qquad \textit{where C is a constant} \tag{25.11}$$

We insert this value (ignoring the constant) into the Higgs Lagrangian from Equation 25.8. For simplicity, only one of the three S_μ fields is shown.

$$\mathscr{L}_H = \frac{1}{2}\,\partial^\mu \phi_H^* \partial_\mu \phi_H + \frac{1}{2}\,g^2\,S^\mu S_\mu\,\phi_H^*\phi_H - \mu^2 H^2 - \frac{1}{4}\,F_{\mu\nu}\,F^{\mu\nu} \tag{25.12}$$

25.6.3 Rotational Fluctuations of the Higgs Field

We now need to consider what I have called *rotational* fluctuations in the Higgs field as portrayed in Figure 25.1. These have nothing to do with rotations in spacetime. They are rotations in SU(2). These variations in the doublet structure of ϕ_H do not change the value of $|\phi_H| = (v + H)$. Therefore they do not change the value of the potential V_H.

We can express the Higgs field in terms of the generator matrices and $|\phi_H|$. I am going to simplify by including the gauge invariance for only *one* of the three SU(2) T matrix generators. The result shown in Equation 25.13 comes directly from Equation 25.7. I also have simplified the symbols to avoid doublets in the equations. The radial value $(v + H)$ indicates the absolute value $|\phi_H|$ of the field. The value $e^{i\alpha T}$ indicates where it sits among all the possible SU(2) ϕ_H doublets.

$$\phi_H = e^{i\alpha T}\begin{bmatrix} 0 \\ v + H \end{bmatrix} = e^{i\alpha T}\,(v + H) \tag{25.13}$$

The next step is to get rid of those pesky Goldstone bosons by using the fact that our Higgs Lagrangian is SU(2) local gauge invariant (as shown in Equation 25.8). We simplify the expression for ϕ_H enormously by applying what is called a *unitary gauge change*. We change the flavour catalogue by multiplying with $e^{-i\alpha T^*}$. T^* is the inverse matrix of the generator matrix T. This means that T^*T gives the identity matrix. The Higgs scalar field simply becomes $(v + H)$.

$$e^{-i\alpha T^*}\,\phi_H = e^{-i\alpha T^*}\,e^{i\alpha T}\,(v + H) = (v + H) \tag{25.14}$$

This changes the expression for the vector field (it is this offsetting change that creates the gauge invariance). The change is of the same form as for QED (Equation 20.26). The effect on ϕ_H and on the vector field is summarised in Equation 25.15. I have labelled the new expression for the vector field W_μ to distinguish it from the original S_μ. This will become one of the weak force bosons. It now contains the additional *variable* α. This is important because it gives an additional degree of freedom which we will need in a moment. Crucially, this cunning change of flavour catalogue does *not* alter the Higgs Lagrangian (this is what local gauge invariance means).

$$\phi_H \to e^{-i\alpha T^*}\,\phi_H = v + H \qquad\qquad S_\mu \to S_\mu + \frac{1}{g}\,\partial_\mu \alpha T^* = W_\mu \tag{25.15}$$

Let me remind you again that there are actually three weak force vector fields, one for each SU(2) generator. The unitary gauge change that we have made involves all three generators and changes each of the three vector fields into W_1, W_2 and W_3 as shown below:

$$\phi_H \to e^{-i\alpha_1 T_1^*}\,e^{-i\alpha_2 T_2^*}\,e^{-i\alpha_3 T_3^*}\,\phi_H = v + H \tag{25.16}$$

$$W_1 = S_1 + \frac{1}{g}\,\partial_\mu \alpha_1 T_1^* \qquad W_2 = S_2 + \frac{1}{g}\,\partial_\mu \alpha_2 T_2^* \qquad W_3 = S_3 + \frac{1}{g}\,\partial_\mu \alpha_3 T_3^*$$

25.6.4 Putting It All Together

The SU(2) invariant Higgs Lagrangian is much easier to calculate after this gauge change. Equation 25.17 shows the Lagrangian before the gauge change as expressed earlier in Equation 25.12. Again, for simplicity, only one of the three S_μ vector fields is shown. The effect of the unitary gauge change is shown on the following line. ϕ_H changes to $(v + H)$ and S_μ changes as in Equation 25.15. We relabel it W_μ to reflect this. The vacuum expectation value v is a constant so, it can be ignored in the derivatives (Equation 25.18). Some simple maths reveals the key terms of the Lagrangian as shown in Equation 25.19:

$$\mathscr{L}_{H/SU2} = \frac{1}{2}\partial^\mu\phi_H^*\partial_\mu\phi_H + \frac{1}{2}g^2 S^\mu S_\mu \phi_H^*\phi_H - \mu^2 H^2 - \frac{1}{4}F_{\mu\nu}F^{\mu\nu} \tag{25.17}$$

$$= \frac{1}{2}\partial^\mu(v+H)\partial_\mu(v+H) + \frac{1}{2}g^2 W^\mu W_\mu (v^2 + 2vH + H^2) - \mu^2 H^2 - \frac{1}{4}F_{\mu\nu}F^{\mu\nu}$$

$$= \frac{1}{2}\partial^\mu H\partial_\mu H + \frac{1}{2}g^2 W^\mu W_\mu (v^2 + 2vH + H^2) - \mu^2 H^2 - \frac{1}{4}F_{\mu\nu}F^{\mu\nu} \tag{25.18}$$

$$= \frac{1}{2}\partial^\mu H\partial_\mu H - \mu^2 H^2 + \frac{1}{2}v^2 g^2 W^\mu W_\mu + 2vW^\mu W_\mu H + W^\mu W_\mu H^2 - \frac{1}{4}F_{\mu\nu}F^{\mu\nu}$$

$$= \frac{1}{2}\partial^\mu H\partial_\mu H - \mu^2 H^2 + \frac{1}{2}v^2 g^2 W^\mu W_\mu - \frac{1}{4}F_{\mu\nu}F^{\mu\nu} + interaction\ terms \tag{25.19}$$

Take a close look at Equation 25.19. This is the SU(2) local gauge invariant form of the Lagrangian of the Higgs scalar field. Lets begin with the first two terms. The first is the radial fluctuation in the Higgs field and the second is a mass term for those fluctuations. This is a *massive* scalar boson associated with radial fluctuations of the Higgs field. It is the *Higgs boson*.

In the third term, note that $v^2 g^2$ is a constant so this is a term for a *massive* vector boson. Don't forget there are three similar interaction terms. The SU(2) local gauge invariance of the Higgs field results in three *massive* weak force bosons. Sharp students might note that a massless spin-1 particle has only two spin states whereas a massive spin-1 particle has three (check back to Section 17.3 if you need to). This extra degree of freedom comes from the variables α_1, α_2 and α_3 that we have incorporated into the expressions for the three W vector bosons (see Equation 25.16).

The terms $W^\mu W_\mu H$ and $W^\mu W_\mu H^2$ are additional interaction terms between these weak force bosons and the Higgs field. I am not going to say any more about these or the H^3 and H^4 Higgs self-interaction terms mentioned in the footnote in Subsection 25.6.2, although they are no doubt of great interest to my good friend Matt von Hippel, who is studying the scattering amplitudes of gauge theories and was a tremendous help editing the advanced sections of this book.

What of the massless Goldstone bosons? There is no sign of them. The $e^{i\alpha T}$ rotational fluctuations in the Higgs field (see Figure 25.1) are included in the expressions for the weak bosons. This may strike you as strange, but it makes sense. These fluctuations involve rotations in SU(2) space and so do the vector boson interaction terms. The gauge change makes things much clearer by gathering together the effect of these SU(2) rotations into the vector boson terms.

Let me illustrate by again comparing with a roulette wheel. It has an inner wheel with the gully and numbers. This inner wheel sits in an outer circular bowl. The croupier spins the inner wheel in one direction and spins the ball around the outer bowl in the opposite direction. Imagine that you try to predict what number the roulette ball will land on. You analyse the maths of the ball's motion around the outer bowl. You analyse the way that the inner wheel's motion is changing the numbers in the gully. Clearly, the simplest approach is to analyse the motion of the ball *relative* to the inner wheel by combining the two motions. This does not change the maths. It simplifies it.

The unitary gauge change is similar. Think of the spin of the roulette ball as the rotational excitations of the Higgs field which are effectively SU(2) shifts in the Higgs doublet. Think of the spin of the inner roulette wheel as the effect of the bosons changing the SU(2) flavour catalogue. Combining the two into the expression for the bosons does not change the maths, but it makes it much easier to interpret, although it took a while for Higgs' idea to get attention (see Box 25.1).

Box 25.1   Of no obvious relevance

In 1964, Peter Higgs sent the original version of his paper outlining the Higgs mechanism for publication by *Physics Letters* in Geneva. It was rejected on the grounds that it did *not warrant rapid publication* being of *no obvious relevance to physics*. Higgs persevered and another version was published later that year, the contents of which led to him winning the 2013 Nobel Prize – proving that persistence pays! Others have had similar experiences. Enrico Fermi's seminal 1933 paper on the weak force was rejected as *too remote from reality*. The work won him the 1938 Nobel Prize.

25.7   The Discovery of the Higgs Boson

The Higgs mechanism predicts the existence of a massive scalar boson associated with what I am calling radial excitation of the Higgs field. The hunt for the *Higgs boson* was afoot! The Higgs field and mechanism were proposed in the 1960s and it took a further fifty years before the existence of the Higgs boson was confirmed at the Large Hadron Collider (LHC). It took that long because there were two major challenges. The first was to make enough Higgs bosons and the second was to build sensitive enough detectors to verify their presence.

The rest mass energy of the Higgs bosons is about 125 GeV, but it takes more energy than you might think to create a decent number of them. For the Higgs discovery, the two LHC beams were each at 4 TeV for a combined total of 8 TeV which is 8,000 GeV. This is because, in order to get energy into the Higgs field, you need the strongest possible interaction with it, which means that ideally you want to use *top quarks* because they have the highest rest mass energy of the elementary fermions at about 170 GeV, almost 1000× greater than that of an up quark. Don't forget that the coupling constant between the Higgs and the elementary fermions depends on their rest mass energy.

The proton beams of the LHC smash together leading to the fusion of gluons. At high enough energy these create *top* quark-antiquark pairs which in turn create Higgs bosons. The Higgs boson decays rapidly and the output is analysed in the LHC detectors. The top diagram in Figure 25.2 is the *4-lepton channel*, nicknamed the *golden pathway.* From left to right, two gluons become a top quark–antiquark pair that becomes a Higgs boson which decays into a Z boson–antiboson pair (labelled Z and Z^*). These decay into four leptons, shown here as electron-positron pairs.

The right of the figure shows an example of the output from the ATLAS detector for the 4-lepton channel (courtesy of CERN), and you can see the clear peak when the combined energy of the four leptons is around 125 GeV. Based on this and the analysis of other pathways, such as the *diphoton channel* which results in two high energy photons (see the lower diagram in Figure 25.2), the LHC was able to confirm the existence of the Higgs boson in July 2012.

25.8   Electroweak Unification

Now we come to one of those elegant twists that make physics exciting. We are going to build U(1) symmetry into the Higgs Lagrangian alongside the SU(2) symmetry. This will unify the weak force with the electromagnetic force.

The first step is to construct the Higgs Lagrangian to be locally gauge invariant with respect to U(1) as well as SU(2). Hopefully you remember U(1) gauge invariance from QED. It is invariant with respect to a change in phase of the wave function. We covered this in detail in Chapter 20. Mathematically this means that the wave function can be multiplied by $e^{i\theta}$ without changing the Lagrangian even if the value of θ varies over space or time. Incorporating this U(1) local gauge invariance requires an additional adjustment to the Klein-Gordon Lagrangian that adds another interaction term. The derivative term is shown in Equation 25.20. It is the same as Equation 20.33 except that I have labelled the vector field B_μ to distinguish it from the electromagnetic field A_μ, and labelled the coupling constant g' to distinguish it from the weak force coupling constant g.

$$\mathcal{L}_{KG/U(1)} = \frac{1}{2}(\partial^\mu \Psi^* - ig' B^\mu \Psi^*)(\partial_\mu \Psi + ig' B_\mu \Psi)...$$

$$= \frac{1}{2}\partial^\mu \Psi^* \partial_\mu \Psi + \frac{1}{2}g'^2 B^\mu B_\mu \Psi^* \Psi... \qquad \textit{highlighting interaction term} \tag{25.20}$$

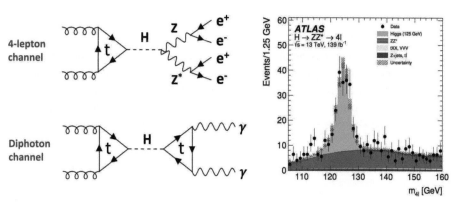

Figure 25.2   Higgs discovery at CERN.

We need to incorporate U(1) local gauge invariance into the Higgs field together with the SU(2) gauge invariance. You might think we can simply add to the Higgs Lagrangian this $B^\mu B_\mu$ term alongside the three $W^\mu W_\mu$ weak force terms that appear in Equation 25.19. It is not that simple.

We can identify each of the massive weak force bosons with their effect on the flavour of the interacting particle. Two of the W terms involve changing its flavour. For example, emitting a W_+ or W_- boson changes the flavour of the particle as shown in Box A of Figure 24.1. Therefore, we can clearly distinguish a W_+ interaction from a W_- interaction. In Feynman diagrams we use the square of the amplitude of each which is $(W_+)^\mu(W_+)_\mu$ and $(W_-)^\mu(W_-)_\mu$.

The third W interaction term which I will label W_3, does *not* involve a change in flavour and nor does the B interaction term that comes from the U(1) symmetry. In terms of Feynman diagrams this makes these two interaction pathways indistinguishable, which means that we *add* the two amplitudes together and square the *sum*. This is a core feature of quantum mechanics. If needed, please refer back to Subsection 21.4.4 which covers how to combine amplitudes in Feynman diagrams, or think back to the interference in the Double-slit experiment (Section 4.4) which comes from adding the amplitudes of two indistinguishable pathways and squaring the result.

Equation 25.21 shows an extract from the Higgs Lagrangian highlighting how the W_3 and B terms combine. Compare this with the expression in Equation 25.19. Do not be confused that the W_3 and B term have different signs when combined. This is convention and, as you will soon see, makes no difference to the underlying argument. The term $(gW_3 - g'B)$ is a linear combination of the W_3 and B vector fields and therefore is in itself a perfectly valid vector field.

$$\mathscr{L}_H = \frac{1}{2}\partial^\mu H \partial_\mu H - \mu^2 H^2 + \frac{1}{2}v^2 g^2 (W_+)^\mu(W_+)_\mu + \frac{1}{2}v^2 g^2 (W_-)^\mu(W_-)_\mu \cdots$$

$$\cdots + \frac{1}{2}v^2(gW_3 - g'B)^\mu(gW_3 - g'B)_\mu - \frac{1}{4}F_{\mu\nu}F^{\mu\nu} + \textit{interaction terms} \tag{25.21}$$

25.8.1 The Z Boson

Let's review things. Equation 25.21 shows the effect on the Higgs Lagrangian of incorporating both SU(2) and U(1) local gauge invariance, typically shown as SU(2) $\otimes$ U(1) in textbooks.[2] The first term contains the partial derivatives of the Higgs field. The second is the potential of the Higgs field (see Equation 25.11), which results in the massive scalar Higgs boson. The third and fourth terms are the massive W_+ and W_- weak force bosons. The fifth term is another massive boson, but it is subtly different from the W_+ and W_- bosons. It couples more strongly to the Higgs field because it incorporates both the W_3 SU(2) coupling *and* the B U(1) coupling. We call this the Z boson.

Equation 25.22 shows how we express this in terms of the Z boson and the two coupling constants. It also shows Z expressed in terms of the W_3 and B fields. The coupling related to SU(2) is called *weak isospin*. The coupling related to U(1) is called *weak hypercharge*. Bear in mind that g, g' and v are all real constants.

$$\frac{1}{2}v^2(gW_3 - g'B)^\mu(gW_3 - g'B)_\mu = \frac{1}{2}v^2(g^2 + g'^2)Z^\mu Z_\mu$$

$$\implies \quad Z = \frac{g}{\sqrt{g^2 + g'^2}}W_3 - \frac{g'}{\sqrt{g^2 + g'^2}}B \tag{25.22}$$

The key terms in the Higgs Lagrangian can be rewritten substituting in Z as shown below. The W_+ and W_- bosons couple identically to the Higgs field so they have the same mass. The Z boson's greater coupling means that it has higher mass. Collider measurements show the Z boson to have rest mass energy of about 91 GeV compared to the W bosons at about 80 GeV (for more on how the Z boson was discovered, refer back to Section 21.5).

$$\mathscr{L}_H = \frac{1}{2}\partial^\mu H \partial_\mu H - \mu^2 H^2 + \frac{1}{2}v^2 g^2 (W_+)^\mu(W_+)_\mu + \frac{1}{2}v^2 g^2 (W_-)^\mu(W_-)_\mu \cdots$$

$$\cdots + \frac{1}{2}v^2(g^2 + g'^2)Z^\mu Z_\mu - \frac{1}{4}F_{\mu\nu}F^{\mu\nu} + \textit{interaction terms} \tag{25.23}$$

$$\textit{Mass W: } m_W = vg \qquad \textit{Mass Z: } m_Z = v\sqrt{g^2 + g'^2} \qquad \implies m_Z > m_W$$

There is a neat way of writing the formula for the Z boson in terms of what is called the *weak mixing angle* or the *Weinberg angle*. This is shown on the left of Figure 25.3. Using this as the definition of the weak mixing angle θ_W, we can express the Z vector field as:

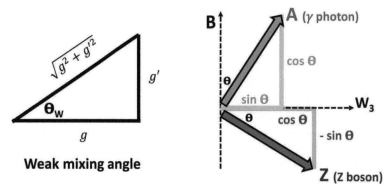

Figure 25.3 Electroweak unification: the combined W_3 and B terms in the Lagrangian result in the massive Z boson and the massless photon.

$$Z = (\cos \theta_W) W_3 - (\sin \theta_W) B \tag{25.24}$$

In summary, to include both U(1) and SU(2) local gauge invariance in the Higgs Lagrangian, we must take into account that neither the W_3 interaction of SU(2) nor the B interaction of U(1) result in a change of flavour, so they are indistinguishable in terms of the outcome of their Feynman diagrams. This means that we must add the interaction amplitudes and square the sum. However, we overcome this complication by incorporating both the W_3 and B couplings with the Higgs field into one term that we call the Z boson. The B component of the Z boson gives it extra coupling to the Higgs field, so the Z boson is more massive than the W_+ and W_- bosons.

25.8.2 The Photon

In the Z boson, we have pulled together those parts of the W_3 and B vector fields that interact with the Higgs field based on the coupling shown in Equation 25.22. What about the parts that do not? Consider the right side of Figure 25.3. The blue arrow shows the Z field in terms of its mix of W_3 (horizontal axis) and B (vertical axis). Check that you see how it matches with Equation 25.24. Defining the Z field is the equivalent of rotating the axis by θ_w away from the W_3 axis towards the B axis. This mixes the W_3 and B components to exactly match their interaction with the Higgs field. Therefore, Z is a *massive vector boson*.

Now, consider the field mix represented by the arrow marked A which is shown in Equation 25.25 alongside Z. It is orthogonal to Z. This is easy to prove by showing that the dot product of Z and A is zero in terms of the W_3 and B coordinates ($Z \cdot A = \cos \theta_W \sin \theta_W - \sin \theta_W \cos \theta_W = 0$). I mentioned earlier that there is no importance to the B field term being negative in Z. You can see in the figure that switching it to positive would change nothing except that we would have to make the B field term in A negative so that A and Z remain orthogonal.

$$Z = (\cos \theta_W) W_3 - (\sin \theta_W) B \qquad A = (\sin \theta_W) W_3 + (\cos \theta_W) B \tag{25.25}$$

Vector field A represents those residual parts of the combined W_3 and B fields that do not interact with the Higgs field. Therefore, the vector boson associated with A is massless. This is our old friend the *photon*! In Chapter 21, I described the photon as the gauge boson associated with U(1) local gauge invariance to phase. While this is true, in reality things are a bit more complicated. The photon is the massless gauge boson associated with the *unified electroweak force*. As such, it also plays a role in SU(2) gauge invariance, alongside the Z, W_+ and W_- bosons.

Redefining the W_3 and B fields in this way does not alter the maths. All we are doing is separating the fields into two orthogonal parts, one of which interacts with the Higgs field and one which does not. We call these respectively the Z boson and the photon. This makes it much easier to interpret the interaction of the W_3 and B vector fields with the Higgs field.

25.9 Summary

I suspect that most readers are exhausted after the mathematical onslaught of the last two chapters. The good news is that you have made it through all of the heavy maths in this book – and survived. Give yourself a well deserved pat on the back!

In the last chapter we considered how the weak force would appear if it were based on an SU(2) symmetry similar to the SU(3) colour symmetry of QCD. In this fantasy scenario, the covariant derivatives of the Dirac Lagrangian would need to include three interaction terms involving three massless weak force bosons in the same way that QCD involves eight massless bosons (the gluons). However, there are multiple problems with this idealised view of SU(2) symmetry, all of which are tied in some way to mass. Different flavours of fermion have different mass (Oops!). The three weak force bosons are massive (Oops!). To everybody's surprise, the weak force interacts only with left-chiral fermions and right-chiral anti-fermions (Oops!).

Left and right-chiral versions of fermions are connected via the Dirac equation to mass. These mass terms result from interactions with the Higgs field that leads to the fermions switching to and fro between left-chiral and right-chiral form as shown on the left side of Figure 25.1.

The Higgs field needs to reflect the fact that mass is a scalar quantity (a number with no direction). It must be a scalar field with a constant non-zero base value through time and space. We model such a scalar field on the Klein-Gordon Lagrangian, the covariant version of which has different interaction terms than the Dirac Lagrangian – interaction terms that give hints of massive vector bosons (as discussed in Section 25.2).

To play a part in SU(2) symmetry, the Higgs field ϕ_H must be a doublet. Its form as two complex scalar values is outlined in Section 25.3. The potential energy of the field shown in Equation 25.5 is such that in its ground state (lowest potential energy) the value of the Higgs field is non-zero and is called the vacuum expectation value v. There are an infinite number of possible values of ϕ_H that have this vacuum expectation value. In Figure 25.1, this is shown as the gully around the potential. Each point in the gully is a different form of ϕ_H for which $|\phi_H| = v$.

The fact that ϕ_H in its ground state takes one of an infinite number of possible values spontaneously breaks the SU(2) symmetry. I compared this to a roulette wheel. The roulette wheel has circular symmetry but when the ball lands on one particular number, some win, some lose. A symmetrical system leads to an asymmetrical result.

We can separate the fluctuations in the Higgs field ϕ_H into what I describe as a *radial* component and a SU(2) *rotational* component. Fluctuations in the radial value (such that $|\phi_H| = v + H$) change the field potential V_H by $\mu^2 H^2$. Changes in the rotational value do not change the potential. This means that H is the only variable that affects the potential V_H in the Higgs Lagrangian.

There was, however, a major headache. The SU(2) rotational fluctuations involve changes in the value of ϕ_H with no change to the potential. These sort of fluctuations are associated with massless particles called Goldstone bosons which would have been experimentally detected.

The insight of Peter Higgs (and others) was that these SU(2) rotational fluctuations could be combined with the associated SU(2) fluctuations coming from the weak force interactions. Combining them gives a clearer view of what is happening. Mathematically this is achieved with a unitary gauge change which incorporates all of the SU(2) fluctuations (none of which affect the potential) into the weak force vector fields. Physicists like to joke that the weak force bosons eat the massless Goldstone bosons. This simplifies the maths and makes it relatively easy to express the full Higgs Lagrangian with SU(2) local gauge invariance. The result is Equation 25.19 shown again below:

$$\mathscr{L}_{H/SU2} = \frac{1}{2}\partial^\mu H \partial_\mu H - \mu^2 H^2 + \frac{1}{2}v^2 g^2 W^\mu W_\mu - \frac{1}{4}F_{\mu\nu}F^{\mu\nu} + \text{interaction terms} \tag{25.26}$$

The first term contains the derivatives of the radial fluctuations of the Higgs field. The second is the mass term for the scalar (spin-0) Higgs boson that Peter Higgs predicted and was confirmed in 2012. The third is actually three terms. These are the weak force bosons W_1, W_2 and W_3 and they have mass because $v^2 g^2$ is a constant. The fourth is also three terms: the Lagrangians of the three weak force vector fields. Suddenly, the weak force and its massive vector bosons make sense.

The final step is to incorporate U(1) symmetry into the Lagrangian. The W_+ and W_- interactions change the flavour of the fermion. They are clearly distinguishable. The W_3 interaction does not and therefore is indistinguishable in outcome from the U(1) B interaction term. The amplitudes of these two must be summed and squared.

Things can be simplified by separating out those components of the combined W_3 and B fields that interact with the Higgs field and are therefore massive, from those that do not interact with the Higgs field and are therefore massless. The former we describe as the Z boson. The latter we call the photon. This neatly unifies the SU(2) symmetry of the weak force and the U(1) symmetry of the electromagnetic force into what is called the electroweak force.

As a final summary, consider how the spontaneous symmetry breaking of the Higgs field has shifted around the degrees of freedom. This is summarised in Table 25.1. Before, the Higgs field has four degrees of freedom (a, b, c, and d in Equation 25.4) and the four massless vector bosons each have two degrees of freedom (two spin possibilities each) for a total of 12 degrees of freedom. After the spontaneous symmetry breaking, the Higgs boson has one degree of freedom (it is a scalar spin-0 particle), the three massive weak force bosons each have three degrees of freedom (three possible spin states each) and the massless photon has two so, the total is also 12.

Table 25.1 Degrees of freedom and the Higgs mechanism.

Before		After	
Higgs field	4	Higgs boson	1
Massless SU(2) bosons	$3 \times 2 = 6$	Massive W_+ W_- Z bosons	$3 \times 3 = 9$
Massless U(1) boson	2	Massless photon	2
	12		12

That then is the story of the weak force and the Higgs field. In the next chapter we will review the Standard Model Lagrangian and discuss some of the things we don't know – the journey ahead.

Notes

1 The H^3 and H^4 terms in the potential indicate that there are possible self-interactions involving 3 or 4 Higgs bosons, but we can ignore these for our purposes here.

2 Technical point: mathematics students may wonder why SU(2) $\otimes$ U(1) is not shown as U(2). This is because the SU(2) symmetry affects only left-chiral particles whereas the U(1) symmetry affects right-chiral particles too.

Chapter 26

The Standard Model and Beyond

CHAPTER MENU
26.1 The Standard Model Lagrangian
26.2 From Einstein and de Broglie to Higgs
26.3 Questions and Problems
26.4 General Relativity and Quantum Mechanics
26.5 Supersymmetry (SUSY)
26.6 String Theory
26.7 Loop Quantum Gravity (LQG)
26.8 That's All Folks!
26.9 Module Memory Jogger

That concludes the overview of the three forces of the Standard Model – there are only three because gravity is not yet included. Each force gives the Standard Model's Lagrangian local gauge invariance linked with some symmetry. For the electromagnetic force, it is U(1) (change of phase). For the strong force it is SU(3) (change of colour). For the weak force it is SU(2) (change of flavour) although this symmetry is obscured by the spontaneous symmetry breaking of the Higgs field. Taken together these symmetries are described as: SU(3) ⊗ SU(2) ⊗ U(1).

The elementary particles are summarised in Figure 26.1. On the left are the three generations of fermions giving a total of 12 elementary fermions plus their antiparticles. They are spin-half and each generation is composed of two quarks (shaded grey) and two leptons (shaded blue). The right side of the figure shows the bosons that should by now be familiar friends. The number of spin-1 vector gauge bosons (shaded grey) depends on the generators linked to each group symmetry (these are discussed in Subsection 23.3.2). U(1) has one generator giving us the massless photon. SU(2) has three in the form of the Z, W_+ and W_- massive weak force bosons and SU(3) has the eight massless gluons. Everyone describes the Z as a weak force boson but, for complete accuracy, I should note again that the Z boson and the photon have combined roles in both SU(2) and U(1) gauge invariance due to electroweak mixing (as shown earlier in Figure 25.3).

In addition, there is the massive scalar Higgs boson. I have also highlighted at the bottom of the figure the *graviton*. I have given it a question mark because gravity is not included in the Standard Model. However, the expectation is that the gravitational force is mediated by a spin-2 boson because it is a tensor field.[1]

26.1 The Standard Model Lagrangian

If you want to see the whole Standard Model Lagrangian $\mathcal{L}_{SM}$ in detail, you can find it on the web with versions generally running to about 40 lines of maths. I do not think that is going to be helpful for most students so let me present the version you might find on a T-shirt or coffee cup, a *very* summary form containing only five terms, and then describe each term (Box 26.1).

Elementary Fermions

d	s	b
down	strange	bottom
u	c	t
up	charm	top
e	μ	τ
electron	muon	tau
ν_e	ν	ν
electron neutrino	muon neutrino	tau neutrino

+ antiparticles

Elementary Bosons

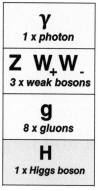

+ graviton?

Figure 26.1 Elementary particles of the Standard Model.

Box 26.1 Abbreviated Standard Model Lagrangian

$$\mathscr{L}_{SM} = i\,\overline{\Psi}\gamma^\mu D_\mu \Psi + y\,\overline{\Psi}\Psi\,\phi_H + \frac{1}{2}D^\mu \phi_H^* D_\mu \phi_H - V(\phi_H) - \frac{1}{4}F_{\mu\nu}F^{\mu\nu} \tag{26.1}$$

The first two terms of Equation 26.1 are the Dirac Lagrangian of the fermions (as shown earlier in Equation 20.34). I have abbreviated savagely by showing only one such Lagrangian. Written in full there is a separate Dirac Lagrangian for each of the 12 elementary fermions plus their antiparticles.

Let's look at the *first* term of Equation 26.1 in more detail. This contains the derivative components of each fermion field. You should recognise the γ^μ Dirac 4 × 4 matrices that mix the four component Dirac wave functions (refer to Subsection 18.3.4). The symbol $D_\mu \Psi$ is the *covariant derivative* which means that it is local gauge invariant to the various symmetries of the Standard Model. Written out in full, it contains interaction terms. U(1) invariance generates the interaction term of QED. SU(2) invariance gives three weak force boson interaction terms although this applies only to the left-chiral components of each fermion as was famously demonstrated by Chien-Shiung Wu. The quarks have colour charge so their derivatives are also SU(3) invariant, which gives the eight strong force gluon interaction terms.

The *second* element in Equation 26.1 is the rest mass for each fermion. It is the Yukawa coupling for each fermion with the Higgs field (refer back to Equation 24.16 if needed). This couples the left-chiral and right-chiral components of the fermion in the zigzag pattern illustrated schematically in Figure 25.1. The mass m of each fermion depends on the coupling to the Higgs field with $m = yv$ where y is the Yukawa coupling constant for each fermion[2] and v is the Higgs vacuum expectation value after spontaneous symmetry breaking.[3]

The structure of the Lagrangian is such that the Action of each fermionic field is minimised when $m^2 = E^2 - p^2$ where m is the fermion's rest mass. The effect is that the fermion is held on-shell although the development of the quantum field can drift briefly off-shell. These off-shell pathways are described as virtual particles and are very important in interactions between particles. For example, they play a role in every QED interaction (as covered in Section 21.3).

The third and fourth terms of Equation 26.1 are the derivatives and the potential of the Higgs field which, being a scalar field, follows the Klein-Gordon Lagrangian. The *third* term contains the *covariant derivatives* of the Higgs field.[4] The SU(2) local gauge invariance gives three interaction terms which, after spontaneous symmetry breaking, can be combined with the rotational SU(2) fluctuations in value of the Higgs field using the unitary gauge change described in Subsection 25.6.3. This results in mass terms for the weak force bosons. The Z boson interaction reflects U(1) as well as SU(2) coupling with the Higgs field, which is why it is more massive than the W_+ and W_- bosons.

The *fourth* term $V(\phi_H)$ is the potential of the Higgs field. After spontaneous symmetry breaking, it can be expressed in terms of the variation in radial excitation H of the Higgs field around the vacuum expectation value as $\mu^2 H^2$ (as in Equation 25.11). As μ is a constant from the expression for the Higgs potential, this is the mass term for the massive Higgs boson. This is a scalar boson so it is spin-0.

The *fifth* and final term in Equation 26.1 is an abbreviated form of the Lagrangians of the spin-1 gauge bosons. These Lagrangians dictate the development of the underlying quantum vector fields. For example, minimising the Action of the Lagrangian of the electromagnetic field leads to Maxwell's equations (see Box 20.6). If written in full, the vector field Lagrangians for both SU(2) and SU(3) contain self-interaction terms. In the case of SU(3) the gluon self-interaction is so strong that quarks always combine in colourless combinations.

I have not bothered to include in Equation 26.1 the equivalent terms for antiparticles such as the positron. You will see these included in some versions of the Standard Model Lagrangian, shown as + h.c. which stands for *Hermitian Conjugate*.

I must emphasise that Equation 26.1 is an abridged form of the Standard Model Lagrangian. However, I still find it amazing that so much physics can be expressed in such a succinct way. This is a powerful demonstration of the symmetries that drive the Standard Model and the very similar mathematical structure that sits behind the electromagnetic, weak and strong forces.

26.2 From Einstein and de Broglie to Higgs

It is worth taking a moment to reflect on the journey that started with Einstein's theory of special relativity and de Broglie's idea of more general wave-particle duality. In the module of this book titled *Essential Quantum Mechanics*, we discussed how these ideas (and experimental results) show physics at the microscopic level to be very different from our day-to-day experience of the world around us. Particles exist in superpositions with related uncertainty between momentum and position such that the result of measurements cannot be precisely predicted, but must be quantified in terms of probability. God plays dice.

In the next module of the book, *Complex Quantum Mechanics*, we delved more deeply into the mathematics with Schrödinger's equation. This uses the relationship of the derivatives of the wave function with energy and momentum to analyse the quantum behaviour of particles, including quantum tunnelling, the quantisation of angular momentum and the stable energy levels of electrons in the Coulomb potential of the atom, which, in turn, decodes the periodic table. In spite of all this progress, there was an underlying problem. Schrödinger's equation equates the first derivative with respect to time $\frac{\partial \psi}{\partial t}$ and the second derivative with respect to space $\frac{\partial^2 \psi}{\partial x^2}$. It treats time and space very differently. It does not conform to special relativity.

This last module on *Relativistic Quantum Mechanics* introduced the Dirac equation, an extraordinary relativistic four component solution for the wave function with $\frac{\hbar}{2}$ of intrinsic angular momentum that we call spin. This matches experimental observations of the electron's spin and more accurately explains the spectral emissions of hydrogen. However, it has other surprising implications including antiparticle states and the creation and destruction of particles.

This led Dirac and others to quantum field theory (QFT) in which particles are quanta of excitation in fields. The behaviour of these quantum fields depends on the path integral which is the sum of all possible development pathways. How these pathways combine depends on the Action of the field which is described by its Lagrangian. The Action holds each quantum of excitation, which we call a particle, closely on-shell, but with enough short-term wiggle room for off-shell pathways, described as virtual particles, to play an important role in particle interactions.

The final huge leap forwards was understanding the role of local gauge invariance. As we have discussed in the last few chapters, the electromagnetic force, weak force and strong force can all be explained, with the help of the Higgs field, as underlying local gauge symmetries. This is pretty much where physics stands. The fit of the Standard Model with experimental data is exceptional, *but*, and it is a big *but*, we know there is still a lot to discover.

26.3 Questions and Problems

In exploring the challenges faced by quantum mechanics and the Standard Model, I want to distinguish between *questions* and *problems*. What is the difference? For me, a *question* is something that we do not know, but if we did, might be incorporated into the Standard Model. A *problem* is something more fundamental that requires a rebuild of the model. We will start by considering a list of questions, some of which have already been discussed in this book. As we go through the list, the questions become more significant leading up to an underlying problem, which is how to integrate gravity with quantum mechanics and the Standard Model.

How do you explain the neutrino? In the Standard Model the neutrinos fit most naturally as massless particles, but experiments show flavour oscillations, which indicate they have a small mass. One possibility is that the neutrino is a completely different type of particle called a Majorana particle (see Box 24.4) or perhaps the neutrino gets its mass from a totally different as yet unknown mechanism. This is an area of active research because the solution may reveal new physics and help us answer some of the other questions in this list.

Where has all the antimatter gone? Most physicists believe the universe started with the Big Bang for which there is considerable supporting evidence. Edwin Hubble's discovery that the universe is expanding, the residual Cosmic Microwave Background Radiation, and the measured abundance of lighter elements all are consistent with the theory. However, a Big Bang should have produced equal amounts of matter and antimatter. Dirac even postulated in his Nobel speech that antimatter galaxies might exist (see Box 18.10), but observations so far suggest not. One possible explanation is that we just need to look harder and further. Another is that there may be differences in the decay patterns of matter and antimatter. However the Standard Model, as currently constructed, does not allow for enough asymmetry to explain the discrepancy.

What is dark matter? Gravitational studies of the rotation of galaxies indicate that they contain a lot more mass than is visible. Note that this mass sits in a halo through the galaxy, not at its centre, so it is not explained by massive central black holes. The data is backed up by research into gravitational lensing that assesses how light bends as it travels around galaxies. The discrepancy is huge and suggests about 85% of the mass of galaxies is dark matter. One explanation is an as yet unknown very Weakly Interacting Massive Particle (known affectionately as a WIMP). Might this be a sterile neutrino (see again Box 24.4), or might it be one of the postulated supersymmetric partner particles that we will discuss in a moment? We don't know.

What about dark energy? This is a tough concept to explain without getting into the details of general relativity (GR), but I will try to give a simplified illustration. Einstein explained gravity as the curvature of spacetime. One feature of his GR equations is that the universe is unstable. Left to it own devices and depending on the amount and type[5] of energy in the universe, the very fabric of space and time might collapse progressively in on itself. Einstein believed the universe to be stable. Therefore, he added a balancing factor to his equations called the *cosmological constant* typically labelled Λ which he believed could offset the shrinking.[6] It can best be described as the energy of the vacuum. When observations revealed that the universe is expanding, Einstein described his effort to stabilise things with the cosmological constant as his *biggest blunder*.

However, modern measurements indicate the rate of expansion of the universe is *accelerating* which suggests Λ does indeed exist. Physicists refer to this Λ as *dark energy*, but have no clear explanation. If they are right, it would represent about 68% of the total energy in the universe. Quantum mechanics includes a possible candidate for dark energy as the quantum fields have ground state energy – but this does not work. Calculations indicate the ground state energy of the quantum fields, if not infinite, should be at least $10^{60} \times$ higher than the value of Λ based on cosmological observations, and that is after applying a high energy cut-off. You may hear this referred to as the *cosmological constant problem* or the *vacuum catastrophe* (see Box 19.5). Oops!

26.4 General Relativity and Quantum Mechanics

By now you should be sensing a major mismatch between quantum mechanics (QM) and general relativity (GR). For a start, taking into account the mysteries of dark energy and dark matter, the Standard Model, which is so exquisitely accurate at very small scales, currently explains less than 5% of the total energy physicists believe may be present in the universe.

On top of this, the two theories have different, arguably incompatible, structures. At the most basic level, QM is, well, quantum. In QM, everything comes in lumps that are discrete. In sharp contrast, GR explains gravity as the curving of spacetime. This is smooth and continuous. Digging further into QM, we see particles in superpositions, with a level of uncertainty in position and/or momentum. The chance that *A* leads to outcome *B* is quantified as a probability. GR is completely the opposite. In the equations of GR, matter precisely and exactly curves spacetime. This curvature precisely and exactly affects how the matter behaves. *A* leads to *B*. There is nothing in the GR equations that speaks of superpositions, uncertainty or probability.

Furthermore, QM and GR describe force through completely different mechanisms. In the Standard Model of QM the three forces are transmitted by the exchange of mediating particles. In GR, the gravitational force is the effect of the curvature of spacetime.

The conundrum is this. QM has been shown again and again to describe accurately the electromagnetic force, the weak force and the strong force. GR has been shown again and again to describe accurately the gravitational force. These two magnificent theories collide majestically at the theoretical level. This is the big *problem* that physics faces. If we want to have a unified theory that includes all four forces, how do we reconcile QM and GR?

The solution is obvious. We need a quantum theory of gravity that fits with QM but, at large scale, gives GR. This may sound a simple task. It is not. Many of the smartest physicists in the world have been grappling with this for decades with limited success. In the following sections, I will touch on two of the best known efforts, *string theory* and *loop quantum gravity*. However, I first want to mention *supersymmetry* which may (and I stress *may*) play an important role in the development of the Standard Model and is key to most versions of string theory.

26.5 Supersymmetry (SUSY)

The central idea of supersymmetry (SUSY) is that there is a major additional symmetry sitting inside the Standard Model – a symmetry between fermions and bosons. If this is the case, then for every fermion there will be a boson superpartner and for every boson there will be a fermion superpartner. Not a single superpartner has been found to date. As a result, I think it fair to say that scepticism is growing. However, the idea of SUSY is compelling enough that many, perhaps even most, physicists still believe it will prove true.

By now it should be clear to you that symmetry underpins the Standard Model. Its Lagrangian is built on the requirement of local gauge invariance to the phase, colour and flavour of the quantum field excitations that we call particles. Let me take the liberty of abbreviating the Standard Model Lagrangian in Equation 26.1 even further to that in Equation 26.2:

$$\mathscr{L}_{SM} = (fermions)_{half} + (bosons)_{integer} \tag{26.2}$$

I have separated out the part of the Lagrangian relating to the half-spin fermions from that of the integer-spin bosons. The half-spin fermions all follow the same basic Dirac Lagrangian. Some of the symmetries of the Standard Model are obscured at low energy by spontaneous symmetry breaking, but the fundamental message is that you can re-categorise the fermions, say, up quark for down quark, without changing the laws of physics. Similarly, we can encapsulate the vector fields of the EM, weak and strong forces with the same vector field structure.[7] However, while the fermions have a symmetry and the bosons have a symmetry, fermions and bosons are *very* different. Their Lagrangians are different. Their spin is different. Their behaviour is different.

SUSY asks how things would be if you could not only re-categorise *within* the fermion group and *within* the boson group (such as the flavour change from up quark with down quark), but you could also re-categorise *across* fermions and bosons. In essence, this is saying that even fermions and bosons are basically the same stuff – that there is an underlying symmetry between *all* excitations of quantum fields.

This sounds a cool idea but, for it to be true, we need to be able to re-categorise the half-spin fermions as integer-spin while re-categorising the integer-spin bosons as half-spin without changing the overall Standard Model Lagrangian. Mathematically this is an easy task. We simply add additional terms to Equation 26.2 to create the symmetry as shown in Equation 26.3. In this form, changing *fermion* and *boson* (or changing *half-spin* and *integer-spin* if you prefer) makes no difference to the total Standard Model Lagrangian.

$$\mathscr{L}_{SUSY} = (fermions)_{half} + (bosons)_{integer} + (fermions)_{integer} + (bosons)_{half} \tag{26.3}$$

Note however that we have added a bunch of additional quantum fields to the Standard Model. Each of these new fields can be excited and that means new particles. For every fermion there would be a bosonic superpartner (named by adding an *s* at the front such as the *selectron* for the electron or the arrestingly named *stop* for the top quark). Similarly for every boson there would be a fermionic superpartner (named by adding *ino* at the end such as the *photino* for the photon or the rather comical *wino* for the *W* boson). None of these superpartners has been discovered to date, but this could be because they are too massive to appear in existing colliders so many physicists are still hoping (see an example in Box 26.2). Beyond the intellectual allure of such a pure symmetry, there are several attractions that fans of SUSY can point to.

One example, is that SUSY seems to better unify the three forces (EM, weak and strong) in the Standard Model. The coupling constant of each is a *running* constant because it varies with energy (see Section 22.6). The general belief is that these forces have a common origin, separating at some stage after the Big Bang. If so, you might expect them to converge to a common value at some higher energy level. Calculating with the Standard Model, this does not happen. However, if you adjust the calculation for SUSY, the values of the three coupling constants neatly converge at about 10^{16} GeV. Another potential benefit of SUSY is that the lightest superpartner particle could be an excellent candidate for dark matter.

Box 26.2   ANITA and the stau

The Antarctic Impulsive Transient Antenna (ANITA) is a balloon-based observatory sitting above the ice sheet. It has detected some unexpected radio emissions from directly below. They appear to be from the decay of tau electrons, but how have they passed through the entire earth? One suggestion is that they originate from stau selectrons, the bosonic superpartner of the tau electron. This is still highly speculative so don't get too excited yet!

And we must not ignore the importance of SUSY to string theory which has been one of the big hopes in the last few decades to unify quantum mechanics and general relativity.

26.6   String Theory

The idea behind string theory is disarmingly simple. The excitations of quantum fields that we call particles are actually tiny vibrating one-dimensional strings. Different vibrations in these strings are associated with different elementary particles. There is an obvious attraction to such a reductionist theory that seeks to describe all the particles and forces as having a common underlying structure.

One of the immediate benefits is that the strings have a length, albeit very small (perhaps 10^{-33} metres compared to the 10^{-10} metres diameter of an atom). This avoids many mathematical problems associated with point particles. In particular, efforts to describe gravity in terms of QFT struggle because the self-interaction in the gravitational field leads to unmanageable infinities as you explore closer and closer to a point particle. String theory neatly sidesteps this difficulty because the strings have a finite size.

Strings can interact by joining together at the ends or splitting apart. Consider the left side of Figure 26.2. It could represent any interaction such as an electron and positron annihilating to form a photon that subsequently decays back into an electron and positron. At the start there are two particles labelled 1 and 2. These are *closed loop* strings. Note that *open loop* strings also play a role in many string theories. From left to right, these closed loops each trace out a two-dimensional surface called a *world sheet*. Their world sheets merge. The loops join to form one larger loop before splitting into loops 3 and 4. The interaction effectively is smeared out over a region of spacetime. At no stage is there a singularity.

26.6.1   Gravity in String Theory

One early attraction of string theory is that it readily accommodates the graviton as a closed loop string. A closed loop can carry a vibration in either of two directions (think clockwise and anticlockwise). Due to symmetry considerations the energy moving in each direction must be the same.[8] As a result excitation comes in units of two, giving a possible description of the graviton that would be expected to be spin-2 in a quantum theory of gravity.

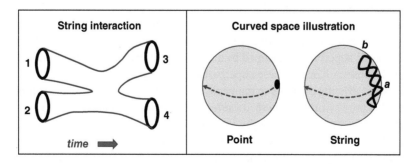

Figure 26.2   String theory illustrations.

Even more compelling is that Einstein's general relativity falls straight out of string theory. Let me give you a flavour of how this happens. Every string is a collection of an infinite number of quantum harmonic oscillators, each of which has a ground state minimum level of excitation (as in Equation 13.16). The volume of space that might be occupied by the points between the ends of the string depends on these oscillators. If you consider a string with, say, five oscillators up to frequency $\omega(5)$ it will occupy a certain volume, but if you add in higher frequencies up to, say, $\omega(10)$ that volume will be larger. Add more and it is larger still. The growth never stops although it slows dramatically. This means that the amount a string is spread out cannot be quantified. It is undefined.

In flat space this has no effect. A string moving through flat space has momentum mv where v is the velocity of its centre of mass. The position of various parts of the string does not change this. However, the picture is very different in *curved* space. Let's consider what happens if space is curved in the shape of the surface of a sphere as shown on the right side of Figure 26.2. A *point* particle moves on the surface of the sphere in a geodesic (great circle). In addition to its mv of linear momentum it has mvr of angular momentum because in this curved space, it is rotating around the vertical axis of the sphere. This is quantised (see Section 14.2 for a reminder).

In the case of a *string*, the space it occupies is undefined. Consider it spread out as shown on the right in Figure 26.2. The part of the string at point a has angular momentum mvr but the part of the string at point b has less angular momentum because it is closer to the axis of rotation. If the volume of space occupied by the string is undefined, its angular momentum is undefined. As a result, you cannot have a consistent string theory in the spherically bent space of Figure 26.2, but there is a solution to this paradox.

The trick is to limit string theory to those spacetimes which have just the right curvature not to create the paradox. The curvature of spacetime is quantified by the metric (see Box 1.1), which is generally labelled $g_{\mu\nu}$. String theorists calculated what limits would need to be placed on $g_{\mu\nu}$ to resolve the string paradox. Can you guess what popped out? The result is Einstein's field equations of general relativity in a vacuum. I find this amazing. In order for string theory to be consistent, spacetime must curve in exactly the way that Einstein showed it does! Furthermore, this relationship is driven by a symmetry in string theory called *conformal symmetry* or *Weyl gauge invariance*. This symmetry is very particular to the worldsheets of strings. There is a neat mathematical fit. This is one reason why everybody was so excited by string theory.

26.6.2 Difficulties with String Theory

String theory is appealing and seductive because it resolves many of the problems related to the singularities that accompany point-like particles. It postulates that all the particles and forces are made up of the same thing – strings. Not only do gravitons fit neatly into the model, but Einstein's field equations for a vacuum are required by it. Sadly, we must now discuss the problems with string theory and why some physicists describe it as a dead end.

The biggest theoretical change is that string theory needs more than just general relativity to be a viable model in curved space. It requires the universe to have 10 dimensions: one of time, the three of space (x, y, z) that we know plus *six additional* spatial dimensions. Obviously we cannot see these six extra dimensions. The idea is that they are curled up to a tiny size (perhaps about 10^{-30} metres) that we cannot sense, but the tiny strings can move in. This is described as *compactification* with the dimensions intertwined in what are called *Calabi-Yau manifolds*.

To be fair, this is not the first time that physicists have considered the possibility of additional dimensions. Shortly after Einstein published his theory of general relativity, Theodor Kaluza proposed that the existence of a fifth dimension might explain electromagnetism. After much initial excitement, this idea fell by the wayside as the details did not work. To date no experiments have given a hint of the existence of additional dimensions. Indeed it is far from clear what experiment might reveal them.

The second challenge and perhaps the biggest experimental disappointment is the lack of evidence for supersymmetry (the SUSY discussed in Section 26.5). Modern string theory relies on SUSY. Not only does it bring fermions into the model, it reduces the number of required dimensions from 26 to ten, and it clears up a number of problems such as troublesome faster-than-light tachyons that appear in earlier versions of the theory. There were high hopes that the Large Hadron Collider might reveal one or more SUSY superpartner particles, but none has been found. String theory can be tweaked to explain the lack of lower energy superpartners, but as some critics say, each tweak makes it that bit uglier.

A third major problem is that string theory can be defined in so many different ways. Over time, five different geometries of string theory developed.[9] There was a leap forwards some 30 years ago when some symmetries were spotted that help tie these theories together. Figure 26.3 illustrates this. The surface of the horizontal cylinder represents a compactified dimension. Think of it like the surface of a straw. From a distance the surface cannot be distinguished. From a distance, the

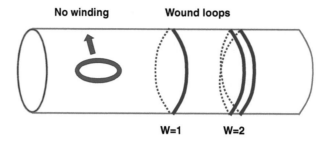

Figure 26.3 T-duality: a symmetry between loops with and without windings.

straw looks like a line. That is our perspective. However, things are different for a tiny ant which can move left and right along the straw *and* around its surface. In string theory the compactified dimensions are much too small for us to sense. However, the tiny strings can move around them.

On the left in Figure 26.3 a loop is moving around a compactified dimension. If it moves with momentum p on the surface, it has angular momentum $L = pr$ so $p = \frac{L}{r}$. Angular momentum is quantised in units of $\hbar$ which means $p = \frac{n\hbar}{r}$ where n is an integer. For a massless particle, $E = p$ giving energy levels proportional to the inverse of the radius of the dimension: $\frac{1}{r}, \frac{2}{r}, \frac{3}{r}$... The right of the figure shows an alternative with the loops wound *around* the dimension. Each loop's energy depends on its length which is $2\pi r W$ where W is the *winding number*. This means the energy levels are proportional to $r, 2r, 3r$...

Suppose your string theory predicts a large gap between energy levels. This might be because the compactified dimension's radius r is *small*, which means large gaps between the energy levels for strings with no winding ($\frac{1}{r}, \frac{2}{r}$...). Or it may be because radius r is *large*, which also means large gaps because of the energy levels of wound strings ($r, 2r$...). It turns out that the two are mathematically equivalent. This is called *T-duality*. There is a similar sort of duality, called *S-duality* between the values of $\frac{1}{g}$ and g for the coupling constants of different string theories.

In 1995 Edward Witten used these dualities to tie together the five versions of string theory into *M-theory* albeit at the cost of adding another dimension increasing the count to eleven.[10] M-theory includes *D-branes* of various dimensions that act as anchor points for strings. While this link sparked new excitement, there is still a vast array of theoretical approaches, especially in the geometries of the six hidden dimensions and how the strings behave in them. The hope must be to explain quantities such as the mass of the elementary particles and the coupling constants of the forces, but after 50 years of study, there are still an estimated 10^{500} different possible versions of string theory, so how do you narrow down the options?

String theory still has many fans and believers. Its mathematical beauty is enticing, but there is no hard evidence and the LHC's failure to find SUSY particles has been a serious blow. Moreover, with so many possible geometries for the compactified dimensions, some go as far as to call it untestable. I think it fair to say that, in recent years, enthusiasm for the subject has slowly declined (hence the joke in Box 26.3).

Box 26.3 Just for laughs:

Is there hard evidence that string theory is right?   *Frayed knot!*

26.7 Loop Quantum Gravity (LQG)

The approach of Loop Quantum Gravity (LQG) is radically different from string theory. Let's compare how the theories address the problem of singularities. In *string theory*, the singularities are avoided because particles are *finite* one-dimensional strings. For a stable model, they must exist in a multi-dimensional spacetime that is curved according to the rules of general relativity. In *Loop Quantum Gravity*, spacetime itself is *quantised*. Singularities cannot exist because there is a limit to how small a chunk of spacetime can be. In layman's terms, spacetime is *granular*.

General relativity's continuous smooth model of spacetime raises a conundrum. There is a limit to how small a volume of space it is possible to measure. The problem is that the smaller the space, the higher the energy needed to probe it. For a photon, $E = \frac{hc}{\lambda}$. The shorter the wavelength, the higher the energy. This energy curves spacetime. In the extreme this creates a black hole. The effect can be calculated using the *Schwarzschild radius*[11]: $R_s = \frac{2GE}{c^4}$ (where G is the gravitational constant, and E is the energy inside the radius). As a result, measuring any length shorter than about 10^{-35} metres creates a black hole. This is the *Planck length*.

$$\text{Planck length:} \quad L_P = \sqrt{\frac{G\hbar}{c^3}} \approx 10^{-35}\, m \tag{26.4}$$

In addition to this challenge, there is a fundamental mismatch in the equations of GR. Einstein's field equations are shown in summary form in Equation 26.5. GR is beyond the scope of this book beyond except to say that the T term on the left side is *energy* and the R and $g_{\mu\nu}$ terms on the right relate to the curvature and metric of spacetime, that is, to the *geometry* of spacetime. One thing that we are absolutely sure of from quantum mechanics is that the energy term on the left side is *quantised*. Therefore, how can the equation be correct if the right side is *continuous*? The answer according to LQG is very straightforward: the geometry of spacetime encapsulated in the right side of Equation 26.5 also must be quantised.

$$\frac{8\pi G}{c^4} T_{\mu\nu} = R_{\mu\nu} - \frac{1}{2} R\, g_{\mu\nu} \qquad \textit{Einstein's GR field equations in summary form} \tag{26.5}$$

The thesis behind LQG is simple. Associated with each force there are fields. For example the EM force has the EM field. LQG argues that the field associated with the gravitational force *is* spacetime. Take a moment to think about this. LQG says that spacetime is nothing more than the gravitational field.

26.7.1 LQG Space as a Quantum Entity

If spacetime is a quantum field then there must be operators associated with measurables such as length, area and volume. Consider an area operator $\hat{A}$ that gives $\hat{A}\Psi = A\Psi$ where A is a spatial area being measured and Ψ is the state of the gravitational field (which is spacetime). LQG introduces quantum effects by assuming that such an *area* operator $\hat{A}$ operates similarly to the *angular momentum* operator $\hat{L}$. I will discuss the logic for this later, but first let's examine how it works.

On the left of Figure 26.4 is a reminder of angular momentum. In classical mechanics, we could set the z axis along the axis of rotation as shown in the figure and measure L_z to give us the total value of angular momentum. Quantum mechanics does not allow this because of the uncertainty relationship between the angular momentum operators. You can have information about the total angular momentum such as $L_{sum}^2 = L_x^2 + L_y^2 + L_z^2$. However, if you know L_z, you cannot know exactly the values of L_x and L_y. This means the plane of rotation cannot be precisely defined. The uncertainty creates a fuzziness.

Now, let's discuss an area of space. The right of Figure 26.4 shows a triangular area. Starting with the classical picture, the vector A_z is perpendicular to the plane of the triangle and its length is set to reflect its *area*. Classically we would say the area of the triangle is $|A_z|$ and the values of A_x and A_y would be zero because they are in the plane of the triangle.

In LQG, the spatial area shares the same uncertainty relationship as angular momentum. The commutator relationship is shown alongside that of angular momentum in Equation 26.6 (α is a constant). If you know the value of A_z, you cannot know the exact value of A_x or A_y. The plane of an area of space is not precisely defined!

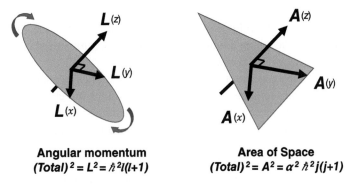

Angular momentum
(Total)^2 = L^2 = $\hbar^2 l(l+1)$

Area of Space
(Total)^2 = A^2 = $\alpha^2 \hbar^2 j(j+1)$

Figure 26.4 Comparison of angular momentum with a spatial area in LQG.

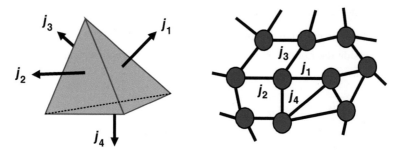

Figure 26.5 From tetrahedrons to a spin network.

$$\text{Ang. momentum: } [\hat{L}_x, \hat{L}_y] = i\hbar \hat{L}_z \qquad \text{LQG area operators: } [\hat{A}_x, \hat{A}_y] = i\alpha\hbar \hat{A}_z \qquad (26.6)$$

The total area of space A is the square root of $A_{sum}^2 = (A_x^2 + A_y^2 + A_z^2)$. This is quantised in the same way as L_{sum}^2 is in the analysis of angular momentum. In LQG the quantum number is labelled j as shown in Equation 26.7 and can take any half-integer value $\frac{1}{2}$, 1, $\frac{3}{2}$, 2... this is just an LQG convention that helpfully acknowledges the half-spin of fermions. The value of α is based around the formulas of GR as $\frac{8\pi G\hbar}{c^3}$. Ignoring some technical details and jumping to the punchline, this gives a minimum possible quantum area of space of about $10^{-70} m^2$.

$$\text{Ang. mom: } \sqrt{L_{sum}^2} = \hbar\sqrt{l(l+1)} \qquad \text{LQG Area: } A_{tot} = \frac{8\pi G\hbar}{c^3}\sqrt{j(j+1)} \qquad (26.7)$$

We can construct a volume of space in the form of tetrahedral-like shapes as shown on the left of Figure 26.5. Each of the four faces has a quantum number j value associated with it, all of which can be known (the uncertainty relationship is between the components A_x, A_y and A_z of each individual face, not between faces). From this, you can calculate an expression for the volume V. As you might expect, it also is quantised.

The total quantum information you can have for the LQG tetrahedron of space is five values which are its volume and the area of each face, that is, (V, j_1, j_2, j_3, j_4). In classical mechanics you need six numbers to fully define a tetrahedron (for example the length of its six sides). This leaves some uncertainty, some quantum fuzziness in the shape of the tetrahedron and the geometry of space.

The minimum size of a quantum of volume in LQG is about $10^{-105} m^3$ which is very very small. There are almost as many of these quanta in the volume of one atom as there are atoms in the known universe. Small it may be, but it is important as, if true, it eliminates the possibility of spacetime singularities and gives a rationale for a high energy cut-off in QFT.

26.7.2 LQG Background Independence: Spin Networks

One attractive feature of LQG is that it is *background independent*. It does not depend on any underlying coordinate structure of spacetime. How then can we piece together spacetime? Consider Figure 26.5. In LQG, space can be built up as adjacent tetrahedrons. Each face of the tetrahedron shown is connected to another tetrahedron face-to-face so the connecting value of j (which embodies the area of the face) matches between adjacent tetrahedrons. This is illustrated on the right of the figure in what is called a *spin network*. The term *spin* is unfortunate. It is because of a historical link to work done by Roger Penrose plus the mathematical association with angular momentum. Don't forget that j is a quantum number related to *area*. It has nothing to do with spin.

In a spin network, every node represents the volume of a particular tetrahedron of space. From each node there are four lines which represent its four faces. The key thing to understand is that, using this sort of network, space is defined by its contiguity to surrounding space, that is, by what is next to what.

Information about the curvature of space is contained in the values of V at each node and the four values of j for its connections. The uncertainty relationship remains (only 5 values versus 6 for full classical definition of a tetrahedron). The structure of the maths from quantum mechanics allows for superpositions. All this is achieved without the need for an underlying coordinate system. This ties much more naturally to GR than is the case with string theory. Everything is relative, just as Einstein required.

An astute reader will have noticed that, so far, we have only discussed LQG *space*. What about *time*? The presence of energy distorts the spin network as you would expect if spacetime is indeed the gravitational field. These changes in the network are what we perceive as *time*. It is quantised with the minimum unit of time being 10^{-43} seconds so the changes are like the clicks of a digital clock. The constantly fluctuating spin network is described as a *spin foam*. The kinematics (change with time) of LQG is beyond our scope, but there has been some notable success, particularly in modelling the entropy of black holes.

Some critics are concerned as to whether this approach can produce a consistent structure that puts time and space on an equal footing as is required by GR. Indeed, there is some disagreement inside the LQG community. While Carlo Rovelli, one of the main architects of LQG, believes that time is an emergent property of the spin foam, his close collaborator Lee Smolin suspects time needs to be built more tightly into the fundamental structure of the theory.

26.7.3 Difficulties with LQG

Before entertaining any criticism of LQG, I want to reflect on its strengths, which are numerous. It represents a fundamental shift in thinking by proposing that spacetime *is* the gravitational field. This is a logical simple message, albeit the mathematical consequences are complicated. And the theory is built on the base of things we know: GR, quantum mechanics and the Standard Model. It does not postulate new concepts like strings, extra dimensions or super-symmetric particles. As some quip, it is a theory with *no strings attached*.

You may wonder what justification there is for applying the maths of angular momentum to space. There are several sensible reasons. It is the maths of the SU(2) symmetry group. For those of you who know some group theory, it covers the SO(3) symmetry of three-dimensional rotations that we are familiar with in space.[12] It is also the maths of spin that drives much of the behaviour of quantum fields and using SU(2) makes it easier to cater for the weird geometry of the spin-half fermions. All this being said, LQG experts such as Carlo Rovelli are the first to admit that it is a *tentative theory* (his words). There is no direct evidence because the quanta of spacetime are much too small to detect.

While it is easy to shoot down a new theory, there are a number of legitimate worries about LQG (see comments in Box 26.4). If there really is a minimum length of space, will that not appear different to different observers (length contraction)? If so, might that allow an observer to know if he or she were moving and drive a stake through the heart of relativity? Another complaint is that LQG is really a theory of space with time added in, and does not address combined spacetime effectively. There are also some questions as to how well LQG leads to GR in the classical limit.

Box 26.4 Are you stringy or loopy?

Like oil and water, the two do not seem to mix:

The reason why I don't work on LQG is the issue with special relativity. If your approach does not respect the symmetries of special relativity from the outset, then you basically need a miracle to happen at one of your intermediate steps. String theorist on LQG

String theory seems to me to have failed to deliver what it offered in the 80s – I do not really understand how people can still have hope in it. LQG theorist on string theory

Conferences have segregated. Loopy people go to loopy conferences. Stringy people go to stringy conferences. They don't even go to 'physics' conferences anymore. I think it's unfortunate that it developed this way. Commented by a researcher in quantum gravity.

The fundamental challenge of LQG is the same as that of string theory. How do you test LQG? There was speculation that in LQG different frequencies of light would travel at different speeds because interaction with the quanta of space would make high energy photons travel very slightly more slowly. Results from the Atmospheric MAGIC telescope gave a hint of this and caused much excitement. It might have provided a shot in the arm for LQG in the same way that the discovery of SUSY particles would bolster string theory, but further analysis revealed no detectable difference.

One theoretical prediction of LQG is that black holes could bounce back as white holes. This would be instantaneous inside the hole, but for an observer outside, might take billions of years because of gravitational time dilation. Might we be able to detect white hole explosions? Researchers toil on in the hope that, through something of this sort, nature might show her hand and give us an indication of the right path forwards. Short of that, it is hard to see how we will know if LQG is reality or just a very pretty idea (to balance the score between strings and loops, I add one final joke in Box 26.5).

Box 26.5 Just for laughs:
Did you hear about the LQG researcher who retired?   *He's out of the loop!*

26.8 That's All Folks!

Congratulations! You have reached the end of this book but, in case you want more, I have added a note on further resources to study string theory and LQG in Box 26.6. I hope that you have learned some things that interest you, but more than anything, I hope that you have enjoyed it.

We have come a long way from the ideas of Newton and Maxwell. In Newton's physics, there was a clear distinction between time and space. In Maxwell's, there was a similarly clear distinction between particles and fields. Then along come Einstein, Heisenberg, Schrödinger, Dirac, Feynman and others to mess things up. Time and space coalesce into spacetime with its curvature incorporating the gravitational field. Particles and fields coalesce into gauge-invariant quantum fields in the Standard Model. The next challenge is to tie the curvature of spacetime together with the quantum fields, but how?

It is frustrating not to have a clearer path forwards, but that is the fascination of physics. Every answer brings further questions. Perhaps this is inevitable. As the boundaries of our knowledge grow, we extend our questions deeper into the unknown. Perhaps there is no end, no theory of everything. Or maybe a major discovery or a new idea is just around the corner. Is the next Einstein already at university starting his or her studies?

Might the next Einstein be you?

Time will tell.

Box 26.6 Other Resources: string theory and Loop Quantum Gravity
I should preface this with a warning. The mathematics behind both string theory and LQG is very advanced. The sources below will give you more on each topic, but will still leave you with many unanswered questions. There is a limit to how much even great teachers such as Susskind and Rovelli can explain in simple terms. For string theory, Leonard Susskind, one of the founding fathers, has a series of lectures on string theory and M theory. These are Stanford university videos available on Youtube. *https://www.youtube.com/watch?v=25haxRuZQUk* (as at December 2022) For LQG, Carlo Rovelli who was and is one of the driving forces in the subject has produced a number of lectures. The first listed below is an introduction. The second is a more complete (and mathematically complicated) lecture series. *https://www.youtube.com/watch?v=IkqHu1BqzDg* (as at December 2022) *https://www.youtube.com/playlist?list=PLwLvxaPjGHxR6zr421tXXlaDGbq8S36Un* (as at December 2022)

26.9 Module Memory Jogger

- *Intrinsic spin of electrons and quarks:* $S = \pm\frac{\hbar}{2}$
- *Rotating fermion wave function by 2π takes $\Psi \to -\Psi$; need 4π rotation for $\Psi \to \Psi$*
- *Pauli exclusion principle: no two fermions can share the same state*
- *Pauli matrices for spin:* $\hat{S}_x = \frac{\hbar}{2}\begin{bmatrix} 0 & 1 \\ 1 & 0 \end{bmatrix}$ $\hat{S}_y = \frac{\hbar}{2}\begin{bmatrix} 0 & -i \\ i & 0 \end{bmatrix}$ $\hat{S}_z = \frac{\hbar}{2}\begin{bmatrix} 1 & 0 \\ 0 & -1 \end{bmatrix}$
- *Massless particles have only two intrinsic spin states*
- *Violation of Bell's inequalities in Aspect experiment: Einstein's spooky action at a distance*
- *The Dirac equation:* $i\gamma^\mu \frac{\partial}{\partial x_\mu} \Psi = m\Psi$
- *Dirac's results: four-component wave function, spin-half particles, antiparticle states*
- *Quantum field theory (QFT): particles are excitations of quantum fields*
- *Short-lived off-shell pathways must be taken into account: virtual particles*
- *Local gauge invariance to phase due to interaction of electron with electromagnetic field*
- *Quantum electrodynamics (QED) interaction term:* $e\,\overline{\Psi}\gamma^\mu A_\mu \Psi$

- *QED Lagrangian:* $\quad \mathscr{L} = i\overline{\Psi}\gamma^{\mu}\partial_{\mu}\Psi - m\overline{\Psi}\Psi - e\overline{\Psi}\gamma^{\mu}A_{\mu}\Psi - \frac{1}{4}F_{\mu\nu}F^{\mu\nu}$
- *Feynman diagrams: experimental outcome is the sum of all possible interaction pathways*
- *Pathway amplitude depends on coupling constant (per interaction vertex) and propagators*
- *QED coupling constant is small so can usually ignore more complicated Feynman diagrams*
- *QED as an effective field theory: adjust values of α and m for a high energy cut-off*
- *Strong force is so strong that standalone particles must be colour-charge neutral: confinement*
- *Strong force local gauge invariance to change in colour: interaction with eight gluon fields*
- *Strong force Lagrangian:* $\quad \mathscr{L} = i\overline{\Psi}\gamma^{\mu}\partial_{\mu}\Psi - m\overline{\Psi}\Psi - g\overline{\Psi}\gamma^{\mu}A_{\mu}^{a}T_{a}\Psi - \frac{1}{4}G_{\mu\nu}^{a}G_{a}^{\mu\nu}$
- *Residual strong force attraction between nucleons due to interchange of virtual mesons*
- *Symmetry and group theory: QED U(1), strong force SU(3), weak force SU(2)*
- *Weak force interaction involves 3 bosons (W^{+}, W^{-}, Z) as expected for SU(2)*
- *Symmetry appears broken: flavours have different mass, bosons are massive, chiral bias*
- *Symmetry restored when interaction with the Higgs field is included in the Lagrangian*
- *Higgs field Mexican hat potential:* $\quad V = -\frac{\mu^{2}}{2}(\phi_{H}^{*}\phi_{H}) + \frac{\lambda}{4}(\phi_{H}^{*}\phi_{H})^{2}$
- *Combined Lagrangian: three massive weak force vector bosons, massive Higgs scalar boson*
- *Electroweak unification of U(1) and SU(2): massive Z boson and massless photon*
- *Standard Model:* $\quad \mathscr{L} = i\overline{\Psi}\gamma^{\mu}D_{\mu}\Psi + y\overline{\Psi}\Psi\phi_{H} + \frac{1}{2}D^{\mu}\phi_{H}^{*}D_{\mu}\phi_{H} - V(\phi_{H}) - \frac{1}{4}F_{\mu\nu}F^{\mu\nu}$
- *Standard Model symmetry:* $\quad$ SU(3) ⊗ SU(2) ⊗ U(1)
- *Major challenge: gravity does not fit in the Standard Model*

Notes

1 Tensors are a step up from vectors. They are important in general relativity, but beyond the scope of this book.

2 Technical point: you may see the Yukawa coupling constant shown as y_{ij}. This reflects the additional subtlety (which is beyond our scope) that, in addition to coupling left-right chiral versions of each fermion, there is some limited coupling between different fermion *generations* such as for example up/charm/top in the case of quarks.

3 This may not be true for the neutrino whose mass is still poorly understood (see Box 24.4).

4 Technical point: the factor of $\frac{1}{2}$ in the Higgs derivatives appears because I am using $(v + H)$ for the Higgs value and v for the v.e.v. Texts using the convention of $\frac{v+H}{\sqrt{2}}$ and the v.e.v as $\frac{v}{\sqrt{2}}$ absorb this $\frac{1}{2}$ factor into the value of the field.

5 The type of energy plays a role because the behaviour of the energy in matter is different from that of radiation-energy.

6 Technical point: in reality his equations remain unstable even with the cosmological constant included.

7 Technical point: I have simplified by including the scalar Higgs field with these vector fields.

8 Technical point: this is called *level matching*. I am told it is due to the symmetries of a loop (using Noether's theorem), but the detailed maths behind string theory is too fiendishly complicated for me.

9 Type 1, Type IIA, Type IIB, SO(32) heterotic, and E8xE8 heterotic.

10 Technical point: the extra dimension of M-theory actually was received positively because it aligned string theory with some 11 dimensional theories of supergravity.

11 Technical point: the Schwarzschild radius is usually shown as $R_{s} = \frac{2Gm}{c^{2}}$. I have switched to energy using $E = mc^{2}$.

12 Technical point: in fact the SU(2) group double covers SO(3).

Index

a

Action   49, 53, 58, 72, 195, 200, 201, 209, 213, 271
Adams, Allan   158
Addition of velocities   16
Al-Khalili, Jim   110
Amplitude   46
Anderson, Carl   184
Angular kinetic energy   129, 156
Angular momentum   126, 156, 161, 180
Angular momentum: quanta   128, 156, 163
ANITA   274
Antimatter   188, 272
Antiparticles   154, 182
Aspect experiment   68, 170
Aspect, Alain   68
Associated Laguerre equation   144, 146
Atlas detector   264
Atomic bomb   24
Atomic orbitals   132, 150, 156

b

Balmer series   136
Balmer, Johann   136
Bell's inequality   170
Bell, John   170
Bernouilli, Johann   130
Bernoulli, Daniel   130
Bhabha scattering   223, 227
Bhabha, Homi   223
Bi-spinors   174
Bohmian interpretation   66
Bohr, Niels   66, 77, 138, 139
Born rule   46, 72
Bosons   165, 193, 235, 269

c

Cardano, Gerolamo   84
Cartesian geometry   47
Chirality   253
Collapse of the wave function   61, 65, 68
Colour charge   154, 236

Commutators   100, 128, 166, 180
Compactification   275
Compton scattering   216
Cooper pairs   165
Copenhagen interpretation   66, 69
Coulomb potential   136, 156, 185, 232
Coupling constant   221, 228, 237, 255
Covalent bonds   153, 217
Covariant versus contravariant   208
Cox, Brian   50, 78
Creation/Annihilation operators   119, 123, 133, 156, 163, 192, 215
Cross section   224
Curl   205

d

Dark energy   272
Dark matter   272
De Broglie, Louis   31, 36, 71
Decoherence   66
Degeneracy   147
Dielectric effect   228
Diphoton channel   225, 264
Dirac delta function   62, 88
Dirac equation   162, 173, 271
Dirac equation: chiral   254
Dirac Lagrangian   204, 212, 247, 252, 258, 270
Dirac matrices   177, 270
Dirac sea   187
Dirac, Paul   119, 173, 174, 177, 188, 189, 202, 227
Divergence   205
Double-slit experiment   32, 50, 65, 71, 154–155
Dressed electron   228
Dyson, Freeman   216

e

Eddington, Arthur   230
Effective Field Theory (EFT)   227
Eigenvalues   97
Eigenvectors   99
Einstein summation   207, 248

Quantum Untangling: An Intuitive Approach to Quantum Mechanics from Einstein to Higgs, First Edition. Simon Sherwood.
© 2023 John Wiley & Sons Ltd. Published 2023 by John Wiley & Sons Ltd.

Einstein, Albert   21, 46, 48, 66, 67, 72, 77, 167, 170, 206
Einstein: $E = mc^2$ 20
Electromagnetic Lagrangian   208
Electromagnetic vector potential   206
Electromagnetism refresher   205
Electron in a box   63, 69, 73
Electron self-energy   229
Electroweak unification   264
Emission spectra   65, 69, 71, 136, 186, 271
Energy-momentum space   24, 190
Entanglement   67
EPR paradox   170
Ether   3
Euler, Leonhard   85
Euler-Lagrange   55, 72, 199, 209
Everett, Hugh   66, 69
Expectation value   101

f
Fermi, Enrico   224, 263
Fermions   165, 193, 234, 269
Feynman diagrams   57, 216, 220
Feynman rules   221
Feynman, Richard   34, 49, 51, 72, 77, 78, 154, 165, 216, 226, 227
Flavour charge   247
Forshaw, Jeff   50, 78
Fourier limitations   44
Fourier transform   43, 61, 87, 93, 111, 156
Fourier, Joseph   44
Free particle wave function   37, 72, 82, 85, 104
Fuzziness   63, 68, 73, 108

g
Galois, Evariste   241
Gauge invariance: global   84, 200, 202, 247
Gauge invariance: local   202, 238, 247, 270, 271
Gauss, Carl   90, 92
Gaussian distribution   89, 93, 111, 113
Gell-Mann matrices   242
Gell-Mann, Murray   237
General relativity   18, 272
Gerlach, Walter   163
Germain, Sophie   92
Golden pathway (4-lepton)   264
Goldstone bosons   260, 263, 267
Goudsmit, Samuel   172
Graviton   235, 269
Green, George   229
Gribbin, John   29, 78
Group theory   240

h
Hadron jet   237
Hafele-Keating   18, 20
Hamiltonian   99
Heisenberg's uncertainty principle   41, 44, 56, 61, 89, 92, 130
Heisenberg, Werner   72, 110, 187, 202

Helicity   253
Hidden variable   46, 66, 170
Higgs boson   175, 199, 225, 269
Higgs field   252, 258, 270
Higgs mechanism   258
Higgs, Peter   246, 263, 267
Hooke, Robert   116
Hund rule   151
Hydrogen energy levels   146, 185
Hydrogen equation: angular   141, 146, 156
Hydrogen equation: radial   144, 146, 156

i
Indices, upper and lower   208
Indistinguishability   191
Interference   50, 83
Interstellar travel   19
Invariant interval equation   5, 21, 70
Ionic bonds   152

k
Kaluza, Theodore   275
Klein-Gordon equation   74, 173, 175
Klein-Gordon Lagrangian   199, 203, 258, 270

l
Lagerstrom, Larry   4
Lagrange, Joseph-Louis   58
Lagrangian   54, 72, 196
Lamb shift   232
Laplace, Pierre-Simon   140
Lavoisier, Antoine   58
LeBlanc, Monsieur   92
Lee, Tsung-Dao   250
Leibniz, Gottfried   123
Lewis, Gilbert   36
Light year   9
Loop Quantum Gravity (LQG)   276
Lorentz invariance   10, 25, 39, 47
Lorentz transformation   10, 26, 36, 71
Lorentz, Hendrik   71
Lyman series   136

m
M-theory   276
Majorana, Ettore   254
Many worlds interpretation   66, 67
Matrix exponential   240
Maxwell, James   3, 206, 210
McFadden, Johnjoe   110
Measurement problem   66
Mendeleev, Dmitri   149, 153
Mesons   154, 244
Metric   6
Millen, James   33
Minkowski spacetime   6, 174, 179, 182
Minkowski, Hermann   12
Mobius strip   164
Molecular bonds   152

Multivariable chain rule   39
Muon   8, 225, 234

n

Neutrino   251, 272
Newton, Isaac   55, 123
Nodes   63
Noether, Emily   195, 201
Non-commutation   91
Normalisation   118
Nuclear bonds   154, 244
Nuclear fission   156
Nuclear fusion   107, 155

o

Off-shell pathways   197, 209
Operation Epsilon   163
Oppenheimer, Robert   66

p

Particle time   20
Paschen series   136
Path integral   49, 72, 155, 190, 195, 216, 227
Pauli exclusion principle   150, 162, 164, 172, 193
Pauli matrices   165, 172, 177, 247, 255
Pauli, Wolfgang   165, 167, 251
Penrose, Roger   66
Periodic table   149
Perturbation theory   219
Photoelectric effect   30, 36
Photon spin   168
Pilot wave   66, 69
Planck energy   232
Planck length   277
Planck's constant   36, 137
Planck, Max   30, 31, 77, 232
Polarised light   169
Positron   57, 183, 187, 201, 218, 229
Potential barriers   108, 113
Potential well: finite   106
Potential well: infinite   63
Probability   46, 50, 60, 72, 73, 84, 89, 101
Propagator   222, 230
Proper distance   10
Proper time   7
Putnam competition   233

q

QCD Lagrangian   241
QED Lagrangian   214, 252
Quantum Biology   110, 113
Quantum chromodynamics   234, 247, 248, 266
Quantum electrodynamics   51, 58, 175, 216, 236
Quantum field theory   117, 125, 189, 205, 271
Quantum Harmonic Oscillator   90, 97, 114, 133, 156, 191, 201
Quantum tunnelling   105, 113, 154, 156, 197, 201, 217, 244

r

Relativistic energy/momentum   22, 71
Renormalisation   227
Residual strong force   244
Resonance   225
Rovelli, Carlo   279, 280
Running constants   231, 273
Rutherford scattering   29
Rydberg, Johannes   136, 137, 146, 186

s

Sakharov, Andrei   113
Schrödinger's cat   34, 71
Schrödinger's equation   73, 95, 104, 117, 139, 156, 173, 185, 271
Schrödinger, Erwin   77, 100, 110, 145
Self-interaction   228
Simple Harmonic Oscillator   114
Spacetime paradoxes   13
Spacetime: leading clocks lag   9, 70
Spacetime: length distortion   8, 70
Spacetime: time distortion   6, 70
Special relativity   3, 70, 74, 84, 156, 174, 190, 271
Spherical harmonics   139, 141, 143, 146, 156
Spin   132, 161, 180, 192, 271
Spin networks (LQG)   278
Spinors   163
Spontaneous symmetry breaking   260, 270
Spooky action   67, 73, 170
Standard Model   66, 234, 269
Standard Model Lagrangian   270
Standing waves   64
Stationary states   102, 156
Stern, Otto   163
Stern-Gerlach experiment   162
String theory   274
Strong force   107, 154, 205, 214, 234, 235
Superposition   42, 60, 72, 86, 96, 121, 166, 271
Supersymmetry   273
Susskind, Leonard   158, 280
Szilard, Leo   156

t

Tamm, Igor   113
Taylor expansion   8, 23, 52, 219
Thomson, George   34
Thomson, Joseph   34
Tokamak   113
Turing, Alan   113
Twin Paradox   16, 70

u

Uhlenbeck, George   172
Ultraviolet catastrophe   30, 71
Unitary gauge change   262
Unitary matrix   239

v

Vacuum energy   192, 272

Valence electrons   149, 153
Valley of Stability   244
Vector cross-product   127
Virtual particles   154, 197, 209, 217, 225, 229
Von Laue, Max   77

w

Wave function: meaning   47, 73
Wave packets   110
Weak force   198, 235, 246, 257
Weyl, Hermann   145, 200

Wigner, Eugene   35
Wilson, Ken   232
Witten, Edward   276
Wu, Chien-Shiung   250, 270

y

Yang, Chen-Ning   250
Young, Thomas   32
Yukawa, Hideki   244, 255

z

Zeno effect   113